The IMA Volumes in Mathematics and Its Applications

Volume 22

Series Editors
Avner Friedman Willard Miller, Jr.

Institute for Mathematics and its Applications
IMA

The **Institute for Mathematics and its Applications** was established by a grant from the National Science Foundation to the University of Minnesota in 1982. The IMA seeks to encourage the development and study of fresh mathematical concepts and questions of concern to the other sciences by bringing together mathematicians and scientists from diverse fields in an atmosphere that will stimulate discussion and collaboration.

The IMA Volumes are intended to involve the broader scientific community in this process.

Avner Friedman, Director
Willard Miller, Jr., Associate Director

* * * * * * * * * *

IMA PROGRAMS

1982-1983	**Statistical and Continuum Approaches to Phase Transition**
1983-1984	**Mathematical Models for the Economics of Decentralized Resource Allocation**
1984-1985	**Continuum Physics and Partial Differential Equations**
1985-1986	**Stochastic Differential Equations and Their Applications**
1986-1987	**Scientific Computation**
1987-1988	**Applied Combinatorics**
1988-1989	**Nonlinear Waves**
1989-1990	**Dynamical Systems and Their Applications**
1990-1991	**Phase Transitions and Free Boundaries**

* * * * * * * * * *

SPRINGER LECTURE NOTES FROM THE IMA:

The Mathematics and Physics of Disordered Media
Editors: Barry Hughes and Barry Ninham
(Lecture Notes in Math., Volume 1035, 1983)

Orienting Polymers
Editor: J.L. Ericksen
(Lecture Notes in Math., Volume 1063, 1984)

New Perspectives in Thermodynamics
Editor: James Serrin
(Springer-Verlag, 1986)

Models of Economic Dynamics
Editor: Hugo Sonnenschein
(Lecture Notes in Econ., Volume 264, 1986)

L. Auslander F.A. Grünbaum J.W. Helton
T. Kailath P. Khargonekar S. Mitter
Editors

Signal Processing

Part I: Signal Processing Theory

Edited by
L. Auslander T. Kailath S. Mitter

With 15 Illustrations

Springer-Verlag
New York Berlin Heidelberg
London Paris Tokyo Hong Kong

Louis Auslander
Center for Large Scale Computation
The Graduate School and University Center, CUNY
New York, NY 10036, USA

Tom Kailath
Information Systems Laboratory
Stanford University
Stanford, CA 94305, USA

Sanjoy K. Mitter
Center for Intelligent Control Systems
Massachusetts Institute of Technology
Cambridge, MA 02139, USA

Series Editors

Avner Friedman
Willard Miller, Jr.
Institute for Mathematics and Its Applications
University of Minnesota
Minneapolis, MN 55455, USA

Mathematics Subject Classification: 22Exx, 43A, 94Axx

Library of Congress Cataloging-in-Publication Data
Signal processing / L. Auslander. . . [et al.], editors.
p. cm.—(The IMA volumes in mathematics and its applications; v. 22)
"Based on lectures delivered during a six week program held at the IMA from June 27 to August 5, 1988"—Foreword.
Contents: pt. 1. Signal processing theory.

1. Signal processing—Mathematics. I. Auslander, Louis. II. University of Minnesota. Institute for Mathematics and Its Applications. III. Series.
TK5102.5.S535 1990
621.382'2—dc20 89-26216

Softcover reprint of the hardcover 1st edition 1990

Camera-ready copy prepared by the IMA using TEX.

9 8 7 6 5 4 3 2 1 Printed on acid-free paper.

ISBN-13: 978-1-4684-6395-8 e-ISBN-13: 978-1-4684-6393-4
DOI: 10.1007/978-1-4684-6393-4

The IMA Volumes
in Mathematics and its Applications

Current Volumes:

Volume 1: Homogenization and Effective Moduli of Materials and Media
Editors: Jerry Ericksen, David Kinderlehrer, Robert Kohn, J.-L. Lions

Volume 2: Oscillation Theory, Computation, and Methods of Compensated Compactness
Editors: Constantine Dafermos, Jerry Ericksen,
David Kinderlehrer, Marshall Slemrod

Volume 3: Metastability and Incompletely Posed Problems
Editors: Stuart Antman, Jerry Ericksen, David Kinderlehrer, Ingo Muller

Volume 4: Dynamical Problems in Continuum Physics
Editors: Jerry Bona, Constantine Dafermos, Jerry Ericksen, David Kinderlehrer

Volume 5: Theory and Applications of Liquid Crystals
Editors: Jerry Ericksen and David Kinderlehrer

Volume 6: Amorphous Polymers and Non-Newtonian Fluids
Editors: Constantine Dafermos, Jerry Ericksen, David Kinderlehrer

Volume 7: Random Media
Editor: George Papanicolaou

Volume 8: Percolation Theory and Ergodic Theory of Infinite Particle Systems
Editor: Harry Kesten

Volume 9: Hydrodynamic Behavior and Interacting Particle Systems
Editor: George Papanicolaou

Volume 10: Stochastic Differential Systems, Stochastic Control Theory and Applications
Editors: Wendell Fleming and Pierre-Louis Lions

Volume 11: Numerical Simulation in Oil Recovery
Editor: Mary Fanett Wheeler

Volume 12: Computational Fluid Dynamics and Reacting Gas Flows
Editors: Bjorn Engquist, M. Luskin, Andrew Majda

Volume 13: Numerical Algorithms for Parallel Computer Architectures
Editor: Martin H. Schultz

Volume 14: Mathematical Aspects of Scientific Software
Editor: J.R. Rice

Volume 15: Mathematical Frontiers in Computational Chemical Physics
Editor: D. Truhlar

Volume 16: Mathematics in Industrial Problems
by Avner Friedman

Volume 17: Applications of Combinatorics and Graph Theory to the Biological and Social Sciences
Editor: Fred Roberts

Volume 18: q-Series and Partitions
Editor: Dennis Stanton

Volume 19: Invariant Theory and Tableaux
Editor: Dennis Stanton

Volume 20: Coding Theory and Design Theory Part I: Coding Theory
Editor: Dijen Ray-Chaudhuri

Volume 21: Coding Theory and Design Theory Part II: Design Theory
Editor: Dijen Ray-Chaudhuri

Volume 22: Signal Processing: Part I
Editors: L. Auslander, F.A. Grünbaum, W. Helton, T. Kailath, P. Khargonekar and S. Mitter

Volume 23: Signal Processing: Part II
Editors: L. Auslander, F.A. Grünbaum, W. Helton, T. Kailath, P. Khargonekar and S. Mitter

Volume 24: Mathematics in Industrial Problems, Part 2
by Avner Friedman

Forthcoming Volumes:

1988-1989: *Nonlinear Waves*

Solitons in Physics and Mathematics

Two Phase Waves in Fluidized Beds, Sedimenation, and Granular Flows

Nonlinear Evolution Equations that Change Type
Computer Aided Proofs in Analysis
Multidimensional Hyperbolic Problems and Computations (2 Volumes)
Microlocal Analysis and Nonlinear Waves

Summer Program 1989: *Robustness, Diagnostics, Computing and Graphics in Statistics*
Robustness, Diagnostics in Statistics (2 Volumes)
Computing and Graphics in Statistics

1989-1990: *Dynamical Systems and Their Applications*
Introduction to Dynamical Systems

FOREWORD

The two volumes of Signal Processing are based on lectures delivered during a six week program held at the IMA from June 27 to August 5, 1988. The first two weeks of the program dealt with general areas and methods of Signal Processing. The problem areas included imaging and analysis of recognition, x-ray crystallography, radar and sonar, signal analysis and 1-D signal processing, speech, vision, and VLSI implementation. The methods discussed included harmonic analysis and wavelets, operator theory, algorithm complexity, filtering and estimation, and inverse scattering. The topics of weeks three and four were digital filter, VLSI implementation, and integrable circuit modelling. In week five the concentration was on robust and nonlinear control with aerospace applications, and in week six the emphasis was on problems in radar, sonar and medical imaging.

Because of the large overlap between the various one-week and two-week segments of the program, we found it more convenient to divide the material somewhat differently. Part I deals with general signal process theory and Part II deals with (i) application of signal processing, (ii) control theory related themes.

We are grateful to the scientific organizers: Tom Kailath (Chairman), Louis Auslander, F. Alberto Grunbaum, J. William Helton, Pramod P. Khargonekar and Sanjoy K. Mitter.

We are also grateful for the generous support given to the IMA program by the Office of Naval Research, the Air Force Office of Scientific Research, the Army Research Office and the National Security Agency.

Signal Processing is undergoing tremendous developments; it is our hope that these two volumes will serve as a source of information and stimulation to mathematical scientists who wish to get acquainted with this field.

Avner Friedman
Willard Miller, Jr.

CONTENTS

CONTENTS

SIGNAL PROCESSING: PART II

I - CONTROL THEORY

II - APPLICATIONS OF SIGNAL PROCESSING

WIDE-BAND AMBIGUITY FUNCTION AND $a \cdot x + b$ GROUP

LOUIS AUSLANDER* AND IZIDOR GERTNER*†

Abstract. In this paper a wide-band ambiguity function is defined. Conditions when the narrow-band approximation is valid are stated. It is shown that the wide-band ambiguity function is a coefficient function of the unitary representation of the $a \cdot x + b$ group. The relation to the "wavelet" expansion is demonstrated. Computational algorithm for the wide-band ambiguity function is presented.

1. Introduction. The classical narrow-band ambiguity function is defined for signals where the Doppler effect is approximated by a frequency shift that is constant across the bandwidth. Therefore the signal spectral shape is not changed. Such signals are called quasiharmonic. Let $s(t)$ be a quasiharmonic signal:

$$s(t) = | s(t) | e^{i\phi(t)} \tag{1}$$

$| s(t) |$ can be directly interpreted as the envelope of the real signal and the quantity

$$\frac{1}{2\pi} \cdot \frac{d\phi(t)}{dt} \tag{2}$$

as its instantaneous frequency.

Let $S(f) = \mathcal{F}\{s(t)\}$ be the Fourier transform of $s(t)$ with f_0 as its central frequency defined as:

$$f_0 = \int_{-\infty}^{\infty} f \mid S(f) \mid^2 df$$

The Doppler shift is then:

$$f_d = 2\pi \frac{v}{c} \cdot f_0 \tag{3}$$

$$\begin{aligned} where: c - &\quad the\ signal\ propagation\ velocity \\ v - &\quad the\ target\ velocity \end{aligned}$$

The spectra of the signal is linearly shifted by the quantity f_d. This approximation is valid when the energy is concentrated near the central frequency f_0 . We call such signals narrow-band. The use of these signals has been widely spread in classical radar and communication.

Unfortunately, in many cases such as signals arising in sonar, seismology, oceanography the narrow-band approximation is not valid. The use of the narrow band ambiguity function may cause an error in the estimation of a velocity and

*Center for Large Scale Computation, The Graduate School and University Center, CUNY, 25 West 43rd street , suite 400, New York , N.Y. 10036

†Supported by DARPA-ACMP 770997 and in part by IMA with funds provided by NSF

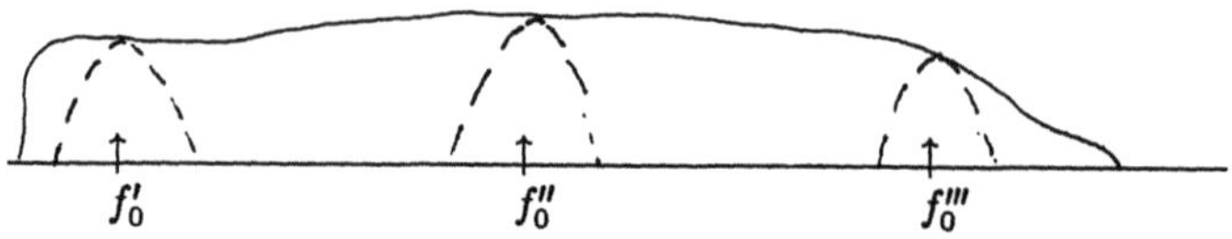

Figure 1: Wide-band signal spectra.

distance. For these signal the deviation from the central frequency is large. Thus when the target is moving there is no uniform Doppler shift for the entire spectra. We can illustrate this heuristically in the following figure 1.

Let f_0' *and* f_0'' be two frequencies in the signal spectra. Consider the neighborhood of the frequencies f_0' and f_0'' According to (3) the Doppler shift will be different in each neighborhood thus causing the signal spectra to strech or to compress. Therefore a more general than narrow-band signal model is required. In [SW66] a signal model that is independent of bandwidth and central frequency was proposed. We will describe this model below.

Let $s_1(t) \in L^2(\Re)$ represent a signal with a constant propogation velocity c . Let $s_2(t)$ be an echo of $s_1(t)$ from a target moving with velocity v . Then

$$s_2(t) = As_1(a \cdot t - \tau) \tag{4}$$

$$\begin{aligned} where: A & \quad - The\ energetic\ parameter\ of\ the\ signal \\ a &= \frac{c-v}{c+v} \quad - The\ Doppler\ strech - compress\ factor \\ \tau &= \frac{2R_0}{c+v} \quad - The\ signal\ delay\ at\ time\ t = 0 \\ R &= R_0 + vt \end{aligned}$$

For our purpose we will ignore signal propogation attenuation and we will normalize to one the signal energy.

$$\begin{aligned} 1 &= \int_{-\infty}^{\infty} | s_2(t) |^2 \, dt \\ &= A^2 \int_{-\infty}^{\infty} | s_1(a \cdot t - \tau) |^2 \, dt \\ &= \frac{A^2}{a} \int_{-\infty}^{\infty} | s_1(t) |^2 \, dt \\ &= \frac{A^2}{a} \end{aligned}$$

Thus the conservation of energy requires that $A = \sqrt{a}$ and the received signal can be represented as:

$$\begin{aligned} s_2(t) &= \sqrt{a} \cdot s_1(a \cdot t - \tau) \\ &= \sqrt{a} \cdot s_1(a(t - \tau')). \qquad (5) \end{aligned}$$

$$\tau' = \frac{2R_0}{c-v}. \tag{6}$$

The problem is to estimate the Doppler parameter a and the delay τ. The maximum likelihood estimator of these parameters is the *max* of the crosscorrelation function :

$$\max_{a,\tau} \quad |\sqrt{a}\int_{-\infty}^{\infty} s_1(t)\cdot s_1^*(a(t-\tau))dt|^2 \tag{7}$$

This suggests the definition of the wide-band ambiguity function:

$$WA_{f,g}(a,\tau) = \sqrt{a}\int_{-\infty}^{\infty} f(t)\cdot g^*(a(t-\tau))dt \tag{8}$$

The maximum likelihood estimator requires one to find the maximum of the wide -band ambiguity function surface between the received signal and the reference signals.

In this paper we will present properties of the wide-band ambiguity function. We will show that wide-band ambiguity function is a unitary representation of the $a\cdot x+b$ group. We will present a computational algorithm to compute the wide-band ambiguity function.

2. Wide-band ambiguity function properties and the $a\cdot x+b$ group.

In this section we will show how to connect the wide-band ambiguity function with a representation of the affine or $a\cdot x+b$ group. Recently, the concept of wavelet expansion has been introduced. A good survey has been presented in [ID88]. We will also show how to relate the wide-band ambiguity function with the wavelets. First we will recall the representation of the $a\cdot x+b$ group. Reader who is familiar with the subject may skip the subsection.

2.1. Relation of the narrow-band and wide-band ambiguity functions. In this section we will show when the wide-band ambiguity function can be approximated by the narrow-band ambiguity function. To do this, let $s(t)$ be a signal and $S(f)$ the Fourier transform of $s(t)$. Then

$$\begin{aligned} s(a(t-\tau)) &= \int_{-\infty}^{\infty} S(f)e^{2\pi i f a(t-\tau)}df \\ &= \int_{-\infty}^{\infty} e^{-2\pi i f a\tau}\cdot S(f)e^{2\pi i f a t}df. \end{aligned} \tag{9}$$

Let f_0 be the central frequency of the spectra $S(f)$. Also denote

$$\begin{aligned} \Delta f &= f-f_0 && -\mathit{the\ frequency\ deviation,} \\ \delta &= \frac{2v}{c+v} && -\mathit{the\ relative\ velocity,} \\ a &= 1-\frac{2v}{c+v} = 1-\delta && -\mathit{the\ Doppler\ factor.} \end{aligned}$$

Then

$$s(a(t-\tau)) = e^{-2\pi i f_0 \delta(t-\tau)} \cdot \int_{-\infty}^{\infty} S(f) e^{2\pi i f(t-\tau)} \cdot e^{-2\pi i \Delta f \delta(t-\tau)} df \tag{10}$$

There are two factors affecting the signal behavior in time

$$e^{-2\pi i f_0 \delta(t-\tau)} \; and \; e^{-2\pi i \Delta f \delta(t-\tau)}$$

The first one shifts the signal spectra ; the second one stretches/compresses the signal envelope. If the target-source velocity $v \ll c$ then $\delta \ll 1$ and thus

$$e^{-2\pi i \Delta f \delta(t-\tau)} \approx 1 \tag{11}$$

In this case

$$s(a(t-\tau)) = e^{2\pi i f_d (t-\tau)} s(t-\tau), \tag{12}$$

where

$$f_d = -f_0 \delta = f_0(a-1)$$

is the classical Doppler shift. From (8) and (5) we get that the wide-band ambiguity function can be written as:

$$\begin{aligned} WA_{h,g}(a,\tau) = & \sqrt{a} \int_{-\infty}^{\infty} h(t) e^{-2\pi i f_0 (a-1)(t-\tau)} \cdot g^*(t-\tau) dt \\ = & \sqrt{a} e^{2\pi i f_0 (a-1)\tau} \int_{-\infty}^{\infty} h(t) \cdot g^*(t-\tau) e^{-2\pi i f_0 (a-1) t} dt \\ & \sqrt{a} e^{2\pi i f_0 (a-1)\tau} \cdot NA_{h,g}(a,\tau), \end{aligned} \tag{13}$$

where

$$NA_{h,g}(a,\tau) = \int_{-\infty}^{\infty} h(t) \cdot g^*(t-\tau) e^{-2\pi i f_0 (a-1) t} dt \tag{14}$$

is the narrow-band ambiguity function.

The approximation error is given as:

$$WA_{h,g}(a,\tau) - NA_{h,g}(a,\tau) = NA_{h,g}(a,\tau) \left\{ \sqrt{a} \cdot e^{2\pi i f_0 (a-1)\tau} \right\} \tag{15}$$

An upper bound on the relative approximation error was obtained by Swick[SW69] in terms of J - the Woodward signal duration and β the Woodward signal bandwidth:

$$| WA(\tau, a) | - \sqrt{a} \, | NA(\tau, a) | \leq \frac{\sqrt{a}}{2\pi} \cdot | \delta | \cdot J \cdot \beta \tag{16}$$

In any case from the above , when the narrow-band approximation is acceptable, we can determine the limiting strech factor a .

From the above discussion we can conclude that factors affecting the narrow-band approximation are:

(1) δ - the target velocity,

(2) $\Delta f = f - f_0$ the signal frequency deviation,

(3) $t - \tau$ - the observation interval.

For sonar signals we usually have the deviation Δf very large, the observation time is usually large also, since one wants to receive more energy in order to improve signal to noise ratio. In this case the narrow band ambiguity function is not valid and one should use the wide-band ambiguity function.

2.2. The $a \cdot x + b$ group.

The $a \cdot x + b$ group is a set $\Re \setminus \{0\} \times \Re$ equipped with the group multiplication law:

$$(a, b) \circ (a', b') = (a \cdot a', b + a \cdot b') \tag{17}$$

The unity of the group is $(1, 0)$:

$$\begin{aligned} (1,0) \circ (a,b) &= (a,b) \\ (a,b) \circ (1,0) &= (a,b) \end{aligned} \tag{18}$$

The right and left inverse of (a, b) is $(\frac{1}{a}, -\frac{b}{a})$:

$$(a,b) \circ \left(\frac{1}{a}, -\frac{b}{a}\right) = (1,0)$$
$$\left(\frac{1}{a}, -\frac{b}{a}\right) \circ (a,b) = (1,0)$$

The group is non-commutative:

$$(a,b) \circ (a',b') = (aa', b + ab')$$
$$(a',b') \circ (a,b) = (aa', b' + a' \cdot b).$$

2.3. Unitary representation. A representation of the group $\Re \setminus \{0\} \times \Re$ in a space $L^2(\Re)$ is a pair:

$$\{\mathcal{U}, L^2(\Re)\}$$

where $\mathcal{U}$ is a mapping which assigns to every element $(a, b) \in \Re \setminus \{0\} \times \Re$ a linear mapping

$$\mathcal{U}_{(a,b)} : L^2(\Re) \Rightarrow L^2(\Re)$$

such that

$$\mathcal{U}_{(1,0)} = 1_{L^2(\Re)} \tag{19}$$

$$\mathcal{U}_{((a,b)\circ(a',b'))} = \mathcal{U}_{(a,b)} \circ \mathcal{U}_{(a',b')}, \quad \forall\, (a,b), (a',b') \in \Re \setminus \{0\} \times \Re \tag{20}$$

In particular

$$\mathcal{U}_{(a,b)} \circ \mathcal{U}_{(\frac{1}{a}, -\frac{b}{a})} = \mathcal{U}_{(1,0)}$$

If $L^2(\Re)$ is a Hilbert space, a linear representation $\{\mathcal{U}, L^2(\Re)\}$ is said to be unitary if the automorphism $\mathcal{U}_{(a,b)}$ forms a unitary operator of

$$L^2(\Re) \quad \forall (a,b) \in \Re \setminus \{0\} \times \Re$$

$\mathcal{U}$ defines a morphism $(a,b) \to \mathcal{U}_{(a,b)}$ of the group $\Re \setminus \{0\} \times \Re$ into a unitary group $\underline{\mathcal{U}}(L^2(\Re))$ of $L^2(\Re)$ operators such that

(22) $$\mathcal{U}_{(a,b)^{-1}} = \left(\mathcal{U}_{(a,b)}\right)^* \quad \forall\, (a,b) \in \Re \setminus \{0\} \times \Re$$

Unitarity is equivalent to:

(23) $$\left\|\mathcal{U}_{(a,b)} f\right\| = \|f\| \quad \forall\, (a,b), \in \Re \setminus \{0\} \times \Re;\ f \in L^2(\Re)$$

or

(24) $$\langle \mathcal{U}_{(a,b)} f, \mathcal{U}_{(a,b)} g\rangle = \langle f, g\rangle \quad \forall\, (a,b), \in \Re \setminus \{0\} \times \Re \quad f, g \in L^2(\Re)$$

Define the translation representation

(25) $$\left(T_{(1,s)}\right)(f) = f(t-s) \qquad (1,s) \in \Re \setminus \{0\} \times \Re.$$

It is easy to see that the representation T is unitary and commutative

(26) $$\begin{aligned} T_{(1,s),(1,s')} &= T_{(1,s+s')} \\ &= T_{(1,s)} \circ T_{1,s')} \end{aligned}$$

The unitarity of the translation representation follows from:

(27) $$\begin{aligned} \langle T_{(1,s)} h, T_{(1,s)} g\rangle &= \int_{-\infty}^{\infty} h(t-s) \cdot g^*(t-s) dt \\ &= \langle h, g\rangle \end{aligned}$$

Define the dilation representation by

(28) $$\left(D_{(\lambda,0)} h\right)(t) = \mid \lambda \mid^{-\frac{1}{2}} \cdot h\left(\frac{t}{\lambda}\right) \qquad \forall (\lambda, 0) \in \Re \setminus \{0\} \times \Re.$$

It is easy to see that representation D is commutative

(29) $$\begin{aligned} D_{(\lambda,0)} \circ D_{(\lambda',0)} &= D_{(\lambda,0)\circ(\lambda',0)} \\ &= D_{(\lambda\cdot\lambda',0)} \end{aligned}$$

and

(30) $$\begin{aligned} D_{(\lambda',0)} \circ D_{(\lambda,0)} &= D_{((\lambda',0)\circ(\lambda,0))} \\ &= D_{(\lambda'\cdot\lambda,0)} \end{aligned}$$

It easy to see that the dilation representation is unitary

$$\begin{aligned}\langle D_{(\lambda,0)}h, D_{(\lambda,0)}g\rangle &= \int_{-\infty}^{\infty} \mid \lambda \mid^{-\frac{1}{2}} \cdot h\left(\frac{t}{\lambda}\right) \cdot \mid \lambda \mid^{-\frac{1}{2}} \cdot g^*\left(\frac{t}{\lambda}\right) dt \\ &= \int_{-\infty}^{\infty} h(t) \cdot g^*(t) dt \\ &= \langle h, g\rangle \end{aligned} \tag{31}$$

Consider the product of dilation and translation representations in the set of unitary representations.

$$DT_{\lambda,s)} \triangleq D_{(\lambda,0)} \circ T_{(1,s)} \tag{32}$$

$$\begin{aligned} DT_{(\lambda,s)} &= D_{(\lambda,0)} \circ \left(T_{(1,s)}h\right)(t) \\ &= D_{(\lambda,0)}h(t-s) \\ &= \mid \lambda \mid^{-\frac{1}{2}} \cdot h\left(\frac{t-s}{\lambda}\right) \end{aligned} \tag{33}$$

The translation and the dilation representations are not commutative

$$\begin{aligned} \left(T_{(1,s)} \circ D_{(\lambda,0)}\right) h &= \mid \lambda \mid^{-\frac{1}{2}} \cdot h\left(\frac{t-\lambda s}{\lambda}\right) \\ &= D_{(\lambda,0)} \circ T_{(1,\lambda s)} \end{aligned} \tag{34}$$

We will introduce a different notation for this unitary representation and call it affine coherent states. This notion has been borrowed from quantum mechanics and is used in the wavelet theory[AK68],[ID88]

$$h^{(\lambda,s)}(t) = \mid \lambda \mid^{-\frac{1}{2}} \cdot h\left(\frac{t-s}{\lambda}\right) \tag{35}$$

The inner product of the affine coherent states and any $g() \in L^2(\Re)$ is the wide -band ambiguity function that we have already defined

$$\begin{aligned} WA_{g,h)}(\lambda,\tau) &= \langle g, h\rangle \\ &= \mid \lambda \mid^{-\frac{1}{2}} \cdot \int_{-\infty}^{\infty} g(t) \cdot h^*\left(\frac{t-\tau}{\lambda}\right) dt \end{aligned} \tag{36}$$

In the language of unitary representation theory this is called the matrix coefficient.

2.4. Wide-band ambiguity function properties.

In this subsection we will summarize the wavelet expansion and we will show the relation to the wide-band ambiguity function. We will use the notations as in [ID88]. The wavelet expansion is connected to the representation of the affine group. Define points of the grid from constants λ_0 , τ_0 > 0, $\lambda_0 \neq 1$. The points of the grid are given by:

$$\lambda_m = \lambda_0^m$$
$$\tau_{mn} = n \cdot \tau_0 \cdot \lambda_0^m \; ; m, n \in Z$$

Define a function on a grid by discrete translation and dilution operators

$$h_{mn}(t) = D_{(\lambda_m, 0)} \circ T_{(1, \tau_{mn})} h(t)$$

(37) $$= \lambda_0^{-\frac{m}{2}} h\left(\lambda_0^{-m} t - n\tau_0\right)$$

We will say $h_{mn}(t)$ is a frame, if there exits constants $0 < A \leq B$ such that for any signal $s(t) \in L^2(\Re)$ the following hold

(38) $$A \, || \, s() \, ||^2 \leq \sum_m \sum_n | \langle h_{mn}, s \rangle |^2 \leq B \, || \, s() \, ||^2$$

Estimates for the bounds A, B are given in[ID88]. If h_{mn} such defined is not a frame, then we can construct a frame from h_{mn} by the following procedure [ID88] Define a operator

(39) $$P = I - \frac{2 \sum_m \sum_n h_m n \langle h_{mn}, \cdot \rangle}{A + B}$$

where

$$I - identity\ operator,$$
$$A, B - are\ constants$$

Then

(40) $$\tilde{h}_{0n} = \left(\frac{2}{A+B} \sum_k P^k \right) h_{0n}$$

and the frame $\tilde{h}_{mn}$ is obtained from $\tilde{h}_{0n}$ by applying the dilution operator

$$\tilde{h}_{mn} = D_{(\lambda_m, 0)} \tilde{h}_{0n}(t)$$

(41) $$= \lambda_0^{-\frac{m}{2}} \tilde{h}_{0n}\left(\lambda_0^{-m} t\right)$$

Then the signal $s(t)$ can be expanded

(42) $$s(t) = \sum_m \sum_n \langle h_{mn}, s \rangle \tilde{h}_{mn}$$

The inner product $\langle h_{mn}, s \rangle$ is a wide-band cross-ambiguity function as we have defined previously and sampled at the points of the grid that have been defined.

If $\frac{B}{A}$ close enough to one ,then the expansion can be approximated [ID88]

(43) $$s(t) = \frac{2}{A+B} \sum_m \sum_n \langle h_{mn}, s \rangle h_{mn}$$

In the next section we will present an algorithm to compute the wide-band ambiguity function

3. Wavelets and the wide-band ambiguity function. In this section we will present properties of the wide-band ambiguity function which follow directly from the affine group representation properties.

[P.1]

$$WA_{(g,h)}(1,\tau) = \int_{-\infty}^{\infty} g(t) \cdot h^*(t-\tau)dt$$

$$= \mathcal{R}_{g,h}(\tau) \tag{44}$$

where $\mathcal{R}_{g,h}(\tau)$ -is the cross correlation of two signals.

[P.2]

$$WA_{g,g}(1,0) = \int_{-\infty}^{\infty} | g(t) |^2 \, dt$$

$$= 1. \tag{45}$$

[P.3]

$$WA_{g,h}(\frac{1}{\lambda},\tau) \leq | WA_{g,h}(1,0) |$$

$$= 1. \tag{46}$$

[P.4]

$$WA_{g,h}(\frac{1}{\lambda},\tau) = \sqrt{\frac{1}{\lambda}} \cdot \int_{-\infty}^{\infty} g(t) \cdot h^*(\frac{t-\tau}{\lambda})dt$$

$$= \sqrt{\frac{1}{\lambda}} \cdot \int_{-\infty}^{\infty} G(\frac{f}{\lambda}) \cdot H^*(f)e^{2\pi i f \tau} df$$

$$= \sqrt{\frac{1}{\lambda}} \cdot \int_{-\infty}^{\infty} G(f) \cdot H^*(f\lambda)e^{2\pi i f \frac{1}{\lambda}\tau} df \tag{47}$$

where $G(f), H(f)$ are the Fourier transform of $g(t), h(t)$.

[P.5] The unitary representation corresponding to the inverse $(\lambda,\tau)^{-1} \in \Re \setminus \{0\} \times \Re$ of the element (λ,τ) is

$$h^{((\lambda,\tau)^{-1})}(t) = h^{(\frac{1}{\lambda},-\frac{\tau}{\lambda})}(t)$$

$$= \sqrt{\lambda}h(\lambda t+\tau) \tag{48}$$

The matrix coefficient is then

$$WA_{g,h}(\frac{1}{\lambda},-\frac{\tau}{\lambda}) = \langle g, h^{(\frac{1}{\lambda},-\frac{\tau}{\lambda})}\rangle$$

$$= \langle h^{(\lambda,\tau)}, g\rangle$$

$$= WA^*_{g,h}(\lambda,\tau) \tag{49}$$

or

$$| WA_{g,h}(\frac{1}{\lambda}, -\frac{\tau}{\lambda}) |^2 = | WA_{g,h}(\lambda, \tau) |^2 \tag{50}$$

[P.6]

$$\begin{aligned}
\int_{-\infty}^{\infty} WA_{g,h}(\frac{1}{\lambda}, \tau) e^{-2\pi\nu\tau} d\tau &= \sqrt{\frac{1}{\lambda}} \cdot \int_{-\infty}^{\infty} \int_{-\infty}^{\infty} G(\frac{f}{\lambda}) \cdot H^*(f) d\tau df \\
&= \sqrt{\frac{1}{\lambda}} \cdot G(\frac{\nu}{\lambda}) \cdot H^*(\nu) \\
&= \sqrt{\frac{1}{\lambda}} \cdot G(\nu) \cdot H^*(\nu\lambda) \qquad (51)
\end{aligned}$$

4. Computation of the wide-band ambiguity function.

We have seen in previous paragraphs the use of wide-band ambiguity function in wavelet expansion and in maximum likelihood estimator of strech/compress parameter a and delay parameter τ in signals arising in sonar. The maximum likelihood estimator is

$$\max_{a,\tau} | WA_{h,g}(a, \tau) |^2$$

where $h()$ is a received signal, $g()$ is a reference signal. In passive sonar case $h(t), g(t)$ may be the received signals at different locations and the locations and the parameters a, τ are relative compress/strech and relative delay factors. From the computation point of view it is the same problem. In the wavelet case the parameters are specified by the discrete grid. We will demonstrate the algorithm for the sonar signal case.

Let $[\tau_{min}, \tau_{max}]$ be reasonable delay interval and $[a_{min}, a_{max}]$ strech/compress factor interval. From the desired accuracy choose the quantization level:

$$\tau_i = \frac{\tau_{max} - \tau_{min}}{N} \cdot i \qquad i = 1, 2 \cdots N$$

and

$$a_j = \frac{a_{max} - a_{min}}{M} \cdot j \qquad j = 1, 2 \cdots M$$

Thus we want to compute for each a_j

$$\max_{\tau} | WA_{h,g}(a_j, \tau) |^2,$$

The above computation assumes that there is only one target in the estimation range and that the noise level is such that after matched correlation processing it is possible, with high certainty, to determine the existence of the signal and locate its maxima with prespecified accuracy. Property P.4 of the wide-band ambiguity

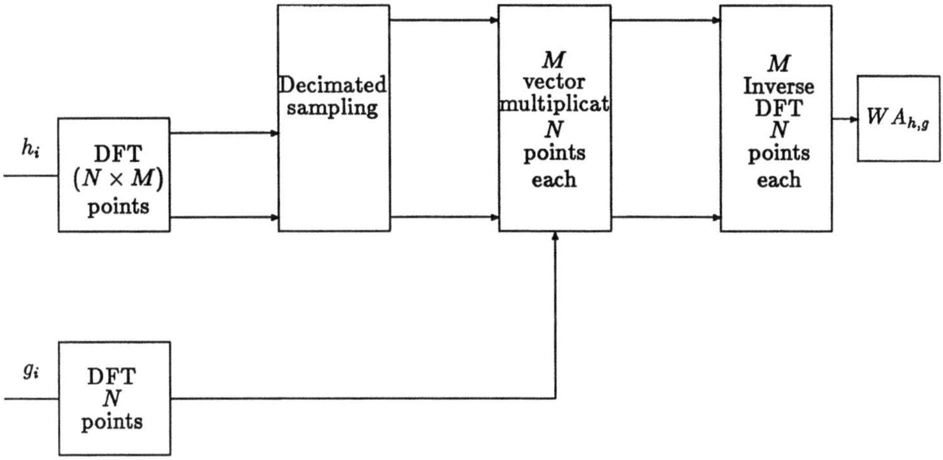

Figure 2: The computational flowchart for wide-band ambiguity function

function will be used for computation. The computational flowchart is presented in the figure 2 above.

The most computationally intensive is the second stage. It includes N-point DFT and another $N \times M$ DFT. From these points M sequencies of length N generated corresponding to different strech/compress factors. The algorithm to compute the wide-band ambiguity function is summarized below:

(1) Precompute for the reference signal its scaled DFT -$N \times M$ points.

(2) Compute for the received signal N point DFT.

(3) Perform M decimated sampling and M vector multiplication of length N each.

(4) Compute M inverse DFT on N points each.

(5) Find maximum on the array of $N \times M$ numbers.

As it is seen from the flowchart the algorithm can be implemented in parallel /pipelined manner. If the narrow-band approximation is valid then more efficient algorithm can be use[AT88], [AG88].

5. Conclusion. The computation of the wide-band ambiguity function is computationally intensive process. It includes the computation of a very long discrete Fourier transform of the reference signal. Then the transformed signal is sampled with the decimated rate determined by the strech/compress Doppler factor. The discrete Fourier transform of the received signal is then vector multiplied by the decimated reference signal. Ultimately the inverse discrete Fourier transform is taken of the product. The computational algorithm is being mapped onto the tree architecture of signal processors .

In the "wavelet case" the computation result is used as a matrix coefficient in the likelihood estimator of the delay and Doppler factors.

REFERENCES

[SW66] DAVID A. SWICK, *An Ambiguity function independent of assumption about bandwidth and carrier frequency*, NRL Report 6471 , Dec. 15 1966..

[SW69] DAVID A. SWICK, *An Review of Wide-band ambiguity functions*, NRL Report 6994 ,Washington , D.C. Dec. 1969.

[ID88] INGRID DAUBECHIES, *The wavelet transform time-frequency localization and signal analysis*, submitted for publication.

[AK68] ASLAKSEN E.W. AND KLAUDER J.R., *Unitary representation of the affine group*, J.Math.Physics, 9 (1968), pp. 208–211.

[AT88] L. AUSLANDER, R. TOLIMIERI, *Computing decimated finite cross-ambiguity functions*, IEEE Transactions on Acoustics, Speech, and Signal Processing, vol. 36, No (March 1988), pp. 359–363.

[AG88] L. AUSLANDER, I.GERTNER, *Computing the ambiguity function on various domains*, submitted to publication.

ON FINITE GABOR EXPANSION OF SIGNALS*

L. AUSLANDER† AND R. TOLIMIERI‡

Abstract. An algorithm is produced computing the coefficients of the finite Gabor expansion from sample values of the resulting signal. The Weil-Zac transform is used to place the cross-ambiguity function at the center stage in the computation where the algorithm found in the authors 'Computing Decimated Finite Cross-ambiguity Functions' can be applied.

1. Introduction. Time-frequency analysis can be found in many branches of mathematics, physics and signal processing. We begin by giving a brief account of several fundamental results which have helped from our understanding of this analysis.

In 1975, we published 'Abelian Harmonic analysis, Theta Functions and Function algebra on a Nilmanifolds' [1]. An important part of this work was devoted to describing functions on a compact Nilmanifold coming from the 3-dimensional Heisenberg group. As an application of this theory, we gave conditions under which a signal $f \in L^2(\mathfrak{R})$ could be expanded as

$$f = \sum_m \sum_n a_{mn} g_{mn}, \quad a_{mn} \in C \tag{1}$$

where

$$g_{mn}(t) = e^{-\pi(t-n)^2} \cdot e^{2\pi imt}, \quad m, n \in Z. \tag{2}$$

In a fundamental work, published in 1946, Gabor [2] suggested such expansions for the simultaneous analysis of signals in time and frequency. Although appearing nearly thirty years before, we were not aware of Gabor's achievement. A transform introduced by A. Weil [3] and studied by J. Brezen [4] plays a crucial part in our work. For functions $f \in L^2(\mathfrak{R})$, the Weil transform of f is defined by the formula

$$H(f)(x, y, z) = e^{2\pi i(z + \frac{xy}{2})} \sum_m f(y + m) e^{2\pi imx}. \tag{3}$$

This transform permits results in $L^2(\mathfrak{R})$ to be translated to results about functions on the compact manifold and is in fact an intertwining operator between two unitary representations of the Heisenberg group. In [1], this transform was systematically studied with special emphasis on its intimate role in the Poisson summation formula and the Plancheral Theorem. The role of this transform in certain number theory applications was pursued in [5] and [6]. A major result proved in 1975 states that if $H(f)(x, y, z)$ is continuous then it must vanish at some point. As shown in this

*This work supported by DARPA ACMP # 770997 and in part by the IMA with funds provided by NSF

†Distinguished Professor Mathematics, CUNY Graduate Center, NYC 10036

‡Professor Electrical Engineering, City College of New York, NYC 10031

work, this places a important restraint on the usefulness of the expansion (1) and in particular on the computing of the coefficients of this expansion from sample values of f. The algorithm presented below carries out this computation. The main point is of course that the functions $g_{mn}, m, n \in Z$, need not constitute an orthonormal basis.

The Weil transform was formally introduced into mixed time-frequency analysis by J. Zac in two works [7] and [8] in 1967. During the last ten years the implications of the transform to time-frequency analysis has been studied in great detail in many works including [9], [10], [11] and [12]. An excellent summary of how the transform relates time-frequency computations appears in A.J.E.M. Janssen's [13].

A part of the current activity involves the generalization of the expansion (1) to arbitrary reference signals $g(t)$ rather than the restricted to $g(t) = e^{-\pi t^2}$. In this general setting, the Gabor expansion (1) is related to the Short-time Fourier transform

$$S_F(u,v) = \int F(t)g(t-u)e^{-2\pi ivt}dt, \tag{4}$$

which is used to analyze the local frequency characteristics of a slowly time varying signal $f(t)$ relative to some window $g(t)$. A variety of time-varying signal processing operations such as time-varying filtering, signal separation, dynamic range and bandwidth compression, general frequency distortion correction and time-scale modification are handled by (4). The same formula

$$A(f,g)(u,v) = \int f(t)g(t-u)e^{-2\pi ivt}dt, \tag{5}$$

defines the cross-ambiguity function of signals f and g which is a complex 'bilinear' signal transform that is used to synthesize radar signals with desired accuracy properties. The ambiguity function is also a matched filter operation on time and frequency.

In 1980, we began a study of the properties and applications of the cross-ambiguity function (5). In [14], we studied the relationship of (5) to the unitary representation theory of the Heisenberg group. Several other authors were also active in this direction. We suggest as reference W. Schempp's text 'Harmonic analysis on the Heisenberg nilpotent Lie group' [15]. An important result in [14] gives conditions under which the functions

$$g_{mn}(t) \equiv g(t-n)e^{2\pi imt} \tag{6}$$

determine an orthonormal basis; namely

$$|H(g)(x,y,z)| = 1, \quad \textit{almost everywhere.} \tag{7}$$

This result was used to exhibit orthonormal basis of special kind related to theta function theory.

In this work, we will relate the Gabor representation to the cross-ambiguity function and construct an algorithm which computes the coefficients of the Gabor representation using sample values of the signal. We are motivated by the classical sampling theory which for completeness and to introduce our point of view will be introduced in the next section. A major part of the computation depends on an algorithm designed in [16] 'Computing Decimated Finite Cross-Ambiguity Functions'. The main result is that up to errors discussed in the body of section 3, computing Gabor coefficients from M^2 sample points of the signal requires M, M-point Fourier Transform's and one 2-dimensional $M \times M$ point Fourier transforms and $2M^2$ multiplications.

For the purpose of algorithm construction we will replace the Weil transform (3) by the Zac transform

$$w(f)(x,y) = \sum_n f(x+n)e^{2\pi iny} \tag{8}$$

The fundamental formula proved in (10) of section 3 is

$$\omega(f)(x,y) = w(g)(x,y)\sum_{m,n} a_{mn}e^{2\pi i(mx+ny)} \tag{9}$$

where the Gabor representation is

$$f = \sum_{m,n} a_{mn}g_{mn}. \tag{10}$$

As pointed out above, $\omega(g)(x,y)$ must vanish somewhere if g is sufficiently smooth which restricts the direct application of formula (9). However, by direct sampling of (9) say by the points $\frac{j}{M}$, $y = \frac{k}{M}$, $0 \le j$, $K < M$, the essential requirement is that the values

$$\omega(g)(\frac{j}{M}, \frac{k}{M}), \quad 0 \le j,k < M, \tag{11}$$

never vanish which is easy to arrange. We can then invert (9) to compute the 'Gabor coefficients' from sample values of the cross-ambiguity function, $A(f,g)(m,n)$, $0 \le m,n < M$. The algorithm in [16] computes these samples from samples of f to produce the final algorithm in (27) of section 3. It is important to observe at that time that several computations combine to arrive at the computational count stated above.

2. Review of Samping Theory for the Fourier Transform. The classical theory of sampling depends upon the periodization mapping

$$P(f)(x) = \sum_{m\in Z} f(x+m), \quad -\infty < x < \infty. \tag{12}$$

For simplicity, we assume that the signal $f(x), -\infty < x < \infty$, is in the Schwartz space. In this case $P(f)(x)$ is an infinitely differentiable periodic function:

$$P(f)(x+1) = P(f)(x), \quad -\infty < x < \infty. \tag{13}$$

The main result is the Poisson summation formula:

$$(14) \qquad P(f)(x) = \sum_{n \in Z} \hat{f}(n) e^{2\pi i n x}, \quad -\infty < x < \infty,$$

where $\hat{f}$ denotes the Fourier transform of f and the series on the right-hand side of (3) uniformly converges. By (3), values of the Fourier transform at the integral points can be determined from the periodization $P(f)$.

The Dilation operator

$$(15) \qquad (D_\lambda)(x) = \lambda^{\frac{1}{2}} f(\lambda x), \quad \lambda > 0,$$

can be used to extend these results. First

$$(16) \qquad \widehat{D_\lambda f}(y) = \lambda^{-\frac{1}{2}} \hat{f}(\frac{y}{\lambda}).$$

Applying the Poisson Summation formula to $D_\lambda f$

$$(17) \qquad P(D_\lambda f)(x) = \lambda^{-\frac{1}{2}} \sum_{n \in Z} \hat{f}(\frac{n}{\lambda}) e^{2\pi i n x}, \quad \lambda > 0.$$

The values $\hat{f}(\frac{n}{\lambda})$, $n \in Z$, are determined by the periodization $P(D_\lambda f)$.

In general, the periodization of a signal results in loss of information. However, if f vanishes outside the interval $[0, 1)$ then

$$(18) \qquad P(f)(x) = f(x), \quad 0 \leq x < 1,$$

and we have

$$(19) \qquad f(x) = \sum_{n \in Z} \hat{f}(n) e^{2\pi i n x}, \quad 0 \leq x < 1,$$

showing that the sample values $\hat{f}(n)$, $n \in Z$, completely determine f over $[0, 1)$. More generally, if f vanishes outside the interval $[0, \lambda)$, $\lambda > 0$, then the dilation $D_\lambda f$ vanishes outside the interval $[0, 1)$ and by (8)

$$(20) \qquad (D_\lambda f)(x) = \lambda^{-\frac{1}{2}} \sum_{n \in Z} \hat{f}(\frac{n}{\lambda}) e^{2\pi i n x}, \quad \lambda > 0;\ 0 \leq x < 1.$$

We can rewrite this as

$$(21) \qquad f(x) = \lambda^{-1/2} \sum_{n \in Z} \hat{f}(\frac{n}{\lambda}) e^{2\pi i \frac{n}{\lambda} x}, \quad \lambda > 0;\ 0 \leq x < \lambda,$$

from which we see that the sample values of the Fourier transform on the set $\frac{1}{\lambda} Z$ completely determine f over $[0, \lambda)$.

Since the Fourier transform is an isomorphism of order 4 on the Schwartz space, all the results above can be applied to the Fourier transform of a signal. The standard form Nyquist sampling theorem applies to signals f having finite interval supported Fourier transform.

Suppose f vanishes outside the interval $[0,1)$. Applying (14)

$$f(\frac{a}{N}) = \sum_{n\in Z} \hat{f}(n)e^{2\pi i\frac{na}{N}}, \quad 0 \le a < N. \tag{22}$$

Setting $n = b + cN$, $0 \le b < N$, $c \in Z$, we have

$$f(\frac{a}{N}) = \sum_{b=0}^{N-1} \left(\sum_{c\in Z} \hat{f}(b+cN)\right) e^{2\pi i\frac{ab}{N}}, \quad 0 \le a < N \tag{23}$$

A discrete periodization appears in (23). In general, if $g(n)$, $n \in Z$, is a sequence of data, we set

$$P_N(g)(b) = \sum_{c\in Z} g(b+cN), \quad b \in Z. \tag{24}$$

We assume that the series on the right-hand side of (24) converges. Then $P_N(g)$ is a periodic sequence

$$P_N(g)(b+N) = P_N(g)(b), \quad b \in Z, \tag{25}$$

and can be viewed as N-point data. Formula (23) can now be written as the N-point Fourier transform of the discrete periodization *mod* N of the sample values of $\hat{f}$.

3. Sampling Theory for Time-Frequency Analysis. The Weil-Zak transform of a signal $f(t)$, $t \in \mathfrak{R}$,

$$\omega(f)(x,y) = \sum_{n\in Z} f(x+n)e^{2\pi iny}, \tag{26}$$

can be used to relate several time-frequency computations. We will again assume, for simplicity that f is in the Schwartz space with the consequence that $w(f)$ is infinitely differentiable. Setting $y = 0$ in (26), we reduce to the periodization. In fact, the formulas

$$w(f)(x,y+1) = w(f)(x,y), \tag{27}$$

$$w(f)(x+1,y) = e^{-2\pi iy}w(f)(x,y), \tag{28}$$

show that $w(f)$ is completely determined by its values on the unit square: $0 \le x,y < 1$. In contrast to the periodization, no information is lost in this process. One can show that the Weil-Zac transform extends to an isometry of $L^2(\mathfrak{R})$. We can recover f from $w(f)$ by the inversion formula

$$f(x+n) = \int_0^1 w(f)(x,y)e^{-2\pi iny}dy, \quad 0 \le x < 1. \tag{4}$$

An indication of the crucial role of the Weil-Zac transform is given by the formula

$$w(\hat{f})(x,y) = e^{-2\pi ixy} w(f)(y,-x). \tag{30}$$

Applying (29), we see that 90°-rotation of $w(f)(x,y)$ determines $\hat{f}$.

Consider the problem of computing the coefficients of the Gabor representation of a signal $f(t), t \in \mathfrak{R}$,

$$f = \sum_{m\in Z} \sum_{n\in Z} a_{mn} \cdot g_{mn}, \tag{31}$$

where

$$g_{mn}(t) = g(t-n) \cdot e^{2\pi imt}, \quad m,n \in Z. \tag{32}$$

Applying (8),

$$w(f)(x,y) = \sum_{m\in Z} \sum_{n\in Z} a_{mn} \cdot w(g_{mn}). \tag{33}$$

Applying the formula

$$w(g_{mn}(x,y)) = w(g)(x,y) e^{2\pi i(mx+ny)}, \tag{34}$$

we can rewrite (33) as

$$w(f)(x,y) = w(g)(x,y) \sum_{m\in Z} \sum_{n\in Z} a_{mn} e^{2\pi i(mx+ny)}. \tag{35}$$

Multiplying both sides of (35) by $w(g)^*(x,y)$, we have the fundamental formula

$$w(f)(x,y) \cdot w^*(g)(x,y) = |w(g)(x,y)|^2 \sum_{m\in Z} \sum_{n\in Z} a_{mn} e^{2\pi i(mx+ny)}. \tag{36}$$

It is important to observe that (36) is a formula of doubly periodic functions. To see this use (27) and (28). First we can write

$$w(f)(x,y) \cdot w^*(g)(x,y) = \sum_{m\in Z} \sum_{n\in Z} b_{mn} e^{2\pi i(mx+ny)}. \tag{37}$$

Under the conditions imposed that f and g are Schwartz functions the Fourier series in (37) converges absolutely.

The coefficients are given by

$$b_{mn} = \langle w(f)(x,y) w^*(g)(x,y), e^{2\pi i(mx+ny)} \rangle, \tag{38}$$

the inner product taken over the unit square. Then

$$b_{mn} = \int_0^1 \int_0^1 w(f)(x,y)w^*(g)(x,y)e^{-2\pi i(mx+ny)}dxdy, \tag{39}$$

which by (34) becomes

$$b_{mn} = \int_0^1 \int_0^1 w(f)(x,y)w^*(g_{mn})(x,y)dxdy \tag{40}$$

$$b_{mn} = \int_{-\infty}^{\infty} f(t)g^*_{mn}(t)dt. \tag{41}$$

It follows by the definition of the cross-ambiguity function given in the introduction that the coefficients are sample values of the cross-ambiguity function of f and g:

$$b_{mn} = A(f,g)(m,n). \tag{42}$$

By (41), formula (36) relates the Gabor coefficients to sample values of the cross-ambiguity function. Since $|\omega(g)(x,y)|^2$ is also doubly periodic, it is natural to write

$$|w(g)(x,y)|^2 = \sum_{m\in Z}\sum_{n\in Z} c_{mn}e^{2\pi i(mx+ny)}; \tag{43}$$

the series in (43) is absolutely converging. We can then write the sample values of the cross-ambiguity function as the 2-dimensional convolution of the coefficients of the Gabor representation with the coefficients of the expansion (43). As is standard, we assume these better coefficients are computed in a pre-computation stage. With above notations

$$b = c * a. \tag{44}$$

The inverse problem can not be obtained as simply since, as has been pointed out, $w(g)(x,y)$ must vanish in the unit square and

$$\frac{1}{|w(g)(x,y)|^2}, \tag{45}$$

although doubly periodic does not have a well-behaved Fourier series. We can still use (36) but our approach must be modified. Set

$$x = \frac{j}{M}, \quad y = \frac{k}{M}, \quad 0 \le j,k < M, \tag{46}$$

in (36). As we saw in formula (23) of the previous section, sampling a discrete Fourier series results in a finite Fourier series of a discrete periodization. We have

$$\sum_{m=0}^{M-1}\sum_{n=0}^{M-1} B_{mn}e^{2\pi i\frac{mj+nk}{M}} = |w(g)\left(\frac{j}{M},\frac{k}{M}\right)|^2 \sum_{m=0}^{M-1}\sum_{n=0}^{M-1} A_{mn}e^{2\pi i\frac{mj+nk}{M}}, \tag{47}$$

where

$$A_{mn} = \sum_{j\in Z}\sum_{k\in Z} a_{m+jM,n+kM}, \qquad B_{mn} = \sum_{j\in Z}\sum_{k\in Z} b_{m+jM,n+kM}. \tag{48}$$

Form the M^2-dimensional vector A and B by lexicographically ordering the coefficients A_{mn}; $0 \le m,n < M$, and B_{mn}; $0 \le m,n < M$ and form the diagonal $M^2 \times M^2$ matrix G by lexicographically ordering $|w(g)(\frac{j}{M},\frac{k}{M})|^2$, $0 \le j,k < M$ down the diagonal. Then we can rewrite (36) as

$$(F(M)\otimes F(M))B = G\cdot(F(M)\otimes F(M))A. \tag{49}$$

ASSUMPTION 1.

$$w(g)\left(\frac{j}{M},\frac{k}{M}\right) \neq 0, \quad 0 \le j,k < M. \tag{50}$$

Then G is invertible and we can write the Gabor coefficients in terms of the sample values of the cross-ambiguity function:

$$A = (F(M)\otimes F(M))^{-1}\cdot G^{-1}\cdot(F(M)\otimes F(M))B. \tag{51}$$

Formula (40) is exact but the vectors A and B are formed by periodizing the Gabor coefficients and the samples of the cross-ambiguity function, respectively. In general, as in the classical case, replacing periodized coefficients by actual coefficients is a source of error. We can side-step the error coming from periodizing Gabor coefficients by assuming

ASSUMPTION 2.

$$f = \sum_{m=0}^{M-1}\sum_{n=0}^{M-1} a_{mn}\cdot g_{mn} \tag{52}$$

Then

$$A_{mn} = a_{mn}, \quad 0 \le m,n < M, \tag{53}$$

and formula (40) computes exactly computes the Gabor coefficients from the cross-ambiguity function samples. We still, of course, must periodize these samples to form B.

3.1 Algorithm to compute the Gabor coefficients. Formula (40) computes Gabor coefficients from cross-ambiguity function samples. We will go one step further and describe an algorithm computing the Gabor coefficients from sample values of the signal. To do so we must give an effective means of computing the vector B from these sample of the signal. This is precisely the content of [AT16].

First we assume that

ASSUMPTION 3. *f vanishes outside* $[0, M)$

Then by definition

$$b_{mn} = \int_0^M f(t)g^*(t-n)e^{-2\pi imt}dt. \tag{54}$$

We approximate the integral in (43) by forming the Riemann sum corresponding to the points, $a + \frac{k}{M}$, $0 \leq a, k < M$:

$$b_{mn} \cong \frac{1}{M} \sum_{k=0}^{M-1} \sum_{a=0}^{M-1} f(a + \frac{k}{M})g^*(a + \frac{k}{M} - n)e^{-2\pi im\frac{a+\frac{k}{M}}{M}}. \tag{55}$$

Setting

$$\alpha = M(a + \frac{k}{M})$$

and

$$F(\alpha) = f(a + \frac{k}{M}),$$

we can rewrite (44) as

$$b_{mn} \cong \frac{1}{M} \sum_{\alpha=0}^{M^2-1} F(\alpha)G^*(\alpha + n \cdot M)e^{-2\pi i\frac{m\alpha}{M}} \tag{56}$$

By the algorithm described in paper [AT16], (45) can be computed by the following steps

a. Compute

$$w'(F)(k,b) = \frac{1}{M} \sum_{a=0}^{M-1} F(k + M \cdot a)e^{2\pi i\frac{ab}{M}}, \quad 0 \leq k, b < M,$$

This stage requires M, M-point one-dimensional Fourier transforme.

b. Precompute

$$w(G)(k,b) = \sum_{a=0}^{M-1} G(k + M \cdot a)e^{2\pi i\frac{ab}{M}}, \quad 0 \leq k, b < M.$$

c. Compute

$$K(k,b) = w'(F)(k,b) \cdot w^*(G)(k,b), \quad 0 \leq k, b < M.$$

d. Compute the *two*-dimensional $M \times M$ Fourier transform

$$b_{mn} \cong \sum_{k=0}^{M-1} \sum_{a=0}^{M-1} K(k,b)e^{+2\pi i\frac{mk-nb}{M}}, \quad 0 \leq m, n < M. \tag{57}$$

We assume

ASSUMPTION 4.

$$B_{mn} = b_{mn}, \quad 0 \leq m, n < M.$$

The first step in carrying out factorization (40) is the two-dimensional $M \times M$ Fourier transform of the two-dimensional array formed from the coefficients b_{mn}. Combining this step with step "*d*" of the algorithm, we can find a permutation matrix P satisfying:

$$(F(M) \otimes F(M))B = P \cdot K, \tag{58}$$

where K is the M^2-dimensional vector formed by lexicographically ordering the coefficients $K(k, b)$, $0 \leq k, b < M$. Placing (47) in (40),

$$A = (F(M) \otimes F(M)) \cdot G^{-1} \cdot P \cdot K. \tag{59}$$

The last assumption 4 can be related to properties of the reference function g. Return to formula (43).

ASSUMPTION 5. *If*

$$g(t) \text{ vanishes outside } [0, 1), \tag{60}$$

then

$$b_{mn} = 0 \text{ if } n < 0 \text{ or } n \geq M.$$

ASSUMPTION 6. *If*

$$\hat{g}(t) \text{ vanishes outside } [0, 1), \tag{61}$$

then

$$b_{mn} = 0 \text{ for } m < 0 \text{ or } m \geq M.$$

Although, both assumptions, can't simultaneously hold, in practical applications they can effectively hold.

Conclusion. Under assumptions A1 - A2, the Gabor coefficients A can be computed by formula (48) using M, M-point one-dimensional Fourier transforms's, one two-dimensional $M \times M$-point Fourier transform and $2M^2$ multiplications. The Riemann sum approximation is a source of error. Assumption 1 is essential for the algorithm to proceed. If the remaining assumptions can be violated, error will result.

[1] L. AUSLANDER AND R. TOLIMIERI, *Abelian Harmonic analysis, theta functions and function algebra on nilmanifolds*, Springer-Verlag (1975).
[2] D. GABOR, *Theory of communications*, J.IEE (London), 93, pp. 429–457, Nov. 1946.
[3] A. WEIL, *Sur certains groupes d'operatours unitaires*, Acta Math., 111 (1964), pp. 143-211.
[4] J. BREZIN, *Function theory on metabelian solvmanifolds*, J. Func. Anal, 10 (1972), pp. 33–51.
[5] R. TOLMIERI, *The multiplicity problem for 4-dimensional solvmanifolds*, Bull. American Math.Soc., 83 (1977), pp. 365–366.
[6] L. AUSLANDER AND R. TOLMIERI, *Algebraic structures for* $\oplus \sum_{n \geq 1} L^2(2/n)$ *compatible with the finite Fourier transform*, Trans AMS No. 244 (1978), pp. 263–272.
[7] J. ZAC, Phys. Rev. Lett., 19 (1967), p. 1385.
[8] J. ZAC, Phys. Rev, 168 (1968), p. 686.
[9] M.J. BASTIANS, *Gabor's expansion of a signal onto Gaussian elementary signals*, IEEE Proc., 68 (1980).
[10] M.J. BASTIANS, *The expansion of an optical signal into a discrete set of Gaussian beams*, Optik, 57, No. 1 (1980), pp. 95–102.
[11] A.J.E.M. JANSSEN. *Gabor representation of generalized functions*, J. of Mathematical Analysis and Applications, 83 (1981), pp. 377-394.
[12] A.J.E.M. JANSSEN, *Weighted Wigner distributions vanishing on lattices*, J. of Mathematical Analysis and Applications, 80, No. 1 (1981), pp. 156–167.
[13] A.J.E.M. JANSSEN, *The Zak transform: a signal transform for sampled time-continuous signals*, Philips J. Res., Vol. 43 (1988), pp. 23–69.
[14] L. AUSLANDER AND R. TOLIMIERI, *Radar ambiguity function and group theory*, SIAM J. Math. Anal, 16, No. 3 (1985), pp. 577–601.
[15] W. SCHEMPP, *Harmonic analysis on the Heisenberg nilpotent Lie group*, Longman Scientific and Technal, Pitman Research Notes in Mathematical Sciences, 147, Harlow, Essex, UK (1986).
[16] L. AUSLANDER AND R. TOLIMIERI, *Computing decimated cross-ambiguity functions*, IEEE Trans. on Acoustics, Speech, and Signal Processing, 36, No. 3 (1988), pp. 359–364.

TWO DIMENSIONAL FFT ALGORITHMS ON DATA ADMITTING 90°-ROTATIONAL SYMMETRY*

GERARD BRICOGNE† AND RICHARD TOLIMIERI‡

Abstract. A multiplicative FFT algorithm is designed to operate on redundant data invariant under 90°-rotational symmetry with a savings in computational count: approximately one-quarter computations relative to direct approach. The goal is to produce an algorithm which makes use of the redundancy without losing completely the advantages afforded by the Cooley-Tukey algorithm and having a structured data flow for implementation simplicity and speed.

Introduction. In this talk, we present an algorithm computing the 2-dimensional finite Fourier Transform on data admitting 90°-rotational symmetry. Suppose $f(a_0, a_1)$, $a_0, a_1 \in Z/n$ denotes a 2-dimensional array of data satisfying

$$(1) \qquad F(a_0, a_1) = f(-a_1, a_0), \ a_0, a_1 \in Z/n.$$

Since the data is redundant, it seems reasonable that the 2-dimensional $n \times n$ F.T. of the data can make use of the redundancy to reduce the computational burden. However, the Cooley-Tukey FFT algorithm [1] operates on the complete n^2 data points. Except in certain cases where data redundancy can be interpreted as redundancy along a coordinate axis the $C-T$ algorithm cannot be directly applied. Several important works have made use of this type of redundancy including Ten Eyck [2].

Our approach will follow more closely the multiplicative method introduced by Rader [3] and generalized by Winograd [4]. The symmetry is incorporated into a ring-structure on the indexing set $Z/n \times Z/n$ and essentially leads to a re-coordinization which naturally fits the symmetry. A variant of Winograd worked out in [5] and [6] and highly dependent on tensor product representation will be the bases of attack.

Although we restrict our attention to data of a prime number of points admitting 90°-rotational symmetry, the method easily extends using the methods of [5] and [6] to other transform sizes as well. The extension of our approach to 120°-rotational symmetry and 60°-rotational symmetry follows using the exact same methods. In a second work, the extension will be carried out.

*The work was supported by D.A.R.P.A. ACMP # 770997 and in part by the Institute for Mathematics and its Applications with funds provided by the NSF and in part by a grant from the National Institute of Health.

Mailing address: Center for Larger Scale Computation, 25 West 43rd Street, Room 400, N.Y.C. 10036

†Trinity College, Cambridge, England
‡C.C.N.Y., Department of Electrical Engineering, N.Y.C. 10036

Ring-Structure. The data indexing set $Z/n \times Z/n$ has a natural ring-structure as the direct product of the ring Z/n with itself. Denoting a typical point by $a = (a_0, a_1)$ we have

$$(1) \qquad a + b = (a_0 + b_0, a_1 + b_1)$$

$$(2) \qquad a \cdot b = (a_0 b_0, a_1 b_1).$$

A second ring-structure can be defined by first forming the quotient polynomial ring

$$(3) \qquad Z/n[x]/(x^2 + 1)$$

consisting of all polynomials $a(x) = a_0 + a_1 x$ where addition and multiplication are taken $\bmod (x^2 + 1)$. Identifying the point $a = (a_0, a_1)$ with the polynomial $a(x) = a_0 + a_1 x$ we have a second ring-product on the indexing set

$$(4) \qquad a * b = (a_0 b_0 - a_1 b_1, a_1 b_0 + a_0 b_1).$$

Addition is unchanged. In particular, setting $i = (0, 1)$, we have

$$(5) \qquad i * a = (-a_1, a_0),$$

which corresponds to 90°-rotation.

Denote by $\mathbf{L}(Z/n \times Z/n)$ the set of all complex-valued functions f on $Z/n \times Z/n$. We can view f as a 2-dimensional array of data

$$(6) \qquad f = [f(j,k)]_{0 \le j,k < n}.$$

The 2-dimensional $n \times n$ finite Fourier transform is defined by

$$(7) \qquad F(f)(b) = \sum_{a \in Z/n \times Z/n} f(a) v^{\langle a,b \rangle}, \qquad v = \exp(2\pi i/n), b \in Z/n \times Z/n,$$

and $\langle a, b \rangle = a_0 b_0 + a_1 b_1$. We will replace this computation by one which is related to the ring-structure (4). First set

$$(8) \qquad \phi(a) = 2a_0 \ , \quad a = (a_0, a_1),$$

and observe that

$$(9) \qquad \phi(a * b) = 2(a_0 b_0 - a_1 b_1).$$

Define

$$(10) \qquad G(f)(b) = \sum_{a \in Z/n \times Z/n} f(a) v^{\phi(a*b)}, \quad b \in Z/n \times Z/n.$$

A direct computation shows that

(11) $$G(f)(b) = F(f)(2b_1, -2b_2), \quad b \in Z/n \times Z/n,$$

which if $(n, 2) = 1$ implies that $G(f)$ and $F(f)$ are related by the output index permutation

(12) $$\begin{bmatrix} 2 & 0 \\ 0 & -2 \end{bmatrix}.$$

Suppose now that f denotes input data and is invariant under 90°-rotation

(13) $$f(i * a) = f(a), \qquad a \in Z/n \times Z/n.$$

We also have

(14) $$f(-a) = f\big((-i) * a\big) = f(a), \qquad a \in Z/n \times Z/n.$$

From (10), direct computation shows

(15) $$G(f)(b) = G(f)(i * b) = G(f)(-b) = G(f)\big((-i) * b\big).$$

Transform Size n a prime $p \equiv 3 \mod 4$. In this case, $x^2 + 1$ is irreducible over Z/p and $Z/p[x]/(x^2 + 1)$ is the field $GF(p^2)$. Denote by $R(p)$ the indexing set $Z/p \times Z/p$ with the field structure. The set $U_2(p)$ of non-zero elements in $R(p)$ is a cyclic group under multiplication. Set

(1) $$a^{m+1} = a * a^m, \quad a \epsilon R(p) \; m \geq 1.$$

A generator w of $U_2(p)$ can be found satisfying

(2) $$i = w^r, \quad r = (p^2 - 1)/4.$$

Order input and output by

(3) $$0; \; 1, w, \dots, w^{p^2 - 2}.$$

Denote the matrix of the mapping G introduced in (11) of section 2 relative to (3) by G, again. Context will make it clear which is meant. Then

(4) $$G = \begin{array}{c|c} 1 & 1 \; 1 \; \dots \; 1 \\ \hline 1 & \\ 1 & C_2(p) \\ \vdots & \\ 1 & \end{array},$$

where

$$C_2(p) = \left[\; {}_v\,\phi(w^{j+k}) \; \right]_{0 \le j,k < p^2-1} \tag{5}$$

We will now make use of data symmetry. First $C_2(p)$ is a skew-circulant matrix and has block skew-circulant form

$$\begin{vmatrix} C_0 & C_1 & C_2 & C_3 \\ C_1 & C_2 & C_3 & C_0 \\ C_2 & C_3 & C_0 & C_1 \\ C_3 & C_0 & C_1 & C_2 \end{vmatrix}, \tag{6}$$

where $C_2 = C_0^*$ and $C_3 = C_1^*$. By $*$ we mean complex conjugation. Assume we are operating on data $f(j,k)$, $0 \le j,k < p$, invariant under 90°-rotation. Then $f(a) = f(i * a)$ which by (2) implies that f is completely determined by its values on the set

$$0; \;\; 1, w, \ldots, w^{r-1}, \quad r = (p^2-1)/4. \tag{7}$$

Setting

$$X = \begin{array}{c|c} 1 & \\ \hline & I_r \\ & I_r \\ & I_r \\ & I_r \end{array}, \tag{8}$$

we have

$$\underline{f} = X \begin{vmatrix} f(0) \\ f(1) \\ f(w) \\ \vdots \\ f(w^{r-1}) \end{vmatrix}, \tag{9}$$

where $\underline{f}$ denotes input data linearly ordered by (3).

Output data $g = G(f)$ is also invariant under 90°-rotation which reduces the computation to

$$(10)\qquad \begin{vmatrix} g(0) \\ 4g(1) \\ 4g(w) \\ \vdots \\ 4g(w^{r-1}) \end{vmatrix} = X^tGX \begin{vmatrix} f(0) \\ f(1) \\ f(w) \\ \vdots \\ f(w^{r-1}) \end{vmatrix}$$

Direct computation shows that

$$(11)\qquad X^tGX = \left|\begin{array}{c|c} 1 & 4\,1_r^t \\ \hline 1_r & H_0 \end{array}\right|,$$

when 1_r is the r-tuple of all ones and H_0 is the real skew-circulant matrix

$$(12)\qquad H_0 = C_0 + C_1 + C_2 + C_3.$$

EXAMPLE $p = 3$

$$X^tGX = \left|\begin{array}{c|cc} 1 & 4 & 4 \\ \hline 1 & 1 & -2 \\ 1 & -2 & 1 \end{array}\right|$$

4. Transform Size n a prime $p \equiv 1 \bmod 4$. In this case, there exists $u \in U(p)$ such that

$$(1)\qquad u^2 \equiv -1 \mod p,$$

and the polynomial $x^2 + 1$ is not irreducible over Z/p. In fact

$$(2)\qquad x^2 + 1 = (x - u)(x + u),$$

and $Z/p[x]/x^2+1$ is not a field. Denote by $R(p)$ the indexing set $Z/p \times Z/p$ with this ring-structure. By the Chinese Remainder theorem, we have a ring-isomorphism Ψ

$$(3)\qquad R(p) \cong Z/p \times Z/p \qquad \text{(direct product)},$$

given by

$$\Psi(a) = (a_0 + a_1 u, a_0 - a_1 u), \quad a \in R(p). \tag{4}$$

Denote the unit group of $R(p)$ by $U_2(p)$. Under Ψ, we have the group-isomorphism

$$U_2(p) \cong U(p) \times U(p). \tag{5}$$

Denote by D_1 and D_2 the subsets of $R(p)$ which under Ψ correspond to $(0) \times U(p)$ and $U(p) \times (0)$, respectively. Then, a partition of $R(p)$ is given by the subsets

$$0\ ;\ D_2\ ,\ D_1\ ;\ U_2(p). \tag{6}$$

The group $U(p)$ is cyclic having a generator, say, z. Then, $U(p)$ consists of the elements

$$1, z, \ldots, z^{p-2}, \tag{7}$$

where $z^{p-1} = 1$. Order $U(p)$ by (7). The ordering induces an ordering of D_1 and D_2 which will be assumed throughout. We could order $U_2(p)$ by taking the lexiographic order on $U(p) \times U(p)$. However, we will order $U_2(p)$ consistent with the 90°-rotation. Set

$$\alpha_1 = (z, z^{-1})\ ,\ \alpha_2 = (1, z). \tag{8}$$

Then we have the group direct product

$$U(p) \times U(p) = \left\{ \alpha_1^j \alpha_2^k : 0 \leq j, k < p-1 \right\} \tag{9}$$

and every element in $U(p) \times U(p)$ can be written uniquely as $\alpha_1^j \alpha_2^k$, $0 \leq j, k < p-1$. Set

$$A_1 = \left\{ \alpha_1^j : 0 \leq j < p-1 \right\}, \tag{10}$$

$$A_2 = \left\{ \alpha_2^k : 0 \leq k < p-1 \right\}, \tag{11}$$

and observe that

$$U(p) \times U(p) = A_1 \times A_2 \qquad \text{(direct product)}. \tag{12}$$

We order $U(p) \times U(p)$ by

$$A_2, \alpha_1 A_2, \ldots, \alpha_1^{p-2} A_2, \tag{13}$$

where A_2 is ordered by the powers of α_2 and order $U_2(p)$ relative to the ordering under Ψ. Finally, we order $R(p)$ by (6). Relative to this ordering, on input and output, the matrix G is

(14)

1	$1^t_{p'}$	$1^t_{p'}$	$q^t_{p'^2}$
$1p'$	$C(p)$	$I(p')$	$C(p) \otimes 1^t_{p'}$
$1p'$	$I(p')$	$C(p)$	$(1^t_{p'} \otimes ((p))\mathcal{S}$
$1p'^2$	$C(p) \otimes 1_{p'}$	$\mathcal{S}^{-1}(1p/ \otimes C(p))$	$\mathcal{S}^{-1}(C(p) \otimes C(p))\mathcal{S}$

$, \quad p' = p - 1,$

where $C(p)$ is the skew-circulant Winograd core

$$(15) \qquad C(p) = \left[\; v^{\phi(z^{j+k})} \; \right]_{0 \le j,k < p^2 - 1}, \quad v = \exp(2\pi i/p),$$

and $\mathcal{S}$ is the matrix direct sum

$$(16) \qquad \mathcal{S} = I_{p'} \oplus S_{p'} \oplus \cdots \oplus S_{p'}^{p-2},$$

with $S_{p'}$ the $p' \times p'$ cyclic shift matrix.

Assume the data f indexed by $R(p)$ is invariant under 90°-rotation. By abuse of notation, we write f for $f_0 \Psi^{-1}$ Then, since z can be chosen satisfying

$$(17) \qquad \Psi(i) = \alpha_1^r \quad , \quad r = (p-1)/4,$$

we have

$$(18) \qquad f(z^r a_0, z^r a_1) = f(a_0, a_1), \quad a = (a_0, a_1).$$

The action of i on D_1 corresponds to the action of α_1^r on $(0) \times U(p)$:

$$(19) \qquad \alpha_1^r(0, z^j) = (0, z^{j-r}), \quad 0 \le j < p - 1.$$

It follows that on $(0) \times U(p)$

$$(20) \qquad f(0, z^j) = f(0, z^{j+r}) = f(a, z^{j+2r}) = f(0, z^{j+3r}).$$

Denoting by $\underline{f}_1$ the r-dimensional vector

$$(21) \qquad \left| \begin{matrix} f(0,1) \\ f(0,z) \\ \vdots \\ f(0, z^{r-1}) \end{matrix} \right| ,$$

we have

$$(22) \qquad \begin{vmatrix} f(0,1) \\ f(0,z) \\ \vdots \\ f(0,z^{p-2}) \end{vmatrix} = (1_0 \otimes I_r)\underline{f}_1$$

Arguing in the same way, if $\underline{f}_2$ is the vector formed by the values of f on the points

$$(23) \qquad (1,0),\ (z,0),\ldots,\ (z^{p-2},0),$$

and if $\underline{f}_3$ is the vector formed by the values of f on the points

$$(24) \qquad A_2, \alpha_1 A_2, \ldots, \alpha_1^{r-1} A_2,$$

we have

$$(25) \qquad \underline{f} = X \begin{vmatrix} f(0) \\ \underline{f}_1 \\ \underline{f}_2 \\ \underline{f}_3 \end{vmatrix} ,$$

where

$$(26) \qquad X = 1 \oplus\ (1_4 \otimes I_r) \oplus (1_1 \otimes I_r) \oplus (1_r \otimes I_s), \quad s = p'r ,$$

and $\underline{f}$ is the vector formed from f by the linear ordering of $R(p)$. In particular, f is determined by its values on the points.

$$(27) \qquad 0; (0,1), \ldots, (0, z^{r-1}); (1,0), \ldots, (z^{r-1}, 0); A_2, \ldots, \alpha_1^{r-1} A_2.$$

The output data $g = G(f)$ also can be written as in (25). It follows that the computation reduces to the computation of the action of the matrix

$$(28) \qquad X^t G\, X.$$

Scalar multiplication by $\frac{1}{4}$ on the output, except at 0, of the action of (28) is necessary. We will now describe the matrix (28). Set

$$(29) \qquad \mathcal{S}_0 = I_{p'} \oplus S_{p'} \oplus \cdots \oplus S_{p'}^{r-1},$$

$$(20) \qquad \mathcal{S}_1 = I_r \otimes I_{p'} \oplus I_r \otimes S_{p'}^r \oplus I_r \otimes S_{p'}^{2r} \oplus I_r \otimes S_{p'}^{3r} ,$$

and observe $\mathcal{S} = \mathcal{S}_1(I_4 \otimes \mathcal{S}_0)$. Then

$$(31) \qquad X^t G X = \mathcal{S}_0^{-1}(x^t \mathcal{S}_1^{-1} G_1 \mathcal{S}_1 X)\mathcal{S}_0,$$

where $G = \mathcal{S}^{-1}G_1\mathcal{S}$. Direct computation shows that $X^t\mathcal{S}_1^{-1}G_1\mathcal{S}_1X$ is the matrix

(32)

1	$4\ 1_r^t$	$4\ 1_r^t$	$4\ 1_{rp'}$
1_r	H_0	$rI(r)$	$H_0 \otimes 1_{p'}^t$
1_r	$4\ I(r)$	H_0	$1_{p'}^t \otimes H_0$
$1_{rp'}$	$H_0 \otimes 1_{p'}$	$1_{p'} \otimes H_0$	J_0

where

(33) $$H_0 = C_0 + C_1 + C_2 + C_3$$

(34) $$J_0 = C_0 \otimes C(p) \oplus C_1 \otimes C(p)S_{p'}^r \oplus C_2 \otimes C(p)S_{p'}^{2r} \oplus C_3 \otimes C(p)S_{p'}S_{p'}^{3r}$$

with

(35) $$C(p) = \begin{vmatrix} C_0 & C_1 & C_2 & C_3 \\ C_1 & C_2 & C_3 & C_0 \\ C_2 & C_3 & C_0 & C_1 \\ C_3 & C_0 & C_1 & C_2 \end{vmatrix}$$

5. Variant. The matrix (32) of section 4 admits a factorization

(1) $$C\ A$$

which decomposes the computation into the action of a pre-addition matrix A followed by the action of a 'multiplicative' matrix C. We use the formula

(2) $$H_0\ 1_r = -1_r,$$

(3) $$J_0\ 1_{rp'} = 1_{rp'},$$

(4) $$J_0(I_r \otimes 1_{p'}) = -H_0 \otimes 1_{p'},$$

(5) $$J_0(1_{p'} \otimes I_r) = -1_{p'} \otimes H_0,$$

to write

(6) $$C = 1 \oplus H_0 \oplus H_0 \oplus J_0$$

and

(7) $A =$

1	$4\ 1_r^t$	$4\ 1_r^+$	$4\ 1_{rp'}^t$
-1_r	I_r	$-4\ I(r)$	$I_r \otimes 1_{p'}^t$
-1_r	$-4\ I(r)$	I_r	$1_{p'}^t \otimes I_r$
$1_{rp'}$	$-I_r \otimes 1_{p'}$	$-1_{p'} \otimes I_r$	$I_{rp'}$

To produce a second variant we set

$$(8)\qquad \begin{vmatrix} H_0 \\ H_1 \\ H_2 \\ H_3 \end{vmatrix} = (F(4) \otimes I_r) \begin{vmatrix} c_0 \\ c_1 \\ c_2 \\ c_3 \end{vmatrix}$$

and

$$(9)\qquad F_0 = I_r \otimes F(4) \otimes I_r,$$

As above

$$(10)\qquad F_1\big(C_0 \otimes C(p)\big)F_0 = 4C_0 \otimes (H_0 \oplus H_1 \oplus H_2 \oplus H_3),$$

$$(11)\qquad F_0\big(C_1 \otimes C(p)S_{p'}^{r}\big)F_0 = 4C_1 \otimes \big(H_0 \oplus (-H_2) \oplus (-iH_3)\big),$$

$$(12)\qquad F_0\big(C_2 \otimes C(p)S_{p'}^{2r}\big)F_0 = 4C_2 \otimes \big(H_0 \oplus (-H_1) \oplus (-H_3)\big),$$

$$(13)\qquad F_0\big(C_3 \otimes C(p)S_{p'}^{3r}\big)F_0 = 4C_3 \otimes \big(H_0 \oplus (-iH_1) \oplus (-H_2) \oplus (iH_3)\big).$$

Denote by P_0 the permutation matrix of size rp' defined by

$$(14)\qquad P_0\ (\underline{a} \otimes \underline{b}) = \underline{b} \otimes \underline{a},$$

for all r-dimensional vectors $\underline{a}$ and p'-dimensional vectors $\underline{b}$. Direct computation shows that the matrix CA can be written as

$$(15)\qquad F^{-1}P^{-1}\ H\ PF^{-1}A,$$

where $F = I_{2r+1} \oplus F_0$, $P = I_{2r+1} \oplus P_0$ and

$$(16)\qquad H = 1 \oplus H_0 \oplus H_0 \oplus (4H_0 \otimes H_0) \oplus (4H_1 \otimes H_1) \oplus (4H_2 \otimes H_2) \oplus (4H_3 \otimes H_3).$$

Finally, we can write the matrix CA as

$$(17)\qquad F^{-1}\ P^{-1}\ H\ B\ P\ F^{-1},$$

where the matrix $B = BF^{-1}A\ F\ P^{-1}$ is

$$(18)\qquad \begin{array}{c|c|c|c} 1 & 4\ 1_r^t & 4\ 1_r^t & 16\ \alpha_4^t \otimes 1_{r^2}^t \\ \hline -1_r & I_r & -4\ I(r) & 4\ \alpha_4^t \otimes 1_r^t \otimes I_r \\ \hline -1_r & -4\ I(r) & I_r & 4\ \alpha_4^t \otimes I_r \otimes 1_r^t \\ \hline \alpha_4 \otimes 1_{r^2} & -\alpha_4 \otimes 1_r \otimes I_r & -\alpha_t \otimes I_r \otimes 1_r & I_{rp'} \end{array}$$

where $\alpha_4^t = [1\ \ 0\ \ 0\ \ 0]$,

REFERENCES

[1] J.W. Cooley and J.W. Tukey, *an algorithm for machine calculation of complex Fourier series*, Math of Comput., 19 (1965), pp. 297–301.

[2] L.F. Ten Eyck, *Crystallographic Fast Fourier Transforms*, Acta Cryst. A29 (1973), pp. 183–191.

[3] C.M. Rader, *Discrete Fourier transforms when the maker of data samples is prime*, Proc. IEEE 56 (1968), pp. 1107–1108.

[4] S. Winograd, *On computing the discrete Fourier transform*, Math. of Comput., 32 (1978), pp. 175–199.

[5] R.W. Johnson, C. Lu and R. Tolimieri, *FFT Algorithms for Transfer Size* $N = p.q$, submitted for publication.

[6] R.W. Johnson,C. Lu and R. Tolimieri, *Modified Winograd FFT Algorithm for Transform Size* $N = p^k$, accepted advances in applied Mathematics (1988).

DISPLACEMENT STRUCTURE FOR HANKEL- AND VANDERMONDE-LIKE MATRICES*

J. CHUN† AND T. KAILATH†

Abstract. We introduce some generalized concepts of displacement structure for structured matrices obtained as products and inverses of Toeplitz, Hankel and Vandermonde matrices. The Toeplitz case has already been studied at some length, and the corresponding matrices have been called near-Toeplitz or Toeplitz-like or Toeplitz-derived. In this paper, we shall focus mainly on Hankel- and Vandermonde-like matrices and in particular show how the appropriately defined displacement structure yields fast triangular and orthogonal factorization algorithms for such matrices.

1. Introduction. Many signal processing problems require solving large systems of linear equations, either directly or via (weighted) least squares. The basic solution tools are *triangular factorization and QR factorization.* These factorizations require $O(n^3)$ or $O(mn^2)$ flops (floating point operations) for an $m \times n$ matrix, which can often be excessively large. Therefore, attention focuses on *structured matrices*, with an eye both to computational reductions and to implementability in special purpose (parallel) hardware). Structured matrices arise in various problems in coding theory, interpolation, control, signal processing and system theory.

Very common examples of structured matrices in the above areas are Toeplitz, Hankel and Vandermonde matrices:

$$T = (t_{i-j}) = \begin{bmatrix} t_0 & t_{-1} & \cdot & t_{n+1} \\ t_1 & t_0 & \cdot & t_{n+2} \\ \cdot & \cdot & \cdot & \cdot \\ t_{m-1} & t_{m-2} & \cdot & t_{-n+m} \end{bmatrix} \in \mathbf{R}^{m\times n}, \tag{1}$$

$$H = (h_{i+j-2}) = \begin{bmatrix} h_0 & h_1 & \cdot & \cdot & h_{n-1} \\ h_1 & h_2 & \cdot & \cdot & h_n \\ \cdot & \cdot & \cdot & \cdot & \cdot \\ h_{n-1} & h_n & \cdot & \cdot & h_{2n-2} \end{bmatrix} \in \mathbf{R}^{n\times n}, \tag{2}$$

$$V \equiv V(\mathbf{c}, K) = \begin{bmatrix} \mathbf{c}^T \\ \mathbf{c}^T\mathbf{K} \\ \mathbf{c}^T\mathbf{K}^{n-1} \end{bmatrix} \in \mathbf{R}^{n\times n}, \tag{3}$$

$$K = \mathrm{diag}(k_1, k_2, \cdot\cdot, k_n),\ k_i \neq 0, \quad \mathbf{c} = [1, 1, \cdot\cdot, 1]^T .$$

*This work was supported by the U.S. Army Research Office, under Contract DAAL03-86-K-0045, and by the SDIO/IST, managed by the Army Research Office under Contract DAAL03-87-K-0033. This manuscript is submited for publication with the understanding that the U.S. Government is authorized to reproduce and distribute reprints for Government purpose notwithstanding any copyright notation therein.

†Information Systems Laboratory, Stanford University, Stanford, CA 94305

These matrices have certain shift invariant properties: The (infinite) Toeplitz matrices are diagonally shift-invariant, Hankel matrices are "reverse diagonally" shift-invariant, and Vandermonde matrices are vertically shift-invariant apart from K.

However, it is easy to see that, for example, Toeplitz structure is not invariant under various frequently occurring operations such as multiplication, inversion, triangular and orthogonal factorization. What is often invariant is a suitably defined notion called the *displacement rank*. For example, let

$$\nabla A \equiv A - Z_n A Z_n^T, \quad A \in \mathbf{R}^{n\times n}, \tag{4}$$

where Z_n is the $n \times n$ lower-shift matrix with ones on the first subdiagonal and zeros everywhere else, and define the displacement rank of A by *rank* (∇A). Then it can be shown that a Toeplitz matrix T and its inverse T^{-1} have the same displacement rank,

$$rank(\nabla T) = rank(\nabla T^{-1}) = 2. \tag{5a}$$

Moreover if T is Hermitian, then

$$inertia(\nabla T) = inertia(\nabla T^{-1}). \tag{5b}$$

Displacement structure as just defined has by now been studied in some detail (see [12] for a recent survey, and also [13], [17]), with many results for closely related definitions in [8] and [10] among others.

In this paper, we shall briefly study an extended form of (4),

$$\nabla_{(F^f,F^b)} A \equiv A - F^f A F^{bT}, \tag{6}$$

and especially another definition

$$\nabla_{(F^f,F^b)} A \equiv F^f A - A F^{bT}, \tag{7}$$

where the matrices $\{F^f, F^b\}$ can be fairly general subject to certain restrictions described in Sec 2.

For reasons that will soon appear we shall call the matrices $\nabla_{(F^f,F^b)} A$ and $\nabla_{(F^f,F^b)} A$ the *Toeplitz displacement*, and the *Hankel displacement* of A *with respect to displacement operators* F^f, F^b}), and denoted as $\alpha_{(F^f,F^b)} A$. The rank of $\Delta_{(F^f,F^b)} A$ will be called the *Hankel displacement rank* of A and denoted as $\beta_{(F^f,F^b)} A$.

We shall say, roughly, that a matrix is "structured" (e.g., "close-to-Toeplitz" or "close-to-Hankel"), if the matrix has a displacement rank (with respect to some $\{F^f, F^b\}$), independent of the size of matrix. The interesting fact, which enables fast algorithms for triangular and orthogonal matrix factorization and matrix inversion, is that such structure is inherited under inversion, multiplication and Schur complementation. This paper will demonstrate this fact for various types of structured matrices.

First in Sec 2, we establish the claims just made about inversion, etc. Based on these results, we present a general factorization algorithm in Sec 3, which will be further specialized in later sections. Thus in Sec 4, we shall show how to compute the triangular factorization of Vandermonde matrices. Triangular factorizations of close-to-Hankel matrices will be presented in Sec 5, and QR factorizations of Vandermonde matrices in Sec 6. References to prior work on each of these applications can be found in the corresponding sections. This paper complements our recent work on Toeplitz and close-to-Toeplitz matrices [5]. On a first reading, readers can skip all "Remarks" in this paper without loss of continuity.

2. Some General Properties of Displacement Operators. In this section, we shall examine useful choices for the *displacement operators* $\{F^f, F^b\}$ in the general definitions (6)-(7), and derive some results on Schur complements that will allow us to easily study the displacement structure of matrix products and inverses.

One of the criteria for choosing displacement operators is to make the corresponding displacement ranks of A as small as possible, because as will be seen presently the displacement rank determines the complexity of various operations on A. The special structures in the matrices (1)-(3), and the results (5) naturally suggest the shift matrix Z_n as an important candidate. In fact, for a Toplitz matrix $T \in \mathbf{R}^{m\times n}$, a Hankel matrix $H \in \mathbf{R}^{n\times n}$ and a Vandermonde matrix $V = V(\mathbf{c}, K) \in \mathbf{R}^{n\times n}$, we can readily see that

(8a)

$$\nabla_{(Z_m,Z_n)}T = T - Z_m T Z_n^T = \begin{bmatrix} t_0 & t_1 & \cdot & \cdot & t_{n-1} \\ t_1 & & & & \\ \cdot & & & & \\ t_{m-1} & & & & \end{bmatrix} \quad \text{has rank 2,}$$

(8b)

$$\Delta_{(Z_n,Z_n)}H = Z_n H - H Z_n^T = \begin{bmatrix} 0 & -h_0 & -h_1 & \cdot & -h_{n-1} \\ h_0 & & & & \\ h_1 & & & & \\ \cdot & & & & \\ h_{n-1} & & & & \end{bmatrix} \quad \text{has rank 2,}$$

(8c)

$$\Delta_{(Z_n,K^{-1})}V = Z_n V - V K^{-1} = \begin{bmatrix} & -\mathbf{c}^T K^{-1} & \\ & & \\ & & \end{bmatrix} \quad \text{has rank 1.}$$

But how about non-Toeplitz, non-Hankel or non-Vandermonde matrices, but ones that are known to be the inverses of some Toeplitz, Hankel or Vandermonde matrices, respectively? The following lemma indicates that these matrices also have displacement structure, a fact first noted in [13] for Toeplitz matrices.

LEMMA 2.1 DISPLACEMENT RANK OF INVERSES. *For any nonsingular matrix* A,

$$\alpha_{(F^f,F^b)}A = \alpha_{(F^{bT},F^{fT})}A^{-1}, \quad \beta_{(F^f,F^b)}A = \beta_{(F^{bT},F^{fT})}A^{-1}.$$

Proof.

$$\alpha_{(F^f,F^b)}A = rank[(A - F^f\ AF^{bt})A^{-1}] = rank[I - F^f\ AF^{bT}\ A^{-1}],$$
$$\alpha^{bT}, F^{fT})A^{-1} = rank[(A^{-1} - F^{bT} - F^{bT}A^{-1}F^f)A] = rank[I - F^{bT}A^{-1}F^fA].$$

But the nonzero eigenvalues of $F^fA)(F^{bT}A^{-1}$ and $F^{bT}A^{-1})(F^fA)$ are identical. Therefore, $\alpha_{(F^f,F^b)}A = \alpha_{(F^{bT},F^{fT})}A^{-1}$. Next, note that

$$\beta_{(F^f,F^B)}A = rank[(F^fA - AF^{bT})A^{-1}] = rank[(F^f - AF^{bT}A^{-1})],$$
$$\beta_{F^{bT},F^{fT})}A = \beta_{(F^{bT},F^{fT})}A^{-1}. \qquad \square$$

In particular, we see immediately from Lemma 2.1 and (8) that

(9a) $$rank[T^{-1} - Z_n^TT^{-1}Z_n] = rank[T - Z_nTZ_n^T = 2,$$

(9b) $$rank[Z_n^TH^{-1} - H^{-1}Z_n] = rank[Z_nH - HZ_n^T] = 2,$$

(9c) $$rank[K^{-1}V^{-1} - V^{-1}Z_n] = rank[Z_nV - VK^{-1}] = 1.$$

Similar results can be obtained for matrix products. However, it will be useful to first consider the displacement properties of the so-called matrix Schur complements. The formation of Schur complements is the heart of triangularization procedures for matrices. Moreover we shall see that working with Schur complements of appropriately defined block matrices leads immediately to results on the displacement structure of matrix products (and inverses). The following lemma generalizes a result first given in [19].

LEMMA 2.2 DISPLACEMENT RANK OF SCHUR COMPLEMENTS. *Let* $\overline{S} \in \mathbf{R}^{(m-i)\times(n-i)}$ *be the Schur complement of* $A \in \mathbf{R}^{i\times i}$ *in* $M \in \mathbf{R}^{m\times n}$, *i.e., let*

$$M \equiv \begin{bmatrix} A & B \\ C & D \end{bmatrix}, \quad S \equiv \begin{bmatrix} 0 & O \\ O & \overline{S} \end{bmatrix} \in \mathbf{R}^{m\times n}, \quad \overline{S} \equiv D - CA^{-1}B, \quad A: \text{nonsingular}.$$

If $F^f \in \mathbf{R}^{m\times n}$ *and* $F^b \in \mathbf{R}^{n\times n}$ *are block lower triangular matrices, i.e.,*

(10) $$F^f = \begin{bmatrix} F_1^f & 0 \\ F_3^f & F_2^f \end{bmatrix}, \quad F^b = \begin{bmatrix} F_1^b & 0 \\ F_3^b & F_2^b \end{bmatrix}, \quad F_1^f \in \mathbf{R}^{i\times i}, \quad F_1^b \in \mathbf{R}^{i\times i},$$

then

(11a) $$\alpha_{(F^f,F^b)}S = \alpha_{(F_2^f,F_2^b)}\overline{S} \le \alpha_{(F^f,F^b)}M,$$

(11b) $$\beta_{(F^f,F^b)}S = \beta_{(F_2^f,F_2^b)}\overline{S} \le \beta_{(F^f,F^b)}M.$$

Proof. It is not hard to check that M^{-1} has the form

$$M^{-1} = \begin{bmatrix} * & * \\ * & \overline{S}^{-1} \end{bmatrix}.$$

Therefore

$$\text{(12a)} \quad \alpha_{(F^f,F^b)}M = \alpha_{(F^{bT},F^{fT})}M^{-1} \geq \alpha_{(F_2^{bT},F_2^{fT})}\overline{S}^{-1} = \alpha_{(F_2^f,F_2^b)}\overline{S} = \alpha_{(F^f,F^b)}S,$$

$$\text{(12b)} \quad \beta_{(F^f,F^b)}M = \beta_{(F^{bT},F^{fT})}M^{-1} \geq \beta_{(F_2^{bT},F_2^{fT})}\overline{S}^{-1} = \beta_{(F_2^f,F_2^b)}\overline{S} = \beta_{(F^f,F^b)}S,$$

where the first and third equalities in (12a) and (12b) follow from Lemma 2.1. The second inequalities follow from the block lower triangularity of F^f and F^b, and from the fact that a submatrix has smaller rank than the matrix. □

REMARK 2.1. Without the triangularity assumption on F^f and F^b, the Schur complements may have larger displacement rank than the matrix itself. See Remark 2.4 below.

Applications. Judicious use of Schur complements will allow us to easily derive the displacement properties of matrix inverses and matrix products. For example, we could alternatively have obtained the results in (9) as follows. Consider the (*extended*) matrices

$$\text{(13a)} \qquad M_1 \equiv \begin{bmatrix} T & I \\ I & O \end{bmatrix}, \quad M_2 \equiv \begin{bmatrix} H & I \\ I & O \end{bmatrix}, \quad M_3 \equiv \begin{bmatrix} V & I \\ I & O \end{bmatrix},$$

and define the displacement operators
(13b)

$$F_1 \equiv \begin{bmatrix} Z_n^T & 0 \\ O & Z_n^T \end{bmatrix}, \quad F_2 \equiv \begin{bmatrix} Z_n & O \\ O & Z_n^T \end{bmatrix}, \quad F_3^f \equiv \begin{bmatrix} Z_n & O \\ O & K^{-1} \end{bmatrix}, \quad F_3^b \equiv \begin{bmatrix} K^{-1} & O \\ O & Z_n^T \end{bmatrix}.$$

Then by using Lemma 2.2 one can check that

$$\text{(13c)} \qquad \alpha_{(Z_n^T Z_n^T)} = \alpha_{(F_1,F_1)}M_1 = 2,$$

$$\text{(13d)} \qquad \beta_{(Z_n^T,Z_n^T)}H^{-1} = \beta_{(F_2,F_2)}M_2 = 2,$$

$$\text{(13e)} \qquad \beta_{(K^{-1},Z_n^T)}V^{-1} = \beta_{(F_3^f,F_3^b)}M_3 = 1.$$

Lemma 2.2 is also useful in determining the displacement ranks of products of matrices. For example, let

$$M_1 \equiv \begin{bmatrix} I & T \\ T^T & O \end{bmatrix}, \quad F \equiv \begin{bmatrix} Z_n & O \\ O & Z_n \end{bmatrix}.$$

Then $-T^TT$ is the Schur complement of I in M_1; and therefore,

$$\text{(14a)} \qquad \alpha_{(F,F)}M_1 = 4 = \alpha_{(Z_n,Z_n)}T^TT.$$

Similarly, let

$$M_2 \equiv \begin{bmatrix} I & H \\ H^T & O \end{bmatrix}, \quad F \equiv \begin{bmatrix} Z_n & O \\ O & Z_n^T \end{bmatrix}.$$

Then, we can see that $H^T H$ has low Hankel displacement with respect to $\{Z_n^T, Z_n^T\}$ (instead of $\{Z_n, Z_n\}$) because

(14.b) $$\beta_{(F,F)} M_2 = 4 = \beta_{(Z_n^T, Z_n^T)} H^T H.$$

Finally, for the product of Vandermonde matrices, consider the matrix

$$M_3 \equiv \begin{bmatrix} I & V^T \\ V & O \end{bmatrix}, \quad F \equiv \begin{bmatrix} K^{-1} & O \\ O & Z_n \end{bmatrix}.$$

Then we can see that VV^T has rank-2 Hankel displacement

(14c) $$\beta_{(F,F)} M_3 = 2 = \beta_{(Z_n, Z_n)} VV^T.$$

This is not surprising because VV^T is, in fact, a Hankel matrix. Also some experiments will show that $V^T V$ does not seem to have a low-displacement rank with respect to "simple" displacement operators.

REMARK 2.2. Referring to the result in (14b), we may note that not $H^T H$, but rather $H^T \tilde{I} H$, where $\tilde{I}$ is the *reverse-identity* (ones on the antidiagonal) matrix has low Hankel displacement rank with respect to the usual Hankel displacement operators $\{Z_n, Z_n\}$. To see this consider the following matrix M_2 and displacement operator;

(15) $$M_2 \equiv \begin{bmatrix} \tilde{I} & H \\ H^T & O \end{bmatrix}, \quad F \equiv \begin{bmatrix} Z_n & O \\ O & Z_n \end{bmatrix}.$$

Then

$$\beta_{(F,F)} M_2 = 4 = \beta_{(Z_n Z_n)} H^T \tilde{I} H.$$

REMARK 2.3. By considering 3×3 or higher order block matrices, one can determine the displacement rank of other *composite* matrices. For instance, the matrix $\tilde{I}(H_4 - H_3 H_1^{-1} H_2)^{-1} \tilde{I}$, where H_i is a Hankel matrix, has Hankel displacement rank 4 with respect to $\{F, F\}$, where

$$F \equiv \begin{bmatrix} Z_n & O & O \\ O & Z_n & O \\ O & O & Z_n \end{bmatrix}$$

REMARK 2.4. One might have noticed that the Hankel matrix H also has a low *Toeplitz displacement* rank, viz, that

$$\nabla_{(Z_n, Z_n^T)} H = H - Z_n H Z_n, \quad \alpha_{(Z_n, Z_n^T)} H = 2.$$

The only difficulty of using this displacement in the context of the fast algorithm in Sec 2, is that the displacement operator Z_n^T is not a block lower triangular matrix. Therefore, the Schur complements of H can have larger displacement ranks than that of H. With a slight modification of Lemma 2.2, one can show that all Schur complements of a Hankel matrix with respect to this displacement operator have rank 3.

REMARK 2.5. A displacement of matrix A can *characterize* the matrix A if we can solve the equations (6) and (7) for A *uniquely* (the equations (6) and (7) are necessarily *consistent* in our context). It is well known that the (consistent) equation (6) uniquely determines the solution A, regardless of what the displacement operators $\{F^f, F^b\}$ are. However, the (consistent) equation (7) has non-unique solutions (see, e.g., [11], [15]) if (and only if) there exists a pair of eigenvalues $\lambda_i(F^f)$ and $\lambda_j(F^b)$ such that

$$\lambda_i(F^f) - \lambda_j(F^b) = 0. \tag{16}$$

This condition holds for most of the displacement operators for close-to-Hankel matrices that we are interested in, which is unfortunate because we are concerned how to find L and U such that $A = LU$ (in a fast way), given *only* the displacement of A. We shall circumvent this non-uniqueness problem by imposing an additional constraint (see Sec 3).

3. Fast Partial Triangular Factorization using Hankel Generators. In the rest of this paper, we shall only consider Hankel displacement structure. Corresponding results on Toeplitz displacement can be found in [5]. Also it is important to mention at this point that we consider only *strongly nonsingular matrices* for triangular factorization, and *full column rank matrices* for QR factorization.

General Schur Reduction. We note (see, e.g., [9], also [5], [17], [18], [23], [24]) that the standard triangular factorization procedure can be regarded as arising from the recursive computation of the Schur complements $\{S_i\}$ of the leading principal submatrices of

$$A_{i-1} = \begin{bmatrix} \\ \mathbf{l}_i \\ \\ \end{bmatrix} d_i \begin{bmatrix} \quad \mathbf{u}_i^T \quad \end{bmatrix} + A_i, \qquad A_i \equiv \begin{bmatrix} O_{i\times i} & O_{i\times(n-i)} \\ O_{(n-i)\times i} & S_i \end{bmatrix}. \tag{17a}$$

Given A_{i-1}, we can determine the quantities in (17a) as

$$\mathbf{l}_i = A_{i-1}\mathbf{e}_i, \quad \mathbf{u}_i = A_{i-1}^T\mathbf{e}_i, \quad d_i = 1/[\mathbf{l}_i]_i = 1/[\mathbf{u}_i]_i, \tag{17b}$$

where $\mathbf{e}_i$ is the unit vector with one at the ith position and zero's elsewhere, and $[\mathbf{v}]_i$ denotes the ith element of the vector $\mathbf{v}$. The computation of the reduced matrix A_i from A_{i-1} will be called (one step) *Schur reduction* [9], [17]. Using $\mathbf{r}$ Schur reduction steps, we can obtain the *(r-step) partial triangular factorization*,

(18)

$$A = \sum_{i=1}^{r} \mathbf{l}_i d_i \mathbf{u}_i^T + A_r$$

$=$ 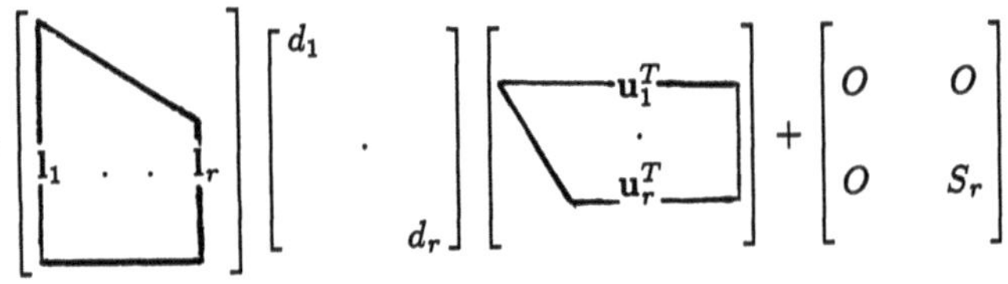

$$\equiv LDU^T + A_r.$$

The "trapezoid" matrices L, U and the diagonal matrix D will be called *(r-step) partial triangular factors* of A. Notice that r Schur reduction steps take $O(rn^2)$ flops.

Schur Reduction using Hankel Displacements. The displacement rank of a matrix A can be much smaller than the rank of the matrix A itself. Also if the chosen displacement operators $\{F^f, F^b\}$ for a given matrix $A \in \mathbf{R}^{n\times n}$ satisfy the condition in Lemma 2.2 *for all* $1 \le i \le r$, viz, that

(19a)

the $r \times r$ leading principal submatrices of F^f and F^B are lower tringular, or pictorially

(19b) $F =$ 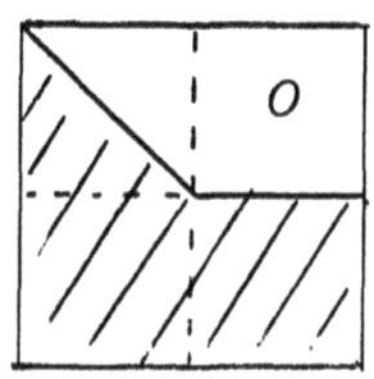

then the displacement ranks of the reduced matrices $(A_i : 0 \le i \le r)$ do not increase:

(20) $$\beta_{(F^f,F^b)} A_i \le \beta_{(F^f,F^b)} A_{i-1}, \qquad 1 \le i \le r.$$

Note that $\Delta_{(F^f,F^b)} A_i$ is determined only by $O(\beta n)$ parameters, whereas the matrix A_i itself needs $O(n^2)$ parameters. Therefore, we can hope that the Schur reduction procedure in (17) can be done more efficiently with $O(r\beta n)$ flops (instead of $O(rn^2)$ flops) if we successively compute $\Delta_{(F^f,F^b)} A_i$ rather than A_i. This is indeed the case as we shall see shortly. For the rest of this paper, we shall restrict ourselves only to displacement operators that have the form (19), because (20) is a nice property to have.

Recalling the definition of displacement,

$$\Delta_{(F^f,F^b)}A_{i-1} = F^f A_{i-1} - A_{i-1}F^{bT}, \qquad F^f \text{ and } F^b \text{ have the form (19)},$$

we can check that

(21a)
$$[\Delta_{(F^f,F^b)}A_{i-1}]\mathbf{e}_i = (F^f A_{i-1} - A_{i-1}F^{bT})\mathbf{e}_i = (F^f - [F^b]_{i,i}I)\mathbf{l}_i,$$

(21b)
$$[\Delta_{(F^f,F^b)}A_{i-1}]^T\mathbf{e}_i = (-F^b A_{i-1}^T + A_{i-1}^T F^{fT})\mathbf{e}_i = -(F^b - [F^f]_{i,i}I)\mathbf{u}_i,$$

where $[F]_{ij}$ denotes the (i,j)th element of F. Therefore, $\mathbf{l}_i$ and $\mathbf{u}_i$ can be obtained by

(22)
$$\mathbf{l}_i = (F^f - [F^b]_{i,i})^{\#}[\Delta_{F^f,F^b})A_{i-1}]\mathbf{e}_i, \quad \mathbf{u}_i = i(F^b - [F^f]_{i,i}I)^{\#}[\Delta_{(F^f,F^b)}A_{i-1}]T\mathbf{e}_i,$$

where the superscript # denotes an *appropriate* generalized inverse. After determining $\mathbf{l}_i$, $\mathbf{u}_i$ and d_i, we can obtain the displacement of the reduced matrix A_i as

$$\begin{aligned}
&(23a) \quad & \Delta_{(F^f,F^b)}A_i &= F^f A_i - A_i F^{bT} \\
&(23b) \quad & &= F^f A_{i-1} - A_{i-1}F^{bT} - F^f\mathbf{l}_i d_i \mathbf{u}_i^T + \mathbf{l}_i d_i \mathbf{u}_i^T F^{bT} \\
&(23c) \quad & &= \Delta_{(F^f,F^b)}A_{i-1} - F^f\mathbf{l}_i d_i \mathbf{u}_i^T + \mathbf{l}_i d_i \mathbf{u}_i^T F^{bT}.
\end{aligned}$$

If $(F^f - [F^b]_{i,i}I)$ and $(F^b - [F^f]_{i,i}I)$ are singular, then there are many $\mathbf{l}_i$'s and $\mathbf{u}_i$'s that would satisfy (21). Let us consider separately the nonsingular and singular cases.

A. Nonsingular cases. If $(F^f - [F^b]_{i,i}I)$ and $(F^b - [F^f]_{i,i}I)$ are nonsingular, i.e., if $[F^b]_{i,i}$ and $[F^f]_{i,i}$ are not eigenvalues of F^f and F^b, respectively, then $\mathbf{l}_i$, $\mathbf{u}_i$ and d_i and therefore A_i will be determined uniquely by taking the ordinary inverse in (22).

Example 3.1. Let $F^f = Z_n$ and $F^b = K^{-1}$. *These displacement operators are useful for the Vandermonde matrix* $V = V(\mathbf{c}, K) \in \mathbf{R}^{n\times n}$ *because they give the smallest displacement rank. Note that* $(Z_n - [K^{-1}]_{i,i}I)$ and $(K^{-1} - [Z_n]_{i,i}I) = K^{-1}$ *are nonsingular for* $1 \le i \le n$.

B. Singular cases. If $(F^f - [F^b]_{i,i}I)$ and/or $(F^b - [F^f]_{i,i}I)$ are singular then the Schur reduction using (22) and (23) is ambiguous. Note that we are not completely free in choosing F^f and F^b, because the structure of a given matrix dictates appropriate F^f and F^b that give the smallest displacement rank. Instead, we shall overcome this difficulty by using the following two approaches. The key observation behind these approaches is the fact that only the projection of $\mathbf{l}_i$ and $\mathbf{u}_i$ lying in *kernel*$(F^f - [F^b]_{i,i}I)$ and *kernel*$F^b - [F^f]_{i,i}I)$ are not uniquely determined.

If we have additional information about such ambiguous components of $\mathbf{l}_i$ and $\mathbf{u}_i$, then we can determine $\mathbf{l}_i$ and $\mathbf{u}_i$ correctly. We shall use this approach for the triangularization of the inverse of Vandermonde matrices in Sec 4.

The other approach is to extend the matrix $A \in \mathbf{R}^{n\times n}$ to a larger one, say $\overline{A} \in \mathbf{R}^{m\times m}$, such that A is a leading principal submatrix of $\overline{A}$. Let $\overline{L} \in \mathbf{R}^{m\times n}$ and $\overline{U} \in \mathbf{R}^{n\times m}$ be n-step *partial* triangular factors of $\overline{A}$. Displacement operators $\overline{F}^f$ and $\overline{F}^b$ for $\overline{A}$ are chosen such that the following is true.

$$\text{(24)} \qquad \overline{L} \perp \mathit{kernel}(\overline{F}^f - [\overline{F}^b]_{i,i}I) \quad \text{and} \quad \overline{U}^T \perp \mathit{kernel}(\overline{F}^b - [\overline{F}^f]_{i,i}I),$$

where $A \perp B$ denotes $A^T B = O$. Now, we can perform n-step partial triangularization of $\overline{A}$ using (22) and (23) unambiguously, because we can compute $\overline{\mathbf{l}}_i$ (the $i$the column of $\overline{L}$) and $\mathbf{u}_i$ (the ith column of $\overline{U}^T$) by taking the Moore-Penrose inverse (pseudo-inverse) in (22). After finding the n-step *partial triangular factors* of $\overline{A} \in \mathbf{R}^{m\times m}$, can obtain the *triangular factors* L and U of A simply by deleting the $m-n$ rows of $\overline{L}$ and $\overline{U}$.

Example 3.2. Let $H \in \mathbf{R}^{n\times n}$ *be a Hankel matrix in (2). for Hankel matrices, the (desirable) displacement operators are* $\{Z_n, Z_n\}$. *Note that* $(Z_n - [Z_n]_{i,i}I) = Z_n$ *is singular. However, if we define*

$$\text{(25a)} \qquad \overline{H}_1 = \begin{bmatrix} H & 0 \\ 0^T & 0 \end{bmatrix} \in \mathbf{R}^{n+n\times(n+1)},$$

then $\beta_{(Z_{n+1},Z_{n+1})}\overline{H}_1$ *is still small (in fact, 4), and partial triangular factors of* H_1 *and the displacement operators* $\{Z_{n+1}, Z_{n+1}\}$ satisfy (24).

Example 3.3. *Let* $H \in \mathbf{R}^{n\times n}$ *be a Hankel matrix in (2). Define*

$$\text{(25b)} \qquad \overline{H}_2 = \begin{bmatrix} H & U_R \\ U_R & O \end{bmatrix} \in \mathbf{R}^{2n\times 2n},$$

where U_R *is the "reverse upper triangular" Hankel matrix (with zero elements in the lower-right corner) such that* $\overline{H}_2$ *is Hankel. As an example, for a* 3×3 *Hankel matrix,* U_R *has the form,*

$$H = \begin{bmatrix} h_0 & h_1 & h_2 \\ h_1 & h_2 & h_3 \\ h_2 & h_3 & h_4 \end{bmatrix}, \quad U_R = \begin{bmatrix} h_3 & h_4 & 0 \\ h_4 & 0 & \\ 0 & & \end{bmatrix}.$$

Now the partial triangular factors of $\overline{H}_2$ *and the displacement operators* $\{Z_{2n}, Z_{2n}\}$ *satisfy (24).*

Generators of Matrices. For a given matrix $A \in \mathbf{R}^{m\times n}$, any matrix pair, $\{X, Y\}$ such that

$$\Delta_{(F^f,F^b)}A = XY^T, \quad X \equiv [\mathbf{x}_1, \mathbf{x}_2, \ldots, \mathbf{x}_\beta] \in \mathbf{R}^{m\times\beta}, \quad Y \equiv [\mathbf{y}_1, \mathbf{y}_2, \ldots, \mathbf{y}_\beta] \in \mathbf{R}^{n\times\beta}$$

is called a *(Hankel) generator of* A *with respect to* $\{F^f, F^b\}$. The numbers β are called the *length* of the generator (with respect to $\{F^f, F^b\}$). A generator with respect to $\{F^f, F^b\}$ with its length equal to the displacement rank is called a *minimal* generator (with respect to $\{F^f, F^b\}$).

Example 3.2 (continued). Generator of $\overline{H}_1$. The matrix $\overline{H}_1$ in (25a) has the displacement,

$$\Delta_{(Z_{n+1},Z_{n+1})}\overline{H}_1 = \begin{bmatrix} 0 & -h_0 & \cdot & -h_{n-2} & -h_{n-1} \\ h_0 & & & & -h_n \\ & & \cdot & & \cdot \\ h_{n-2} & & & & -h_{2n-2} \\ h_{n-1} & h_n & \cdot & h_{2n-2} & 0 \end{bmatrix},$$

and therefore has a generator , $\{X_1, Y_1\}$ where

$$\text{(26a)} \qquad X_1 = \begin{bmatrix} 1 & 0 & 0 & 0 \\ 0 & \cdot & -h_n & h_0 \\ \cdot & \cdot & \cdot & \cdot \\ \cdot & 0 & -h_{2n-2} & \cdot \\ 0 & 1 & 0 & h_{n-1} \end{bmatrix}, \quad Y_1 = \begin{bmatrix} 0 & 0 & 0 & 1 \\ -h_0 & h_n & \cdot & 0 \\ \cdot & \cdot & \cdot & \cdot \\ \cdot & h_{2n-2} & 0 & \cdot \\ -h_{n-1} & 0 & 1 & 0 \end{bmatrix}.$$

Example 3.3 (continued). Generator of $\overline{H}_2$. The matrix$\overline{H}_2$ in (25b) has the displacement,

$$\Delta_{(Z_{2n},Z_{2n})}\overline{H}_2 = \begin{bmatrix} 0 & -h_0 & \cdot & \cdot & -h_{2n-2} \\ h_0 & & & & \\ \cdot & & & & \\ \cdot & & & & \\ h_{2n-2} & & & & \end{bmatrix},$$

and therefore has a generator, $\{X_2, Y_2\}$, where

$$\text{(26b)} \qquad X_2 = \begin{bmatrix} 1 & 0 & \cdot & \cdot & 0 \\ 0 & h_0 & \cdot & \cdot & h_{2n-2} \end{bmatrix}^T, \quad Y_2 = \begin{bmatrix} 0 & -h_0 \cdot & \cdot & -h_{2n-2} & \\ 1 & 0 & \cdot & \cdot & 0 \end{bmatrix}^T.$$

Fast Schur Reduction using Hankel Generators. Now, let

$$\Delta_{(F^f,F^b)}A_{i-1} = X^{(i-1)}Y^{(i-1)T}.$$

Notice that the matrix products involving $\Delta_{(F^f,F^b)}A_{i-1}$ in (22) can be done more efficiently as

(27a)

$$[\Delta_{(F^f,F^b)}A_{i-1}]\mathbf{e}_i = [X^{(i-1)}[Y^{(i-1)}\mathbf{e}_i]], \quad [\Delta_{(F^f,F^b)}A_{i-1}]T\mathbf{e}_i = [Y^{(i-1)}[X^{(i-1)}[X^{(i-1)T}\mathbf{e}_i]$$

where matrix-vector products are performed in the sequence as shown with the square-brackets. Furthermore, a generator of A_i can be obtained as

$$\text{(27b)} \qquad X^{(i)} = [X^{(i-1)}, -F^f\mathbf{l}_i d_i, \mathbf{l}_i d_i], \quad [Y^{(i-1)}, \mathbf{u}_i, F^b\mathbf{u}_i],$$

because of (23c). Although the generator given in (27b) is not minimal, it is possible to delete the two redundant columns in $X^{(i)}$ and $Y^{(i)}$ in an efficient way [16].

However, the above Schur reduction procedure is still not efficient, because of the matrix inversions required in (22), and the matrix-vector multiplications $F^f\mathbf{l}_i$ and $F^b\mathbf{u}_i$ in (27b). Nevertheless, for structured matrices A (e.g., Hankel, block-Hankel, Hankel-block, Vandermonde (etc), displacement operators, F^f and F^b are extremely simple so that such operations are trivial.

4. Fast Triangular Factorization of Vandermonde Matrices. The problem of finding the coefficients of the n th degree interpolating polynomial can be formulated as a problem of solving an $(n+1) \times (n+1)$ Vandermonde matrix equation. Bjorck and Pereyra first noted [2] that the *divided-difference scheme* (which needs $O(n^2)$ flops for finding *Newton's form* of the interpolating polynomials, in fact, solves the Vandermonde matrix equations. They also presented an algorithm that needs $O(n^2)$ flops for the factorization $V^{-1} = UL$, along with other extensions. Recently, Gohberg, Kailath and Koltracht [8] obtained the algorithm of Bjorck and Pereyra by a different route and gave different extensions. In this section, we shall present two fast algorithms for computing the factorizations $V = LU$ as well as $V^{-1} = UL$. We believe that our approach is more fundamental and provides richer insight.

Consider the Vandermonde matrix

$$V = V(\mathbf{c}, K) = \begin{bmatrix} \mathbf{c}^T \\ \mathbf{c}^T K \\ \\ \mathbf{c}^T K^{n-1} \end{bmatrix} \in \mathbf{R}^{n\times n}, \tag{28a}$$

where

$$K = \operatorname{diag}(k_1, k_2, \cdot\cdot, k_n), \quad \mathbf{c}^T = [1,\ 1, \cdot\cdot, 1]. \tag{28b}$$

Note that

$$\Delta_{(Z_n,k^{-1})} V = \mathbf{x}^{(0)}\mathbf{y}^{(0)T},\ \mathbf{x}^{(0)} \equiv \mathbf{e}_1 \in \mathbf{R}^{n\times 1},\ \mathbf{y}^{(0)} \equiv -K^1\mathbf{c} \in \mathbf{R}^{n\times 1}. \tag{29}$$

Therefore, V has a generator $\{\mathbf{x}^{(0)}, \mathbf{y}^{(0)}\}$ of length 1. Now the Schur reduction steps specialized for Vandermonde matrices can be summarized in the following theorem.

THEOREM 4.1. *Let* $\Delta_{(Z_n,K^{-1})} V_{i-1} = \mathbf{x}^{(i-1)}\mathbf{y}^{(i-1)T}$. *Then*

$$\mathbf{l}_i = -k_i[\mathbf{y}^{(i-1)}]_i (I - k_i Z_n)^{-1} \mathbf{x}^{(i-1)}, \tag{30a}$$

$$\mathbf{u}_i = -K\mathbf{y}^{(i-1)}\mathbf{x}^{(i-1)T}\mathbf{e}_i = -K[\mathbf{x}_i^{(i-1)}\mathbf{y}^{(i-1)}, \tag{30b}$$

$$d_i = 1/[\mathbf{l}_i]i \tag{30c}$$

and $\Delta_{(Z_n,K^{-1})} V_i = \mathbf{x}^{(i)}\mathbf{y}^{(i)T}$, *where*

(31a)

$$\mathbf{x}^{(i)} = \left[[\mathbf{y}_{i+1}^{(i-1)}\mathbf{x}^{(i-1)} - d_i[\mathbf{u}_i]_{i+1} Z_n \mathbf{l}_i + (d_i[\mathbf{u}_i]_{i+1}/k_{i+1})\mathbf{l}_i \right] / [\mathbf{x}^{(i)}]_{i+1},$$

(31b)

$$\mathbf{y}^{(i)} = [\mathbf{x}^{(i-1)}]_{i+1}\mathbf{y}^{(i-1)} - d_i[\mathbf{l}_i]_i \mathbf{u}_i + d_i[\mathbf{l}_i]_{i+1} K^{-1}\mathbf{u}_i.$$

Proof. (30) follows immediately from (22). From (27b), we have

$$X^{(i)}Y^{(i)T} = \mathbf{x}^{(i-1)T} - Z_n\mathbf{l}_i d_i \mathbf{u}_i^T + \mathbf{l}_i d_i \mathbf{u}_i^T K^{-1}. \tag{32}$$

The matrix $X^{(i)}Y^{(i)T}$, which is the displacement of the i th reduced matrix, has the null rows and columns from 1 to i. Because it is a rank-one matrix, a minimal generator of $X^{(i)}Y^{(i)T}$ can be obtained simply by taking the $(i+1)$st row and the $(i+1)$st column with an appropriate normalization;

$$\mathbf{x}^{(i)} = X^{(i)}Y^{(i)T}\mathbf{e}_{i+1}/[\mathbf{x}^{(i)}]_{i+1}, \quad \mathbf{y}^{(i)} = Y^{(i)}X^{(i)T}\mathbf{e}_{i+1}. \tag{33}$$

The generator $\{\mathbf{x}^{(i)}, \mathbf{y}^{(i)}\}$ in (31) follows from (33), after inserting (32) for $X^{(i)}Y^{(i)T}$.

Now we shall summarize the algorithm.

Algorithm 4.1. Fast Triangular Factorization of a Vandermonde Matrix

Input: A generator $\{\mathbf{x}^{(0)}, \mathbf{y}^{(0)}\}$ in (29) of $V \equiv V_0$;

Output: Tringular factorization, $V = LU$;

for i := 1 **to** n **do begin**

Compute $\mathbf{l}_i, \mathbf{u}_i, d_i$ using (30); /* $O(n)$ flops (see Remark 4.1 below) */

Obtain a generator of V_i using (31); /* $O(n)$ flops */

end

$L := [\mathbf{l}_1, \mathbf{l}_2, \cdots, \mathbf{l}_n]$; $U^T := [\mathbf{u}_1, \mathbf{u}_2, \cdots, \mathbf{u}_n]$; $D := diag(d_1, d_2, \cdots, d_n)$

return $(\{L, U, D\})$;

REMARK 4.1. The matrix-vector multiplication, $\mathbf{p} \equiv (I - K_i Z_n)^{-1}\mathbf{x}^{(i-1)}$ in (30a) needs only $O(n)$ flops, because it is essentially the back-substitution procedure solving the bi-diagonal system, $(K - k_i Z_n)\mathbf{p} = \mathbf{x}^{(i-1)}$.

REMARK 4.2. The above algorithm can be applied to any matrix V of the form (28a) with an arbitrary lower triangular matrix K and any vector c (rather than the $\mathbf{c}$ in (28b)). However, for such cases the algorithm may take greater than $O(n^2)$ flops.

Triangularization of the inverse of Vandermonde Matrices. Consider the matrix,

$$M \equiv \begin{bmatrix} V & I \\ I & O \end{bmatrix}, \quad V \in \mathbf{R}^{n \times n}, \tag{34}$$

and its partial triangularization:

$$M = \begin{bmatrix} L_1 \\ U_2 \end{bmatrix} D \, [U_1 L_2] + \begin{bmatrix} O & O \\ O & S \end{bmatrix}, \tag{35}$$

where U_2 and L_2^T are necessarily upper-triangular because M is banded. We can see that

(36a) $$V^{-1} = U_2 D L_2,$$

because

(36b) $$V = L_1 D U_1, \quad I = U_2 D U_1, \quad I = L_1 D L_2.$$

Therefore, we can obtain the factorization $V^{-1} = U_2 D L_2$ by the n-step partial-triangularization of the matrix M using the fast Schur reduction. To do so, we first need to find a generator of M with respect to appropriate displacement operators. Our choices of displacement operators are

(37a) $$F^f = \begin{bmatrix} Z_n & O \\ O & K^{-1} \end{bmatrix}, \quad F^b = \begin{bmatrix} K^{-1} & O \\ C & C_n^T \end{bmatrix},$$

where C_n is the *circular shift-down* matrix, i.e.,

(37b) $$C_n = \begin{bmatrix} 0 & & & 1 \\ 1 & 0 & & \\ & & & \\ & & 1 & 0 \end{bmatrix} \in \mathbf{R}^{n \times n}.$$

Note that $\Delta_{(F^f, F^b)} M = \mathbf{x}^{(0)} \mathbf{y}^{(0)T}$ where

(38) $$\mathbf{x}^{(0)} = \mathbf{e}_1 \in \mathbf{R}^{2n \times 1}, \ \mathbf{y}^{(0)} = [-\mathbf{c}^T K^{-1}, -\mathbf{e}_n^T]^T \in \mathbf{R}^{2n \times 1}, \ \mathbf{e}_n \in \mathbf{R}^{n \times 1}.$$

Note that F^f and F^b in (37a) have the form in (19). Also $(F^b - [F^f]_{i,i} I)$ is nonsingular for $1 \le i \le n$, and therefore, $\mathbf{u}_i$ can be obtained by taking the ordinary inverse in (22). However, $(F^f - [F^b]_{i,i} I)$ is singular for $1 \le i \le n$, and

(39) $$kernel(F^f - [F^b]_{i,i} I) = span \begin{bmatrix} \mathbf{z} \\ \mathbf{e}_i \end{bmatrix}, \quad \mathbf{z} \in \mathbf{R}^{n \times 1}, \quad \mathbf{e}_i \in \mathbf{R}^{n \times 1}, \quad 1 \le i \le n,$$

where $\mathbf{z}$ is the null-vector, i.e., $[\mathbf{z}]_i = 0$ for all i. Therefore, if we take the pseudo-inverse in (22) for $\mathbf{l}_i$, then only the element $[\mathbf{l}_i]_{n+i}$ is not determined. However, note that this element $[\mathbf{l}_i]_{n+i}$ can be determined by other means. Namely,

$$[\mathbf{l}_i]_{n+i} = 1/(d_i \cdot [\mathbf{u}_i]_i),$$

because $U_2 = (DU_1)^{-1}$ from (36b).

REMARK 4.3. The reason of using the F^b in (37a) rather than F_3^b in(13b) is to make $F^b - [F^f]_{i,i} I$ be nonsingular. We cannot, however, use C_n in the place of Z_n for F^f, because the resulting F^f would not have the form (19).

Now we shall summarize the n-step partial triangularization with the following theorem. The proof is similar to that for Theorem 4.1, and we shall omit it.

THEOREM 4.2. *Let* $\Delta_{(F^f,F^b)}M_{i-1} = \mathbf{x}^{(i-1)}\mathbf{y}^{(i-1)T}$, *where* $M_0 \equiv M$ *and* F^f, F^b *are as in (34) and (37a), respectively. Then*

$$\mathbf{u}_i = -(F^b - [F^f]_{i,i}I)^{-1}\mathbf{y}^{(i-1)}\mathbf{x}^{(i-1)T}\mathbf{e}_i, \tag{40a}$$

$$\mathbf{l}_i = (F^f - [F^b]_{i,i}I)^{+}\mathbf{x}^{(i-1)}\mathbf{y}^{(i-1)T}\mathbf{e}_i + \mathbf{e}_{n+1}/(d_i \cdot [\mathbf{u}_i]), \tag{40b}$$

$$d_i = 1/[\mathbf{l}_i]_i, \tag{40c}$$

where A^+ *denotes the pseudo-inverse of* A. *Also,* $\Delta_{(F^f,F^b)}M_i = \mathbf{x}^{(i)}\mathbf{y}^{(i)T}$, *where*

(41a)

$$\mathbf{x}^{(i)} = \left[[\mathbf{y}^{(i-1)}]_{i+1}\mathbf{x}^{(i-1)} - d_i[\mathbf{u}_i]_{i+1}F^f\mathbf{l}_i + (d_i[\mathbf{u}_i]_{(i-1)}/k_{(i-1)})\mathbf{l}_i\right]/[\mathbf{x}^{(i)}]_{i+1},$$

(41b)

$$\mathbf{y}^{(i)} = [\mathbf{x}^{(i-1)}]_{i+1}\mathbf{y}^{(i-1)} - d_i[\mathbf{l}_i]_i\mathbf{u}_i + d_i[\mathbf{l}_i]_{i+1}F^b\mathbf{u}_i.$$

Note that the computations of (40) take only $O(n)$ flops because

$$(F^b - [F^f]_{i,i}I)^{-1} = \begin{bmatrix} K & O \\ O & C_n \end{bmatrix}.$$

$$(F^f - [F^b]_{i,i}I)^{+} = \begin{bmatrix} -k_i(I - k_iZ_n)^{-1} & O \\ O & K^{(i)} \end{bmatrix}, \quad K^{(i)} \equiv K \quad \text{except } [K]_{i,i} = 0,$$

We shall summarize the algorithm.

Algorithm 4.2. Fast Triangular Factorization of V^{-1}.

Input: A generator $\{\mathbf{x}^{(0)}, \mathbf{y}^{(0)}\}$ in (38) of $M \equiv M_0 \in \mathbf{R}^{2n\times 2n}$;

Output: Tringular factorization, $V^{-1} = UL$;

for i := 1 **to** n **do begin**

Compute $\mathbf{l}_i, \mathbf{u}_i, d_i$ using (40); /* $O(n)$ flops */

Obtain a generator of V_i using (41); /* $O(n)$ flops */

end

$L := [\mathbf{l}_1, \mathbf{l}_2, \cdots, \mathbf{l}_n]$; $U^T := [\mathbf{u}_1, \mathbf{u}_2, \cdots, \mathbf{u}_n]$; $D := diag(d_1, d_2, \cdots, d_n)$;

return (D and the bottom halfs of L, U^T);

5. Fast Triangular Factorization of close-to-Hankel Matrices. In this section, we shall only consider *strictly* lower triangular displacement operators F^f and F^b, i.e., those with zeros on the main diagonal. It will be seen that the use of such displacement operators greatly simplifies finding minimal generators of Schur complements. Such displacement operators can be used for Hankel, Hankel block and block Hankel matrices.

Berlekamp [1], [19] (see also [3], [7]) was perhaps the first to describe a fast $O(n^2)$ algorithm (needs inner-product computations) for solving Hankel matrix equations; the closely related Berlekamp-Massey algorithm [19] is an algorithm of Philips [21]. Rissanen [22] extended the results of Philips to block-Hankel matrices; The Berlekamp-Massey algorithm involves certain inner-product computations, which is a bottle-neck for parallel evaluation. Recently, following earlier work of Kung [14] and Citron [7], Lev-Ari and Kailath [18] presented another fast algorithm that does not need inner-product computation. The results in this section can be regarded as an extension of the results of Lev-Ari and Kailath [18] to Hankel-block and block-Hankel matrices. Furthermore, we shall give a fast algorithm for computing the triangular factorization of the inverse of Hankel matrices.

Let $\{X, Y\}$ be a Hankel generator of a matrix with respect to strictly lower triangular displacement operators $\{F^f, F^b\}$ *that also satisfy the condition (24).* Otherwise, we assume that the matrix has been extended appropriately such that (24) holds.) We say that a Hankel generator is *proper* if, for certain i, all the elements in the ith row of X and above, except for the element $[X]_{i,1}$, are zero, and all elements in the ith row of Y and above, except the element $[Y]_{i,\beta}$ are zero. Thus a proper generator has the form

$$(42)\quad X = [\mathbf{x}_1, \cdot\cdot, \mathbf{x}_\beta = \begin{bmatrix} * & 0 & & 0 \\ * & * & & * \\ & & & \\ * & * & & * \end{bmatrix}, \quad Y = [\mathbf{y}_1, \cdot\cdot, \mathbf{y}_\beta] = \begin{bmatrix} 0 & & 0 & * \\ * & & * & * \\ & & & \\ * & & * & * \end{bmatrix}.$$

Before we show how to convert a non-proper generator to a proper one, we shall summarize the one-step Schur reduction with the following theorem. Often we shall denote a proper generator as $\{X_p, Y_p\}$ for clarity.

THEOREM 5.1. *Let* $\Delta_{(F^f,F^b)}A_{i-1} = X_p^{(i-1)}Y_p^{(i-1)T}$, *where*

$$X_p^{(i-1)} = [\mathbf{x}_1^{(i-1)}, \mathbf{x}_2^{(i-1)}, \cdot\cdot, \mathbf{x}_\beta^{(i-1)}], \quad Y_p^{(i-1)} = [\mathbf{y}_1^{(i-1)}, \mathbf{y}_2^{(i-1)}, \cdot\cdot, \mathbf{y}_\beta^{(i-1)}].$$

Then

$$(43)\quad \mathbf{l}_i = [\mathbf{y}_\beta^{(i-1)}]_i (F^f)^+ \mathbf{x}_\beta^{(i-1)}, \quad \mathbf{u}_i = -[\mathbf{x}_1^{(i-1)}]_i (F^b)^+ \mathbf{y}_1^{(i-1)}, \quad d_i = 1/[\mathbf{l}_i]_i,$$

and $\Delta_{(F^f,F^b)}A_i = X^{(i)}Y^{(i)T}$, *where*

$$(44a)\quad X^{(i)} \equiv [(\mathbf{x}_1^{(i-1)} - d_i[\mathbf{x}_1^{(i-1)}]_i \mathbf{l}_i), \mathbf{x}_2^{(i-1)}, \cdot\cdot, \mathbf{x}_\beta^{(i-1)}],$$

$$(44b)\quad Y^{(i)} \equiv [\mathbf{y}_1^{(i-1)}, \mathbf{y}_2^{(i-1)}, \cdot\cdot, (\mathbf{y}_\beta^{(i-1)} - d_i[\mathbf{y}_\beta^{(i-1)}]_i \mathbf{u}_i)].$$

Proof. (43) is immediate from (22). Using (43), we have

$$F^f \mathbf{l}_i d_i = [\mathbf{y}_\beta^{(i-1)}]_i d_i \mathbf{x}_\beta^{(i-1)}, \quad F^b \mathbf{u}_i = -[\mathbf{x}_1^{(i-1)}]_i \mathbf{y}_1^{(i-1)}.$$

Therefore, from (27b)

$$(45)\ X^{(i)} = [X^{(i-1)}, -[\mathbf{y}_\beta^{(i-1)}]_i d_i \mathbf{x}_\beta^{(i-1)}, \mathbf{l}_i d_i], \quad Y^{(i)} = [Y^{(i-1)}, \mathbf{u}_i, -[\mathbf{x}_1^{(i-1)}]_i \mathbf{y}_1^{(i-1)}]$$

Now, (44) follows from (45). □

We can obtain the triangular factorization of close-to-Hankel matrices by applying Theorem 5.1 repeatedly as described below:

Algorithm 5.1. r-step Partial Triangular Factorization of Close-to-Hankel Matrices.

Input: A generator $\{X^{(0)}, Y^{(0)}\}$ of $A \equiv A_0 \in \mathbf{R}^{r \times n}$;

Output: Tringular factorization, $L \in \mathbf{R}^{n \times r}$ and $U \in \mathbf{R}^{r \times n}$;

for i := 1 **to** r **do begin**

Construct a proper generator of A_{i-1}; /* See below. */

Compute $\mathbf{l}_i, \mathbf{u}_i$, d_i using (43);

Obtain a generator of A_i by (44);

end

$L := [\mathbf{l}_1, \mathbf{l}_2, \cdots, \mathbf{l}_r]$; $U^T := [\mathbf{u}_1, \mathbf{u}_2, \cdots, \mathbf{u}_r]$; $D := diag(d_1, d_2, \cdots, d_r)$;

return (L, U, D);

Example 5.1 Hankel matrices. *One can use Algorithm 5.1 to find the n-step partial triangularization of the extended matrices in (25) with any of the two generators in (26); both generators will need the same amount of computation. Triangular factors of Hankel matrices are obtained from the partial triangular factors of the extended matrices. Also note that $Z^+ = Z^T$.*

Example 5.2 Block-Hankel matrices. *The block Hankel matrix,*

$$(46) \qquad H \equiv \begin{bmatrix} B_0 & B_1 & \cdot & \cdot & B_{n-1} \\ B_1 & B_2 & \cdot & \cdot & B_n \\ \cdot & \cdot & \cdot & \cdot & \cdot \\ B_{n-1} & B_n & \cdot & \cdot & B_{2n-2} \end{bmatrix} \in \mathbf{R}^{nb \times nb}, \quad B_i \in \mathbf{R}^{b \times b},$$

has a low-rank displacement with respect to the block shift displacement operator Z_{nb}^b. However, this displacement operator and the block Hankel matrix (46) do not satisfy

(24). We first add a block of null arrows and a block of null columns to (46) to get the extended matrix $\overline{H} \in \mathbf{R}^{(n+1)b\times(n+1)b}$. *Note that* $\overline{H}$ *also has low displacement rank with respect to* $\{Z^b_{(n+1)b}, Z^b_{(n+1)b}\}$. *One can easily find a generator of* $\overline{H}$ *with respect to these displacement operators. Now, we can find triangular factors of* H *from the nb-step partial triangular factors obtained by using the algorithm 5.1.*

Example 5.4 Inverse of Hankel Matrices. *Let* $H \in \mathbf{R}^{n\times n}$ *be a hankel matrix. Define the matrices*

$$(47) \qquad M \equiv \begin{bmatrix} H & \widetilde{I} \\ \widetilde{I} & O \end{bmatrix} \in \mathbf{R}^{2n\times 2n}, \quad \overline{M} \equiv \begin{bmatrix} M & \mathbf{0} \\ \mathbf{0}^T & 0 \end{bmatrix} \in \mathbf{R}^{(2n+1)\times(2n+1)},$$

where $\widetilde{I}$ *is the reverse identity (ones on the antidiagonal) matrix. The matrix* M *is a Hankel-block matrix of the form of* B *in Example 5.3, and the matrix* $\overline{M}$ *has displacement rank 4 with respect to* $\{Z_{2n+1}, Z_{2n+1}\}$. *We use the Algorithm 5.1 to get n-step partial triangular factorization of* $\overline{M}$,

$$\overline{M} = \begin{bmatrix} L_1 \\ C \\ 0^T \end{bmatrix} D[U_1 G\ 0] + \begin{bmatrix} O & O \\ O & S \end{bmatrix}$$

Note that

$$H^{-1} = U_2 D L_2, \quad U_2 \equiv \widetilde{I}C, \quad L_2 \equiv G\widetilde{I},$$

where U_2 *and* L_2 *are upper-triangular, because*

$$H = L_1 D U_1, \quad \widetilde{I} = CDU_1, \quad \widetilde{I} = L_1 D G.$$

Example 5.5 Inverse of Hankel Matrices. *Instead of using the extended matrix* $\overline{M}$ *in (47), one can use the following extended matrix,*

$$(48) \qquad \overline{M} \equiv \begin{bmatrix} \overline{H}_2 & \mathbf{0} & I_{2n} \\ \mathbf{0}^T & 0 & \mathbf{0}^T \\ I_{2n} & \mathbf{0} & O \end{bmatrix} \in \mathbf{R}^{(4n+1)\times(4n+1)},$$

where $\overline{H}_2$ *is as defined in (25b), and* I_{2n} *is the* $2n\times 2n$ *identity matrix. The matrix* $\overline{M}$ *in (48) has displacement rank 2 with respect to* $\{F, F\}$, *where* $F = Z_{2n} \oplus Z^T_{2n+1}$. *Also one can check that* $\Delta_{(F,F)}\overline{M} = XY^T$, *where*

$$(49) \quad X = \begin{bmatrix} 1 & 0 & \cdot & \cdot & 0 & 0 \\ 0 & h_0 & \cdot & \cdot & h_{2n-2} & 1 \end{bmatrix}^T, \quad Y = \begin{bmatrix} 0 & -h_0 & \cdot & \cdot & -h_{2n-2} & -1 \\ 1 & 0 & \cdot & \cdot & 0 & 0 \end{bmatrix}^T.$$

Note that n-step partial triangular factors of $\overline{M}$ *and the displacement operator* F *satisfy (24). Therefore, we can use the Algorithm 5.1 to get n-step partial triangular factorization of* $\overline{M}$, *and obtain triangular factors of* H^{-1}.

Construction of Proper Generators. The basic tool for constructing a proper (Hankel) generator is the use of *elimination* matrices $E_{i,j}(\eta)$, defined as the identity except for the element $[E_{i,j}(\eta)]_{i,j} = \eta$. Notice that $E_{i,j}^{-1}(\eta)$ is also different from the identity, except that $[E_{i,j}^{-1}(\eta)]_{i,j} = -\eta$. Let $\{X, Y\}$ be a non-proper generator of A. Without loss of generality we shall assume that $[X]_{i,1} \neq 0$. If not, we can always interchange (implicitly) columns of X and rwos of Y^T to obtain such a generator of A. We can annihilate all elements in the ith row of X except the element $[X]_{i,1}$ by post-multiplying with the $n-1$ elimination matrices *pivoting with* the element $[X]_{i,1}$,

$$X\ E_{1,2}(\eta_2)E_{1,3}(\eta_3)\cdot\ \cdot E_{1,\beta}(\eta_\beta), \qquad Y\ E_{1,2}^{-T}(\eta_2)e_{1,3}^{-T}(\eta_3)\cdot\ \cdot E_{1,\beta}^{-T}(\eta_\beta),$$

where $\eta_k = -[X]_{i,k}/[X]_{i,1}$.

Again assuming that $[Y]_{i,\beta} \neq 0$, we can similarly annihilate all elements in the ith row of Y except the element $[Y]_{i,\beta}$ by post-multiplying with the $n-1$ elimination matrices;

$$Y\ E_{\beta-1,\beta}(\gamma_{\beta-1})E_{\beta-2,\beta}(\gamma_{\beta-2})\cdot\ \cdot E_{1,\beta}(\gamma_1), \qquad X\ E_{\beta-1,\beta}^{-T}(\gamma_{\beta-1})E_{\beta-2,\beta}^{-T}(\gamma_{\beta-2})\cdot\ \cdot E_{1,\beta}^{-T}(\gamma_1),$$

where $\gamma_k = -[Y]_{i,k}/[Y]_{i,\beta}$. Note that the last annihilation $E_{1,\beta}$ does not destroy the zero at $[X]_{i,\beta}$. This procedure will require $2n$ elimination matrices, and therefore $2\beta n$ flops.

If a matrix A is symmetric, then the matrix $\Delta_{(F,F)}A = XY^T$ is skew symmetric, and therefore, has the same number of positive and negative eigenvalues. Hence the *symmetric* displacement has the form

$$\Delta_{(F,F)}A = XPX^T, \quad P = \begin{bmatrix} O & -\widetilde{I}_\delta \\ \widetilde{I}_\delta & O \end{bmatrix},$$

where $\widetilde{I}_\delta$ is the $\delta \times \delta$ reverse identity matrix (Check the generators (26a) and (26b)). We call the generator $\{X, XP^T\}$ *skew symmetric*. With a skew symmetric generator $\{X, XP^T\}$ note that we only need to apply

$$X\ E_{1,2}(\eta_2)E_{1,3}(\eta_3)\cdot\ \cdot E_{1,\beta}(\eta_\beta),$$

to obtain a proper generator. Also, we only need to compute $\mathbf{I}_i$ in (43) and $X^{(0)}$ in (44a), and Algorithm 5.1 will give the Cholesky factorization $A = LDL^T$.

REMARK 5.1. Algorithm 5.1 for finding the Cholesky factorization of the symmetric Hankel matrix in (2) by using the generator (26b) (choosing alternative pivoting elements) is identical to the Euclidean algorithm for finding

$$GCD(p(x),\ q(x)), \quad p(x) \equiv h_0x^{2n-2} + \cdot\ \cdot + h_{2n-3}x + h_{2n-2}, \quad q(x) \equiv x^{2n-1}.$$

We encourage readers to check this equivalence by using the 3×3 Hankel matrix,

$$H = \begin{bmatrix} 5 & 3 & 2 \\ 3 & 2 & 1 \\ 2 & 1 & 4 \end{bmatrix}$$

REMARK 5.2. Algorithm 5.1 reduces to the algorithm of Sugiyama *et. al.* [24] if we use the generator (49) to find the triangular factorization of the inverse of Hankel matrix. Also the $O(n\log^2 n)$ algorithm of Brent *et. al.* [4] (see also [6]) for solving Toeplitz system of equations is closely related to the divide-and-conquer version of the algorithm by Sugiyama *et. al.* after permuting the rows of a Toplitz matrix to make it Hankel. Furthermore, Berlekamp-Massey algorithm can be regarded as the "Levinson version" of the above procedure, so that one can work with only the bottom part of the algorithm applied to (49) (see [5]).

6. Fast QR Factorization of Vandermonde Matrices. We shall show that the Algorithm 5.1 can be used for the QR factorization of the transpose of the Vandermonde matrix,

$$V^T = [\mathbf{c}, K\mathbf{c}, \cdots, K^{n-1}\mathbf{c}], \quad K = \mathrm{diag}(k_1, k_2, \cdots, k_n).$$

First, notice that the matrix $VV^T \equiv H$ is a Hankel matrix,

$$VV^T = H = (h_{i+j-2}) = \begin{bmatrix} \mathbf{c}^T K\mathbf{c} & \mathbf{c}^T K^2\mathbf{c} & \cdot & \mathbf{c}^T K^n\mathbf{c} \\ \mathbf{c}^T K^2\mathbf{c} & \mathbf{c}^T K^3\mathbf{c} & \cdot & \mathbf{c}^T K^{n+1}\mathbf{c} \\ \cdot & \cdot & \cdot & \cdot \\ \mathbf{c}^T K^n\mathbf{c} & \mathbf{c}^T K^{n+1}\mathbf{c} & \cdot & \mathbf{c}^T K^{2n-1}\mathbf{c} \end{bmatrix} \in \mathbf{R}^{n\times n}.$$

Let us define

$$\overline{M}_1 \equiv \begin{bmatrix} \overline{H}_1 & \overline{V}_1 \\ \overline{V}_1^T & O \end{bmatrix} \in \mathbf{R}^{(2n+1)\times(2n+1)}, \tag{50}$$

where $\overline{H}_1$ is as (25a) and $\overline{V}_1^T = [V, K^n\mathbf{c}]$. Note that

$$\begin{aligned} &\beta_{(F_1,F_1)}\overline{M}_1 = 4, \quad F_1 \equiv Z_{n+1} \oplus K^{-1} \\ &\Delta_{(F_1,F_1)}\overline{M} = XPX^T, \end{aligned} \tag{51a}$$

where

(51b)

$$X = \begin{bmatrix} 1 & 0 & \cdot & \cdot & 0 & 0 & \cdot & \cdot & 0 \\ 0 & \cdot & \cdot & 0 & 1 & 0 & \cdot & \cdot & 0 \\ 0 & -h_n & \cdot & -h_{2n-2} & 0 & 0 & \cdot & \cdot & 0 \\ 0 & h_0 & \cdot & \cdot & h_{n-1} & k_1^{-1} & k_2^{-1} & \cdot & k_n^{-1} \end{bmatrix}^T, \quad P = \begin{bmatrix} & & & -1 \\ & & -1 & \\ & 1 & & \\ 1 & & & \end{bmatrix}.$$

It is easy to check that the displacement operators $\{F_1, F_1\}$ and the n-step partial triangular factors of $\overline{M}_1$ satisfy (24). Therefore, we can use Algorithm 5.1 for the n-step partial triangularization of

$$\overline{M}_1 = \begin{bmatrix} R^T \\ 0^T \\ Q \end{bmatrix} [R \; 0 \; Q^T] + \begin{bmatrix} O & O \\ O & S \end{bmatrix}. \tag{52}$$

Comparing (50) and (52), it is easy to see that

$$VV^T = R^T R, \quad V^T = QR, \quad Q^T Q = I.$$

REMARK 6.1. Instead of using the matirx $\overline{M}_1$ in (50), one may use the extended matrix

$$\overline{M}_2 = \begin{bmatrix} \overline{H}_2 & \overline{V}_2 \\ \overline{V}_2^T & O \end{bmatrix} \in \mathbf{R}^{3n \times 3n},$$

where $\overline{H}_2$ is as in (25b), and

$$\overline{V}_2^T \equiv [V, W], \quad W \equiv K^n V.$$

For the matrix $\overline{M}_2$, one can check that

$$\begin{aligned} &\beta_{(F_2,F_2)} \overline{M}_2 = 2, \quad F_2 \equiv Z_{2n} \oplus K^{-1}, \\ &\Delta_{(F_2,F_2)} \overline{M}_2 = XPX^T, \end{aligned}$$

where

$$X = \begin{bmatrix} 1 & 0 & \cdot & \cdot & \cdot & \cdot & \cdot & \cdot & 0 \\ 0 & h_0 & \cdot & \cdot & h_{2n-2} & k_1^{-1} & \cdot & \cdot & k_n^{-1} \end{bmatrix}, \quad P \equiv \begin{bmatrix} 0 & -1 \\ 1 & 0 \end{bmatrix}.$$

7. Concluding Remarks. We introduced some generalized notion of displacement structure and developed some of their properties. The displacement structures associated with Toeplitz and close-to-Toeplitz matrices have been the most studied so far, with some new results in [5]. In this paper we have focused on Hankel and close-to-Hankel matrices, and presented a general algorithm for triangular factorization of such matrices and their inverses. This general algorithm was also extended to obtain the triangular factorizations of Vandermonde and close-to-Vandermonde matrices and their inverses, and the QR factorizations of Vandermonde matrices and close-to-Vandermonde matrices. (The QR factorization of Hankel matrices can be obtained via the QR factorization algorithm for Toeplitz matrices [5]). Relationships with all earlier algorithms for these problems have also been noted. we remark that Algorithm 5.1 can be easily implemented as a divide-and-conquer fashion. The approach taken in [6] can be used for this purpose.

REFERENCES

[1] E. BERLEKAMP, *Algebraic coding theory*, McGraw-Hill, New York (1968).

[2] A. BJORCK AND V. PEREYRA, *Solution of Vandermonde systems of equations*, Math Com., 24 (1980).

[3] R. BLAHUT, *Theory and practice of error control codes*, Addison-Wesley, Reading, MA (1983).

[4] R. BRENT, F. GUSTAVSON AND D. YUN, *Fast Solution of Toeplitz Systems of Equations and Computation of Pade Approximants*, Journal of Algorithms, 1 (1980), pp. 259–295.

[5] J. CHUN AND T. KAILATH, *Generalized displacement structure for block-Toeplitz, Toeplitz block and Toeplitz-derived matrices*, Proc. NATO advanced Study Inst. on Signal Processing (1989).

[6] J. Chun and T. Kailath, *Divide-and-conquer solutions for least-squares problems for matrices with displacement structure*, Proc. Sixth Army Conf. on Applied Math. and Comput., Boulder, CO (June, 1988).

[7] T. Citron, *Algorithms and architectures for error correcting codes*, Ph.D. Thesis, Stanford Univ., (August 1986).

[8] I. Gohberg, T. Kailath and I. Koltracht, *Efficient solution of linear systems of equations with recursive structure*, Linear Algebra and its Appl., 80 (1986), pp. 81–113.

[9] G. Golub and C. Van Loan, *Matrix computations*, Johns Hopkins Univ. Press, Maryland (1983).

[10] G. Heinig and K. Rost, *Algebraic methods for Toeplitz-like matrices and operators*, Akademie-Verlag, Berlin (1984).

[11] T. Kailath, *Linear Systems*, Prentice-Hall, Englewood Cliffs, New Jersey (1980).

[12] T. Kailath, *Signal processing applications of some moment problems*, Proceedings of Symposia in Applied Mathematics, 37 (1987), pp. 71–109.

[13] T. Kailath, S. Kung and M. Morf, *Displacement ranks of matrices and linear equations*, J. Math. Anal. Appl., 68 (1979), pp. 395–407. See also Bull. Amer. Math. Soc., 1 (1979), pp. 769–773.

[14] S. Kung, *Multivariable and multidimensional systems: analysis and design*, Ph.D. Thesis, Stanford Univ. (1977).

[15] P. Lancaster, *Theory of matrices*, Academic Press, New York (1969).

[16] H. Lev-Ari, Personal communication (1988).

[17] H. Lev-Ari and T. Kailath, *Lattice filter parameterizations and modeling of nonstationary process*, IEEE Trans. Inform. Theory, IT-30 (1984), pp. 2–16.

[18] H. Lev-Ari and T. Kailath, *Triangular factorization of structured Hermitian matrices*, Operator Theory, Advances and Applicátions, 18, Birkhäuser, Boston (1986), pp. 301–324.

[19] J. Massey, *Shift-register synthesis and BCH decoding*, IEEE Trans., Information Theory, IT-15 (1969), pp. 122–127.

[20] M. Morf, *Doubling algorithms for Toeplitz and related equations*, Proceedings of the IEEE International Conf. on ASSP, Denver (1980), pp. 954–959.

[21] J. Philips, *The triangular decomposition of Hankel matrices*, Math Comp., 25, (1971) pp. 599–602.

[22] J. Rissanen, *Algorithms for triangular decomposition of block Hankel and Toeplitz matrices with applications to factoring positive matrix polynomial*, Math. Comput., 27, Jan. (1973), pp. 147–154.

[23] I. Schur, *Über Potenzreihen, die im Innern des Einheitskreises beschrankt sind*, J. für die Reine und Angewandte Mathematik, 147 (1917), pp. 205–232.

[24] I. Schur, *On power series which are bounded in the interior of the unit circle. 1*, (English translation of [23]), Operator Theory, Advances and Applications, 18, Birkhauser, Boston (1986), pp. 31–60.

[25] Y. Sugiyama, M. Kasahara and T. Namekawa, *A method for solving key equations for decoding Goppa codes*, Int. J. on Control, 27 (1975), pp. 87–99.

WAVELET ANALYSIS AND SIGNAL PROCESSING

R.R. COIFMAN*

We would like to describe a range of methods used in analysis over the last fifty years, which recently have found a number of applications in signal and image processing as well as in the analysis of linear and nonlinear operator theory. This so-called wavelet analysis can sometimes be used as a flexible substitute to the Fourier transform.

There is evidence that sensory perceptions such as vision, hearing, touch are processed by our brain in a similar fashion.

To be specific, let ψ denote a function with sufficient decay (say $(\psi(x)) \leq \frac{c}{1+x^2}$) with $\int \psi dx = 0$ and let $\psi_\varepsilon(x) = \frac{1}{\varepsilon}\psi\left(\frac{x}{\varepsilon}\right), \psi_\varepsilon^a = \frac{1}{\varepsilon}\psi\left(\frac{x-a}{\varepsilon}\right)$. We'll call such a function a wavelet. The wavelet transform of a signal f will be given by correlating f and ψ_ε^a.

$$u(a,\varepsilon) = \langle f, \psi_\varepsilon^a\rangle = \int f(x)\overline{\psi}_\varepsilon^a(x)dx = (f \star \psi_\varepsilon)(a).$$

Here a, ε could be restricted to some discrete subset; most commonly we'll consider $\varepsilon = 2^{-j}$ and $a = 2^{-j}k$ with $j, k \in \mathbf{Z}$. The choice of ψ will depend on our need.

In general, the functions ψ_ε^a or their discretised samples $\psi_{2^{-j}}^{2^{-j}k} := \psi_j^k$ are not independent implying strong constraints on $u(a,\varepsilon)$. (For example, if $\psi(x) = e^{-x^2}$ then $u(a,\varepsilon^{\frac{1}{2}})$ satisfies the heat equation $\left(\frac{\partial^2}{\partial a^2} - \frac{\partial}{\partial \varepsilon}\right) u(a,\varepsilon^{\frac{1}{2}}) = 0$.)

There is a simple way to recover the function $f(x)$ from the knowledge of $u(a,\varepsilon)$, as

$$f(x) = \int_0^\infty \frac{d\varepsilon}{\varepsilon} \int_{-\infty}^\infty da \langle f, \psi_\varepsilon^a\rangle \tilde{\psi}_\varepsilon^a(x) \tag{1,1}$$

where $\tilde{\psi}$ is any "reasonable" function such that in Fourier variables

$$\int_0^\infty \hat{\psi}(\varepsilon\xi)\hat{\tilde{\psi}}(\varepsilon\xi)\frac{d\varepsilon}{\varepsilon} \equiv 1 \tag{1,2}$$

(In particular, since the above integral depends only on the sign of ξ we can take $\tilde{\psi} = \psi$ whenever ψ is either even or odd.)

A discrete version of this reconstruction will have the following form

$$f(x) = \Sigma\langle f, \psi_j^k\rangle \tilde{\psi}_j^k(x) \tag{1,3}$$

i.e., these formulas behave as if we had a basis ψ_j^k and a dual basis $\tilde{\psi}_j^k$ or even an orthonormal basis (Formula (1,3) can be interpreted either as an approximation to (1,1) or as an exact formula with an appropriate choice of ψ_j^k.)

*Department of Mathematics, Yale University, New Haven, CT 06520

Formula (1,1) is easily proved by taking Fourier transforms of both sides of the equation, leading to

$$\hat{f}(\xi) = \int_0^\infty \frac{d\varepsilon}{\varepsilon} \hat{f}(\xi)\hat{\psi}(\varepsilon\xi)\hat{\tilde{\psi}}(\varepsilon\xi)$$

(here we used the fact that the right hand side of (1,1) can be rewritten as $\int_0^\infty \frac{d\varepsilon}{\varepsilon}(f \star \psi_\varepsilon) \star \tilde{\psi}_\varepsilon)$.

(1,1) is known as the Calderon reproducing formula [1]; it was rediscovered recently by Morlet and Grossman leading the way to a range of applications in signal processing.

Various discrete versions have been obtained in a variety of settings (see [3], [5], [6]), the most remarkable of which involve special choices of ψ such that $2^{j/2}\psi(2^j x - k)$ forms an orthonormal basis of L^2. The construction of a smooth function ψ was obtained by Y. Meyer leading him and his collaborators to develop the general framework known as multiresolution analysis (see §2).

In broad terms wavelet analysis like Fourier analysis is concerned with the relation between properties of the coefficients $\langle f, \psi_\varepsilon^a \rangle$ and features of the function f, resulting in a better understanding of transformations on the functions or their coefficients. Fortunately, although the term wavelet has only been introduced recently, harmonic analysts have, over the last fifty years, developed an extensive well understood theory commonly referred to as Littlewood Paley theory in which wavelet coefficients of most classical spaces have been completely characterized by their size. This situation is in strong contrast to Fourier series and relates directly to some advantages in signal processing applications. More precisely, let $f(\theta)$ be periodic, and $e_k(\theta) = e_k = e^{ik\theta}$ then

(1,4) $$f(\theta) = \sum_{-\infty}^{\infty} \langle f, e_k \rangle e^{ik\theta}$$

The only spaces of functions which can be described by a size condition on the coefficients $\langle f, e_k \rangle$ are $L^2(d\theta)$ and similar Hilbert spaces (say, functions having derivatives in L^2). This is reflected by the fact that such a simple operation as setting some coefficients equal to 0 is frequently an unbounded operation on the space.

In contrast for the expansion (1,3) we can give size condition to characterize most classical spaces. As an example, the condition $|f(x) - f(x')| \le c|x - x'|^\delta$, $0 < \delta < 1$ is equivalent to

$$|\langle f, \psi_j^k \rangle| \le cz^{-j\delta} .$$

Similar, more complicated conditions can be given for most smoothness classes, L^p spaces, $1 < p < \infty$, Soboleff spaces [16], [1].

Another related distinction between the two kinds of expansion involves the possibility of reading local spatial (or time) information from the knowledge of the expansion coefficients. Assume that the signal being analyzed looks as follows

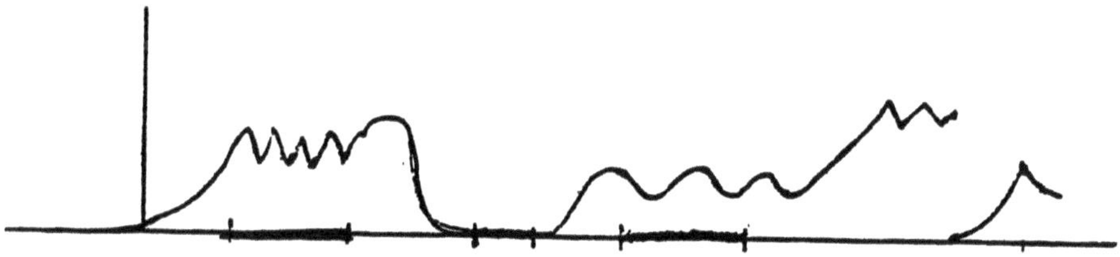

exhibiting different behavior at different time regions A, B, C, D, E. The Fourier coefficients of such a signal involve integrals over the full signal and each coefficient reflects the presence of some frequency somewhere. Moreover, in the reconstruction of $f(\theta)$ say in the region B where $f \equiv 0$ we must have subtle cancellations since many coefficients do not vanish (any modification of coefficients, say by setting some as 0 may drastically affect the cancellation and change the nature of the function.)

To better understand the information in the coefficients in the expansion (1,1) or (1,3) we'll consider two "simple" choices of wavelets, the first by taking

$$\psi(x) = \begin{cases} 1 & 0 \le x < \frac{1}{2} \\ -1 & \frac{1}{2} \le x < 1 \end{cases}$$

$\psi_j^k = 2^{j/2}\psi(2^j x - k)$ or ψ_I where $I = I_j^k$, the support of ψ_j^k, is a dyadic interval $0 < 2^j x - k < 1$. The orthonormal basis ψ_I is well known as the Haar basis. The coefficients $\langle f, \psi_I \rangle$ represent the difference of average values of the function in the two halves of the interval I. Thus the non vanishing of such a coefficient provides us with information on the changes in the function in I on the scale of I. In region B we'll get 0 coefficients for all intervals I contained in B. For the other example we'll take ψ to be the Meyer wavelet having the property that its Fourier transform is supported in $\frac{2\pi}{3} < |\xi| \le \frac{8\pi}{3}$. In space it has rapid decay, permitting us to think that ψ_k^j is localized around the interval I_j^k. This time the fact that $\langle f, \psi_I \rangle$ is not zero tells us simultaneously that the Fourier transform of f has frequencies in the range $\approx 2^j$ and that these frequencies occur around the interval I (distinguishing, for example, between region A and C).

The realization of the function f as a sum

$$f(x) = \sum_j \Delta_j(x)$$

where

$$\Delta_j(x) = \sum_{k=0} \langle f, \psi_j^k \rangle \psi_j^k(x)$$

permits us to distinguish between various features of the functions occuring on different scales. Here, each $\Delta_j(x)$ has frequencies in the range 2^j moreover, each $\Delta_j(x)$ reflects the behaviour of f in a neighborhood of x of size roughly 2^{-j}. This decomposition in "octaves" is the Littlewood Paley decomposition of f.

In applications it is important to realize that we have the flexibility to adapt the wavelet to our problem and to the features we wish to detect.

Whether we choose to consider orthonormal bases or special overdetermined representations in terms of linearly dependent functions may depend on our needs.

In the example which stimulated J. Morlet to come up with a reconstruction like (1,1) he was confronted with the following situation.

In oil exploration one of the methods consists of banging the ground with a large piston attached to a heavy truck. A compression wave $\psi(x)$ having roughly the form

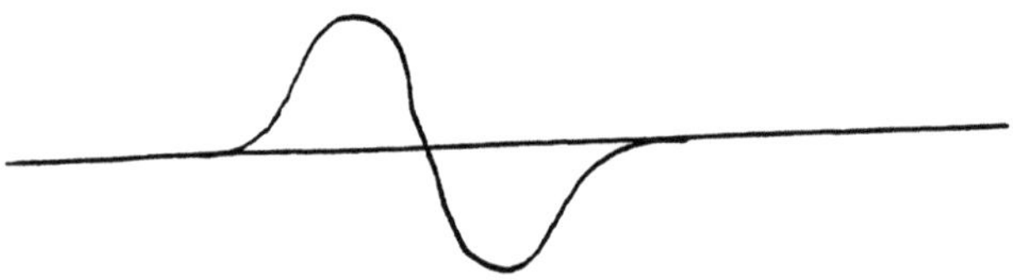

is applied to the surface. This compression travels to depth t on time t and at time $2t$ some reflexion is measured on the surface representing an average of some unknown function r around depth t.

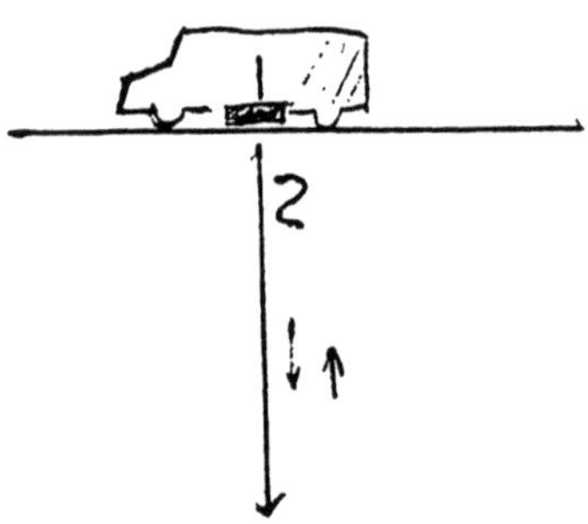

This measurement is repeated at various times T_k and the ground is compressed at different rates dilating the shape of ψ to ψ_ε. This we see the results of the measurements as providing wavelet coefficients of the unknown ground reflectivity function r. The inversion formula (1,1) or its discrete version (1,3) will provide $r(x)$. Of course this simplistic approach is only a first step, but at least it permits us to convert the measured data into the functions $r(x)$ enabling comparison of different test sites.

Other situations where the choice of ψ is forced on us involve various models for sensory perception such as vision, hearing, and touch. As we well know our sensitivity to sound frequencies involves a logarithmic scale we hear by octaves. This corresponds to the way our ear is built. The analysis seem to be similar to wavelet analysis, at least for a certain range of frequency. The membrane in our ear is modulating different frequencies roughly by dilates of a certain wavelet, essentially recording wavelet coefficients, see [15].

As for vision, various authors [14] have proposed the difference of two Gaussians or D.O.G model in which the signal transmitted to the brain by nerve endings correspond to differences of two Gaussian averages around various points of the retina, where the averages correspond almost exactly to rescaling by factors of 2. Here again the theory is that the information stored consists of various wavelet coefficients with $\psi(x) = e^{-|x|^2} - \frac{1}{2}e^{-|x|^2/4}$. The point being that we are conserving differences of shading corresponding to different scales and perceiving contrasts. Of course, in this case the functions ψ_j^k are highly dependent and we are oversampling our picture. The reconstruction formula (1,3), by "projecting" back into the space of "pictures", enables us to correct for possible errors, as well as to process the picture by concentrating say, on changes occurring on a specific scale (say, an outline of an object) and erasing other scales.

Another advantage of this analysis in image or sound processing is that the problem of positioning Gabor windows and matching various sections does not arise, since we work on all scales (allowing a fixed oscillation in each window rather than all frequencies as in Gabor's functions).

Morlet and Grossman have observed that consideration of complex wavelets for real signals permit easy detection of local periodicity or jumps. They consider wavelets like $\psi(x) = e^{i10x}e^{-x^2/2}$ (whose Fourier transform is $e^{-|\xi-10|^2/2}$ and is "practically" 0 at 0) and plot the phase of the coefficients. The equiphase lines tend to be parallel if a given period is present and converge to points where a discontinuity of the signal or its derivative exists.

Of course in situations where we do not want to oversample but rather to condense information, the orthonormal wavelets become advantageous; in fact the numerical algorithms involved are very similar to algorithms devised for image compacting (see [13]).

§**2.** The existence of orthonormal bases of the form $2^{j/2}\psi(2^jx - k)$ can be reformulated and better understood in the following terms (introduced by Mallat and Meyer).

A multiresolution analysis is an increasing sequence V_j of subspaces of L^2 with the following properties.

a) $\bigcap_{j\in\mathbf{Z}_0} V_j = \{0\} \qquad \cup V_j = L^2$

b) $f(x) \in V_j \Longleftrightarrow f(2x) \in V_{j+1}$

c) The space V_0 can be realized as the span of a single function $\Delta(x)$ and its translates, i.e., $f \in V_0 \Leftrightarrow f(x) = \Sigma\alpha_k\Delta(x-k)$. Moreover, we require

$c\|f\|_{L^2(\mathbf{R})} \le (\Sigma|\alpha_k|^2)^{1/2} \le c_1\|f\|_{L^2(\mathbf{R})}$ In other words, $\Delta(x-k)$ is an unconditional basis of V_0.)

For example, consider V_0 to be the span of $\chi(x-k)$ where $\chi(x) = \begin{cases} 1 & 0 < x \le 1 \\ 0 & elsewhere \end{cases}$. Then V_j is the space of simple functions constant on dyadic intervals of the form $(2^{-j}k, 2^{-j}(k+1))$.

A second example is provided by taking

$$\Delta(x) = \begin{cases} 1+x & -1 < x \le 0 \\ 1-x & 0 \le x \le 1 \\ 0 & elsewhere \end{cases}$$

In that case V_0 is the space of piecewise linear functions in $L^2(\mathbf{R})$ with breaks on integer points and V_j are linear splines with breaks on dyadic point $2^{-j}k$.

An orthonormal basis is constructed in two steps. First we find $\varphi \in V_0$ such that $\varphi(x-k)$ forms an orthonormal basis of V_0. We would be tempted to use a Gram Schmidt construction on Δ, but this would not commute with translations. Instead we consider the Gram matrix

$$G = [g_{k-j}] = [\langle\Delta(x-j), \Delta(x-k)\rangle] = [\int \Delta(x-k)\overline{\Delta}(x-j)dx]$$

calculate $G^{-1/2} = [h_{k-j}]$ and let

$$\varphi(x) = \Sigma h_k \Delta_k(x) = \Sigma h_k \Delta(x-k).$$

In our case, calculating Fourier transforms we find

$$\hat{\varphi}(\xi) = (\Sigma h_k e^{ik\xi})\hat{\Delta}(\xi) = H(\xi)\hat{\Delta}(\xi)$$

where $H(\xi) = \frac{1}{G(\xi)^{1/2}}$ and $G(\xi) = \Sigma e^{ik\xi} g_k$. Thus

$$\hat{\varphi}(\xi) = \left(\frac{\sin \xi/2}{\xi/2}\right)^2 \left(1 - \frac{2}{3}\sin^2\frac{\xi}{2}\right)^{-1/2}.$$

(Observe here that in the first example the functions $\chi(x-k)$ are already orthonormal and i.e., $\varphi = \chi$.)

In order to obtain an orthogonal basis for L^2 we write $V_0 = W_{-1} \oplus V_{-1}$ as an orthogonal direct sum and look for a function ψ such that $\frac{1}{\sqrt{2}}\psi(\frac{x}{2} - k)$ is an orthonormal basis of W_{-1}. The functions $\psi_j^k = 2^{j/2}\psi(2^j x - k)$ will be our orthonormal basis for $L^2(\mathbf{R})$.

We start by observing that if $f \in V_0$ and $f = \Sigma\alpha_k\varphi(x-k)$, then

$$\hat{f}(\xi) = (\Sigma\alpha_k e^{ik\xi})\hat{\varphi}(\xi) = m(\xi)\hat{\varphi}(\xi)$$

where

$$\|f\|^2_{L^2(\mathbf{R})} = \frac{1}{2\pi}\int_0^{2\pi} |m(\xi)|^2 dz \ .$$

In particular, $\frac{1}{2}\varphi(\frac{x}{2}) = \Sigma\alpha_k\varphi(x+k)$,

(2,1)
$$\hat{\varphi}(2\xi) = m_0(\xi)\hat{\varphi}(\xi) \ .$$

It follows immediately from the orthogonality of $\varphi(x+k)$ that

(2,2)
$$|m_0(\xi)|^2 + |m_0(\xi+\pi)|^2 = 1$$

We are looking for a function $\psi \in V_0$ such that $\psi \perp V_{-1}$, i.e., $\hat{\psi} = \ell(\xi)\hat{\psi}(\xi)$ and f is orthogonal to functions of the form $m(2\xi)m_0(\xi)$ where m is 2π periodic. (Here we used the fact that if $f \in V_{-1}, \hat{f}(\xi) = m(2\xi)\hat{\varphi}(2\xi) = m(2\xi)m_0(\xi)\hat{\varphi}(\xi)$.) In short, we would like to have

$$\int_0^{2\pi} m(2\xi)m_0(\xi)\bar{\ell}(\xi)d\xi = 0$$

for all 2π periodic functions $m(\xi)$. Therefore, we must have

$$m_0(\xi)\bar{\ell}(\xi) + m_0(\xi+\pi)\bar{\ell}(\xi+\pi) \equiv 0 \ .$$

Clearly, in view of (2,2), all functions of the form

$$e^{i\xi}\overline{m}_0(\xi+\pi)e^{2ki\xi}$$

satisfy this relation, leading us to take

$$\hat{\psi}(2\xi) = e^{-i\xi}\overline{m}_0(\xi+\pi)\hat{\varphi}(\xi),$$

or in the case $\hat{\varphi}$ is real,

$$\hat{\psi}(\xi) = e^{-i\xi/2}[\hat{\varphi}(\xi/2)^2 - \hat{\varphi}^2(\xi)]^{1/2}$$

which, for the example of linear splines, gives

$$\hat{\psi}(\xi) = e^{-i\xi/2}\sin^2\frac{\xi}{4}\left(\frac{\sin\xi/4}{\xi/4}\right)^2\left(\frac{1-\frac{2}{3}\cos^2\frac{\xi}{4}}{1-\frac{2}{3}\sin^2\frac{\xi}{4}}\right)^{1/2}\left(1-\frac{2}{3}\sin^2\frac{\xi}{2}\right)^{-1/2}.$$

Here the function $m_0(\xi)$ can be taken as the fundamental object defining the sequence V_j and φ, ψ. In fact, from (2,1) we deduce that

$$\hat{\varphi}(\xi) = \prod_1^\infty m_0(2^{-j}\xi) \ .$$

This remark has led I. Daubechies to construct compactly supported smooth ψ (by choosing m_0 as an appropriate trigonometric polynomial.)

Meyer's basis is obtained by choosing a smooth nonnegative function $\theta_1(\xi)$ supported on $(\frac{2\pi}{3}, \frac{8\pi}{3})$ such that

$$\theta_1^2(\xi) + \theta_1^2(2\xi) = 1$$

and

$$\theta_1^2(\xi) + \theta_1^2(2\pi - \xi) = 1$$

for $\frac{2\pi}{3} < \xi < \frac{4\pi}{3}$. We then take

$$\hat{\varphi}(\xi) = \begin{cases} 1 & 0 < |\xi| < \dfrac{2\pi}{3} \\ (1 - \theta_1^2(\xi))^{1/2} & \dfrac{2\pi}{3} < \xi| < \dfrac{4\pi}{3} \\ 0 & |\xi| > \dfrac{4\pi}{3} \end{cases}$$

and $\hat{\psi}(\xi) = \theta_1(\xi)e^{-i\xi/2}$.

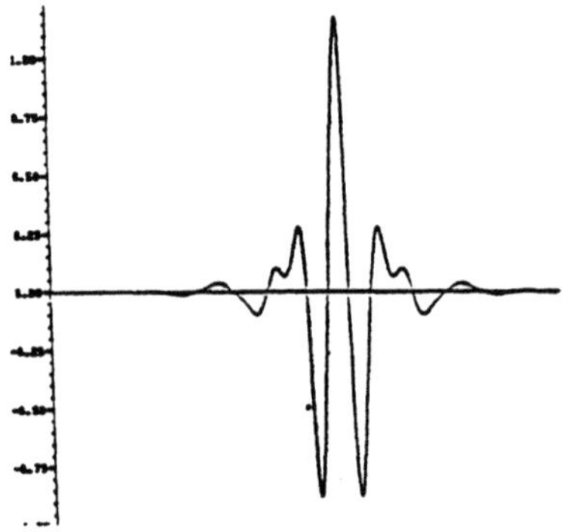

We end this section by observing that using the coefficients of m_0 various fast $(N \log N)$ algorithms have been devised for calculations of wavelet coefficients, see [13] for the orthogonal case and [12] for general wavelets. We also observe that various higher dimensional orthonormal wavelet bases can be obtained, either by constructing a multiresolution analysis or simply by considering tensor products. For example, in $\mathbf{R}^2$ we can take

$$\varphi_{k_1}^j(x_1)\psi_{k_2}^j(x_2), \psi_{k_1}^j(x_1)\varphi_{k_2}^j(x_2), \psi_{k_1}^j(x_1)\psi_{k_2}^j(x_2)$$

localized roughly at distance 2^{-j} from the point $(2^{-j}k_1, 2^{-j}k_2)$.

§3 Linear transformations in the wavelet basis. As we all know, the importance of the trigonometric basis e_k stems from the fact that it diagonalizes all linear transformations commuting with rotations such as differentiation or convolution, permitting simple algebraic manipulations of such operations.

Although algebraically simple the properties of such operations on the coefficient side are usually difficult to translate to the function side. This has motivated Marcinkiewicz in 1938 to break such operations in terms of the Litlewood Paley components of the functions and has subsequently led to a number of generalizations of this analysis to a variety of settings.

A natural question then is to describe the matrix realization (in the wavelet basis) of common classes of operators in mathematics. Although convolution operators lose their simple form, they frequently have a matrix realization which is almost diagonal. This feature is preserved for a large class of operators which behave asymptotically as if they commute with translations and dilations. Consider for example, the Hilbert transform

$$H(f)(x) = \lim_{\varepsilon \to 0} \frac{1}{\pi} \int_{|x-t|>\varepsilon} \frac{f(t)}{x-t} dt.$$

The boundedness in L^2 of this operator can easily be obtained by Fourier transform since

$$\widehat{H(f)}(\xi) = i \operatorname{sgn} \xi \hat{f}(\xi) \; .$$

In terms of an orthonormal wavelet bases ψ_I (here we chose dyadic intervals as an index set)

$$a_{I,J} = \langle H\psi_I, \psi_J \rangle$$

can easily be estimated (or calculated exactly for the case of the Haar basis) and one can easily verify

$$\sum_j |a_{I,J}| < C$$

In fact, if we chose Meyer's basis this matrix decays rapidly away from the diagonal. In any case, the boundedness in L^2 and other spaces become obvious. The point of course, is that we could have taken transformations

$$K(f)(x) = \int k(x,t) f(t) dt,$$

where say, $k(x,t) = -k(t,x)$ $|x-t||k(x,t)| + |x-t|^2 |\partial_x k(x,t)| \leq c$, $\int k(x,t)dt = 0$,[1] and would have obtained essentially the same estimate on the matrix $a_{I,J}$.

Observe also that for integral kernels as above, the usual method to numerically calculate the operation $K(f)$ at N distinct points x_i involves roughly replacing it by an N^2 matrix $K(x_i, t_j)$. The matrix obtained in the wavelet basis is almost diagonal, involving $C N \log N$ entries, reducing substantially the complexity of the problem.

Finally, we would like to conclude by mentioning that wavelet analysis or generalized Littlewood Paley theory has been carried out in a very general non translation

[1] The general case can be reproduced by "standard" paraproducts to this case.

invariant setting in which the geometry and notion of scale is allowed to vary from point to point. More precisely, we can perform such analysis on metric spaces endowed with a measure such that balls of radius r and $2r$ have comparable volume. For example, a compact Riemannian manifold, or a Lipschitz surface in $\mathbf{R}^3$ or, $\mathbf{R}^2$ equipped with the distance $\|(x,t)-(x_1,t_1)\| = (|x_1 - x|^2 + |t - t_1|)^{1/2}$ in which the space and time variable scale differently. [4].

In terms of possible applications we can think of image processing and analysis using a distorted lens and storing various differences of distorted means corresponding to a variable geometry at different points. Various reconstruction formulas like (1,1) and (1,3) exist with potential applications to be explored.

REFERENCES

[1] A.P. Calderón, *Intermediate spaces and interpolation, the complex method*, Studia Math. 24 (1964), pp. 113–190.

[2] R.R. Coifman and Y. Meyer, *The discrete wavelet transform*, preprint, Yale University, Department of Mathematics (1987).

[3] R.R. Coifman and R. Rochberg, *Representation theorems for holomorphic and harmonic functions in L^p*, Asterisque 77 (1980), pp. 11–66.

[4] R.R. Coifman and Guido Weiss, *Analyse harmonique non commutative sur certains espaces homogenés*, Springer-Verlag 242 (1971).

[5] I. Daubechies, *The wavelet transform, time frequency localization and signal analysis*, preprint, Bell Lab., Murray Hill.

[6] I. Daubechies, A. Grossman, and Y. Meyer, *Painless nonorthogonal expansions*, J. Math. Phys. 27 (1986), p. 1271.

[7] P. Goupillaud, A. Grossman, and J. Morlet, *Cycle-octave and related transforms in siesmic signal analysis*, Geoexploration 23 (1984/85), p. 85.

[8] A. Grossman and J. Morlet, *Decomposition of functions into wavelets of constant shape, and related transforms*, Center for Interdisciplinary Research and Research Center Bielefeld-Bochum-Stochastic, University of Bielefeld, Report No. 11, December (1984), in "Mathematics and Physics 2." L. Streit, editor, World Scientific Publishing Co., Singapore.

[9] A. Grossman, J. Morlet, and T. Paul, *Transforms associated to square-integrable group representations, I: General results*, J. Math. Phys. 26 (1985), p. 2473.

[10] A. Grossman, J. Morlet, and T. Paul, *Transforms associated to square-integrable group representations, II: Examples*, Ann. Inst. H. Poincaré 45 (1986), p. 293.

[11] A. Grossman and J. Morlet, *Decomposition of Hardy functions into square integrable wavelets of constant shape*, SIAM J. Math. Anal. 15 (1984), p. 723.

[12] R. Kronland-Martinet, J. Morlet, and A. Grossman, *Analysis of sound patterns through wavelet transforms*, To appear in international Journal of Pattern Recognition and Artificial Intelligence, special issue on Expert Systems and Pattern Analysis.

[13] S. Mallat, *A theory for multiscale decomposition: the scale change representation*, Grasp Laboratory, Department of Computer and Information Science, University of Pennsylvania, Philadelphia, PA (1986), 19104–6389.

[14] D. Marr, *Vision*, W.H. Freeman and Company (1982).

[15] Y. Meyer, S. Jaffard, and O. Rioul, *L'analyse par ondelettes*, Pour la Science (1987), pp. 28–37.

[16] Y. Meyer, *Principe d'incertitude, bases hilbertienne et algebre d'operateurs*, Seminaire Bourbaki 662 (Feb. 1986).

[17] Franz Tuteur, *Wavelet transformations in Signal detection*, Preprint, Elect. Engin., Yale University (1988).

ESTIMATING INTERESTING PORTIONS OF THE AMBIGUITY FUNCTION

EPHRAIM FEIG*

The cross ambiguity function of two functions $f, g \in L^2(\mathbf{R})$ is defined as

$$A_{f,g}(\tau, v) = \int_{-\infty}^{\infty} f(t)\overline{g}(t-\tau)e^{ivt}dt \tag{1}$$

When $f = g$ we have the auto-ambiguity function, which we label as $A_f(\tau, \nu)$. In radar one would like to estimate the values of the absolute value of the ambiguity function on some window $0 \leq \tau \leq R$ and $-D \leq \nu \leq D$ which is much smaller than the support of the function. A typical example taken from [1] requires that the ambiguity function be sampled on a rectangular grid of size 50×70, but the functions f and g have support spanning approximately 2^{14} samples. The function f will correspond to the transmitted pulse, and g will correspond to the received echo. All computations not involving g will be done in advance once and for all.

A straight forward method for estimating the ambiguity function in its window is as follows. For every τ of interest, we first compute the vector of samples $f(t)\overline{g}(t-\tau)$ then take a huge FFT on this product vector and then simply ignore the FFT outputs which are not of interest to us. We could be considerably better using signal processing techniques for spectral windowing. These are based on the following well known result relating the Fourier transform on the Reals and the Discrete Fourier transform [2]. The theorem below deals with functions on the Reals; the extensions to functions on $L^2(\mathbf{R}^2)$ are straightforward.

Let $f(t)$ and $F(w)$ be a Fourier transform pair; that is,

$$F(w) = \int_{-\infty}^{\infty} f(t)e^{-iwt}dt \tag{2}$$

We introduce the periodic functions

$$\widetilde{f}_T(t) = \sum_{n=-\infty}^{\infty} f(t+nT) \qquad \widetilde{F}_W(w) = \sum_{n=-\infty} f(2+nW) \tag{3}$$

where T and W are arbitrary constants.

*Thomas J. Watson Research Center, P.O. Box 218, Yorktown Heights, NY 10598

THEOREM 1. *Let T be an arbitrary constant, N an arbitrary integer, $T_1 = T/N$, $w_0 = 2\pi/T$, and $W = Nw_0$. Then for any integer m,*

$$\widetilde{f}_T(mT_1) = \sum_{n=0}^{N-1} \widetilde{F}_W(nw_0)e^{2\pi imn/N} \tag{4}$$

Theorem 1 is illustrated by the suggestive commuting diagram of operators (Figure 1) [3]. The operators P and $P^{\perp}$ are periodizing operators; the fundamental domain T on the left hand side is dual to that on the right hand side, w_1. The operators S and $S^{\perp}$ are sampling operators; the unit step size T/N on the left hand side is dual to that on the right hand side, w_0. The operators Φ, Φ_T and Φ_d are the Fourier transforms on the Reals, Circle, and $\mathbb{Z}/N\mathbb{Z}$, respectively.

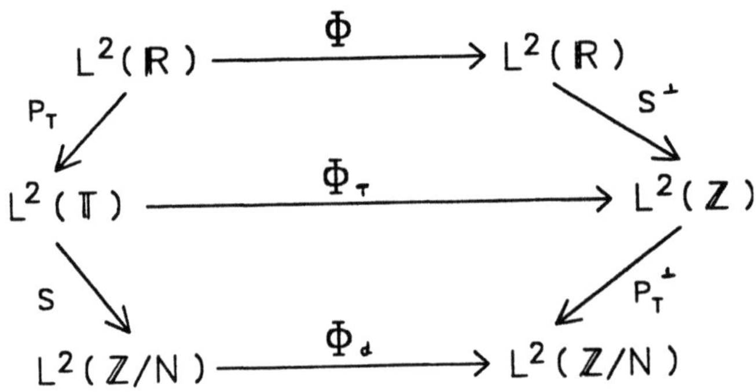

FIGURE 1

We can improve on the naive algorithm for the ambiguity function by speeding up the discrete Fourier transform part. After all, we are only interested in a narrow interval of the output around the origin. Theorem 1 says we can low pass filter first and then apply a smaller FFT on a sparse sampling. This approach was mentioned in [1]. In that same paper, Tolimieri and Winograd introduce a second approach, which globalizes the computation. By this I mean that the second method mixes parts of the computation for the various τ' s of interest together. First we discretize the functions f and g to K samples, where $K = LN$ and N is our desired size for the shortened FFT. We assume that a symmetric low pass filter wich has been designed to meet prescribed aliasing tolerance, and that the filter uses $2M+1$ taps. Let us write the loss pass filtering procedure as

$$\sum_{m=-M}^{M} f_{rN+m}\overline{g}_{rN+m-\tau}h_m \tag{5}$$

Setting

$$\psi_m(r) = f_{rN+m}h_m \qquad \eta_{m-\tau}(r) = \hat{g}_{rN+m-\tau} \tag{6}$$

the filtering equation takes the form of convolutions

$$\sum_{m=M}^{M} \psi_m(r)\eta_{m-\tau}(r) \tag{7}$$

For each r we can compute simultaneously the convolution values for all t' s of interest using a fast algorithm for convolutions. For certain values of K, L, and M of interest to the radar engineer, these new algorithms may be very practical. These algorithms exhibit typical space-time trade-offs. The time saving is at the expense of storing the precomputed products $f_{rN+m}h_m$.

Complete globalization is achieved by Auslander and Tolimieri in [4], but they estimate the values of the ambiguity function on a sparse grid, most of which is outside our window of interest. Here I give an elementary derivation for their algorithm.

THEOREM 2. *The two dimensional Fourier transform equation (1) is*

$$f(s)\hat{\bar{g}}(r)e^{irs} \tag{8}$$

This observation already appeared in [5]. Theorems 1 and 2 together imply that in order to estimate sparsely sampled values of the ambiguity function, we periodize the two dimensional function of equation (8) and then perform a sparse 2-D FFT. The output will be sampled values of a periodized ambiguity function. If the periodization window is large enough to contain the essential support ot the ambiguity function, the aliasing error will be small and the approximation good. Periodizing the sampled value of the Fourier transform yields

$$\sum_a \sum_b f(s+bB)\hat{\bar{g}}(r+aA)e^{i(r+aA)(s+bB)} \tag{9}$$

The discretization parameters a and b and the window parameters A and B should be chosen so that rbB, saA, and abB are all integer multiples of 2π. Then equation (9) becomes

$$e^{irs}\left(\sum_b f(s+bB)\right)\left(\sum_a \hat{\bar{g}}(r+aA)\right) \tag{10}$$

Thus, the two dimensional periodization is achieved by first precomputing and storing the products $e^{irs}\left(\sum_b (f(s+bB)\right)$. The folded one dimensional Fourier transform $\left(\sum_a \hat{\bar{g}}(r+aA)\right)$ is easily computed, again using theorem 1, by first folding $g(t)$ and then doing a small FFT. This folded $\hat{g}$ is multiplied, pointwise, by the precomputed matrix, and the output is then obtained with a 2-D FFT.

The Auslander-Tolimieri algorithm can be used to estimate the ambiguity function in its region of interest. Suppose $W(\tau, v)$ is a window which isolates this region. We would like to compute the sampled values of the periodized product $A_{f,g}(\tau,v)W(\tau,v)$, where the periodization window contains the essential support of the product. Because the Fourier transform of a product equals the convolution of Fourier transform, this could be done by first computing sparsely sampled values of the convolution $\widehat{A}\otimes\widehat{W}$ and then doing a small 2-D FFT. The following was shown in [6].

THEOREM 3. *If* $\widehat{W}(r,s) = W_1(r)W_2(s)\; eA^{-irs}$ *then*

$$(\widehat{A}_{f,g}\otimes\widehat{W})(r,s) = e^{-irs}A_{\widehat{\overline{g}},W_1}(r,-s)A_{f,W_2}(s,-r) \tag{11}$$

The convolution we want could be obtained by computing ambiguity functions. The windows are designed in advance, and so the values of $e^{-irs}\, A_{f,W_2}(s,-r)$ could be precomputed. We then compute sparsely sampled values of $A_{\widehat{\overline{g}},W_1}(r,-s)$ using the Auslander-Tolimieri algorithm, do the pointwise product of equation (11), and finish with a small 2-D FFT.

The windows we must use are of a very special form. They are essentially themselves ambiguity functions. Their construction is indeed an issue, but this should not be as difficult as the classical synthesis problem [7, 8] which asks for best approximation of two dimensional functions by auto-ambiguity ($f = g$) surfaces. Here we have two independent functions to play with. Furthermore, the classical problem searched for surfaces approximating delta functions around the origin. Here we do not require such concentration of volume; we merely require a fast decay at the window boundaries.

Finally, it is often also useful to estimate the aliased cross ambiguity surface. This is particularly true when the transmitted pulse is a train of chirps, so that the auto-ambiguity function is approximately a periodic delta function, and the aliasing reflects this periodicity. We can use the Auslander-Tolimieri algorithm for this task too. We use the following identity from [9].

THEOREM 4.

$$\int\!\!\int |A_{f,g}(\tau,v)|^2 e^{i(x\tau-yv)}\, d\tau\, dv = A_f(x,y)\overline{A}_g(x,y) \tag{12}$$

To estimate the values of the aliased ambiguity surface, we first precompute $A_f(\tau,v)$. We then use the Auslander-Tolimieri algorithm to compute sparse samples of $\widehat{A}_g(\tau,v)$, and do pointwise multiplication with the precomputed matrix. We finish with a small 2-D FFT.

REFERENCES

[1] R. TOLIMIERI AND S. WINOGRAD, *Computing the Ambiguity Surface*, IEEE-ASSP 33, 5 (1985), pp. 1239–1245.

[2] A. PAPOULIS, *Signal Analysis*, McGraw-Hill, New York (1977).

[3] E. FEIG AND F. GREENLEAF, *Norrowband chemical-shift imaging*, Applied Optics, 24, December 1, 1985, pp. 3980.

[4] L. Auslander and R. Tolimieri, *Computing the decimated ambiguity function*, IEEE-ASSP 36 No. 3 (March 1988), pp. 359–364.

[5] E. Feig, *Transformed Ambiguity Function*, IBM RC 9747 (#43082) 12/15 (1982).

[6] E. Feig, *Computational Methods with the Ambiguity Function*, IMB RC 13140 (#58795) 9/17 (1987).

[7] C.H. Wilcox, *The Synthesis Problem for Radar Ambiguity Functions*, MRC Technical Summary Report # 157, Mathematics Research Center, United States Army, The University of Wisconsin, April (1960).

[8] S.M. Sussman, *Least-Squares Synthesis of Radar Ambiguity Functions*, IRE Trans. Information Theory, April (1962).

[9] C.A. Stutt, *Results on Real Part/Imaginary-Part and Magnitude-Phase Relations in Ambiguity Functions*, IEEE Trans. Information Theory, October (1964).

THE BAND METHOD FOR EXTENSION PROBLEMS AND MAXIMUM ENTROPY*

I. GOHBERG†, M.A. KAASHOEK‡ AND H.J. WOERDEMAN‡

Abstract. A review is given of a general scheme to deal with positive and strictly contractive extension problems. The application of the general scheme is explained on nonstationary extension problems.

0. Introduction. In a series of papers [4–10] H. Dym and I. Gohberg developed a method for solving various extension problems, among them the matrix-valued version of the Carathéodory-Toeplitz problem and the Nehari problem and their continuous analogues. The paper [7], which concerned the extension problem for integral operators, presented an abstract approach which could serve as a basis for dealing with extension problems in general. In fact, in this approach all extension problems referred to above appear as examples of a general abstract extension problem. The latter point of view was developed further in the papers [11, 12, 13] of the present authors, which in the setting of the general scheme gave the description of all solutions and the maximum entropy principle. In order to obtain these general results additional structure had to be introduced into the general scheme. In the papers [11, 12, 13] also new applications were given.

The present paper gives a concise review of the final version of the abstract scheme as it appears in [11], 12, 13]. The application of the general scheme is explained on one class of examples, namely the nonstationary positive and strictly contractive extension problems.

CHAPTER I. THE GENERAL SCHEME.

The general scheme for extension problems discussed in this paper is developed in the setting of *-algebras with a special additive and multiplicative structure. The description of this structure and the positive extension theorems are in the first section. In the second the general scheme is used to derive an abstract contractive extension theorem. The third section deals with maximum entropy in the framework of the general scheme.

I.1. Positive extensions. Let $\mathcal{M}$ be an algebra with a unit e and an involution *. Throughout this section we suppose that $\mathcal{M}$ admits a direct sum decomposition of the form

$$\mathcal{M} = \mathcal{M}_1 \dotplus \mathcal{M}_2^0 \dotplus \mathcal{M}_d \dotplus \mathcal{M}_3^0 \dotplus \mathcal{M}_4, \tag{1.1}$$

---

*These results were presented by the second author at the Signal Processing Workshop, Institute for Mathematics and its Applications, University of Minnesota, July, 1988.

†Department of Mathematical Sciences, The Raymond and Beverly Sackler Faculty of Exact Sciences, Tel Aviv University, Ramat Aviv 69978, Israel.

‡Department of Mathematics and Computer Science, Vrije Universiteit, De Boelelaan 1081, 1081 HV Amsterdam, The Netherlands.

where $\mathcal{M}_1, \mathcal{M}_2^0, \mathcal{M}_d, \mathcal{M}_3^0$ and $\mathcal{M}_4$ are linear subspaces of $\mathcal{M}$ and the following conditions are satisfied:

(i) $e \in \mathcal{M}_d, \mathcal{M}_1 = \mathcal{M}_4^*, \mathcal{M}_2^0 = (\mathcal{M}_3^0)^*, \mathcal{M}_d = \mathcal{M}_d^*$;

(ii) the following multiplication table describes some additional rules on the multiplication in $\mathcal{M}$:

(1.2)

	$\mathcal{M}_1$	$\mathcal{M}_2^0$	$\mathcal{M}_d$	$\mathcal{M}_3^0$	$\mathcal{M}_4$
$\mathcal{M}_1$	$\mathcal{M}_1$	$\mathcal{M}_1$	$\mathcal{M}_1$	$\mathcal{M}_+^0$	$\mathcal{M}$
$\mathcal{M}_2^0$	$\mathcal{M}_1$	$\mathcal{M}_+^0$	$\mathcal{M}_2^0$	$\mathcal{M}_c$	$\mathcal{M}_-^0$
$\mathcal{M}_d$	$\mathcal{M}_1$	$\mathcal{M}_2^0$	$\mathcal{M}_d$	$\mathcal{M}_3^0$	$\mathcal{M}_4$
$\mathcal{M}_3^0$	$\mathcal{M}_+^0$	$\mathcal{M}_c$	$\mathcal{M}_3^0$	$\mathcal{M}_-^0$	$\mathcal{M}_4$
$\mathcal{M}_4$	$\mathcal{M}$	$\mathcal{M}_-^0$	$\mathcal{M}_4$	$\mathcal{M}_4$	$\mathcal{M}_4$

where

$$\mathcal{M}_+^0 := \mathcal{M}_1 \dotplus \mathcal{M}_2^0, \mathcal{M}_-^0 := \mathcal{M}_3^0 \dotplus \mathcal{M}_4,$$
$$\mathcal{M}_c := \mathcal{M}_2^0 \dotplus \mathcal{M}_d \dotplus \mathcal{M}_3^0. \tag{1.3}$$

We shall refer to $\mathcal{M}$ as the *algebra with the band structure* (1.1). The space $\mathcal{M}_c$ is called the *band* of $\mathcal{M}$. We need the following additional notations:

$$\mathcal{M}_+ := \mathcal{M}_+^0 \dotplus \mathcal{M}_d, \mathcal{M}_- := \mathcal{M}_-^0 \dotplus \mathcal{M}_d,$$
$$\mathcal{M}_2 := \mathcal{M}_2^0 \dotplus \mathcal{M}_d, \mathcal{M}_3 := \mathcal{M}_3^0 \dotplus \mathcal{M}_d. \tag{1.4}$$

Note that $\mathcal{M}_1$ (resp. $\mathcal{M}_4$) is a two-sided ideal of the subalgebra $\mathcal{M}_+$ (resp. $\mathcal{M}_-$). Furthermore, if $d \in \mathcal{M}_d$ is invertible, then $d^{-1} \in \mathcal{M}_d$.

If $\mathcal{A}$ is an algebra with a unit and an involution $*$, we say that an element $a \in \mathcal{A}$ is *nonnegative definite* in $\mathcal{A}$ (notation: $a \geq_{\mathcal{A}} 0$) if there exists an element $c \in \mathcal{A}$ such that $a = c^*c$. The element $a \in \mathcal{A}$ is called *positive definite* in $\mathcal{A}$ (notation: $a >_{\mathcal{A}} 0$) if there exists an invertible element $c \in \mathcal{A}$ such that $a = c^*c$. We shall write $b \geq_{\mathcal{A}} a$ instead of $b - a \geq_{\mathcal{A}} 0$, and $b >_{\mathcal{A}} a$ instead of $b - a >_{\mathcal{A}} 0$. When $\mathcal{A} = \mathcal{M}$ we shall omit the subscript $\mathcal{M}$.

Let us introduce the following two types of factorizations for positive elements in $\mathcal{M}$. Let $b \in \mathcal{M}$ be positive definite in $\mathcal{M}$. We shall say that b admits a *left spectral factorization* (relative to the decomposition (1.1)) if $b = b_- b_-^*$ for some $b_+ \in \mathcal{M}_+$ with $b_+^{-1} \in \mathcal{M}_+$. We shall say that b admits a *right spectral factorization* (relative to the decomposition (1.1)) if $b = b_b^*$ for some $b_- \in \mathcal{M}_-$ with $b_-^{-1} \in \mathcal{M}_-$. Note that b admits a left spectral factorization if and only if b^{-1} admits a right spectral factorization.

We shall use the symbols P_i $(i = 1,\ldots,4), P_i^0$ $(i = 2,3)$, $P_\pm, P_\pm^0, P_c$ and P_d to denote the natural projections of $\mathcal{M}$ onto the subspaces of the same index along their natural complement in $\mathcal{M}$. Thus, for instance,

$$P_+ = P_1 + P_2, P_- = P_3 + P_4, P_c = P_2 + P_3^0 = P_2^0 + P_d + P_3^0 = P_2^0 + P_3.$$

Let k be a selfadjoint element in the band $\mathcal{M}_c$. An element $b \in \mathcal{M}$ is called a *positive extension* of k if $P_c b = k$ and b is positive definite in $\mathcal{M}$. We are interested in finding all positive extensions of a given k. The following type of a positive extension will play an important role. A positive extension b of k is called a *(positive) band extension of* k if in addition $b^{-1} \in \mathcal{M}_c$. In what follows we will just speak about a band extension and omit the adjective positive.

The next theorem reduces the construction of a band extension of a given $k \in \mathcal{M}_c$ to solving linear equations.

THEOREM I.1.1. *Let $\mathcal{M}$ be the algebra with band structure (1.1), and let $k = k^* \in \mathcal{M}_c$. The element k has a band extension b which admits a left and a right spectral factorization if and only if the equations*

$$P_2(kx) = e\ , P_3(ky) = e\ , \tag{1.5}$$

have solutions x and y with the following properties:

(i) $x \in \mathcal{M}_2, y \in \mathcal{M}_3$,

(ii) *x and y are invertible, $x^{-1} \in \mathcal{M}_+, y^{-1} \in \mathcal{M}_-$,*

(iii) *$P_d x$ and $P_d y$ are positive definite in $\mathcal{M}_d$.*

Moreover, if such an element b exists, then b is unique and given by

$$b = x^{*-1}(P_d x)x^{-1} = y^{*-1}(P_d y)y^{-1}. \tag{1.6}$$

To describe the set of all positive extensions of a given $k \in \mathcal{M}_c$, we need extra requirements on the algebra $\mathcal{M}$. We shall assume that $\mathcal{M}$ is a *-subalgebra of a B^*-algebra $\mathcal{R}$ with norm $\|.\|_\mathcal{R}$, and $\mathcal{R}$ has a unit e which belongs to $\mathcal{M}$. Further, we assume that the following two axioms hold:

AXIOM (A1). If $f \in \mathcal{M}$ is invertible in $\mathcal{R}$, then $f^{-1} \in \mathcal{M}$;

AXIOM (A2). If $f_n \in \mathcal{M}_+, f \in \mathcal{M}$ and $\lim_{n\to\infty} \|f_n - f\|_\mathcal{R} = 0$, then $f \in \mathcal{M}_+$.

Note that if $e - f^*f$ is positive definite in $\mathcal{M}$, then $e - f^*f$ is positive definite in $\mathcal{R}$, and hence $\|f\|_\mathcal{R} < 1$.

The following theorem gives a description of the set of all positive extensions of a (certain type of) given $k \in \mathcal{M}_c$.

THEOREM I.1.2. *Let $\mathcal{M}$ be the algebra with band structure (1.1), and assume that $\mathcal{M}$ is a *-subalgebra of a B^*-algebra $\mathcal{R}$ such that the unit e of $\mathcal{R}$ belongs to $\mathcal{M}$ and Axioms (A1) and (A2) hold. Let $k = k^* \in \mathcal{M}_c$, and suppose that k has a band extension b which admits a left and a right spectral factorization:*

$$b = u^{*-1}u^{-1} = v^{*-1}v^{-1}, u^{\pm 1} \in \mathcal{M}_+, v^{\pm 1} \in \mathcal{M}_-. \tag{1.7}$$

Then each positive extension of k is the form

$$T(g) = (g^*v^* + u^*)^{-1}(e - g^*g)(vg + u)^{-1}, \tag{1.8}$$

where g is an element of $\mathcal{M}_1$ such that $e - g^*g$ is positive definite in $\mathcal{M}$. Furthermore, formula (1.8) gives a 1-1 correspondence between all such g and all positive extensions of k. Alternatively, each positive extension of k is of the form

$$S(f) = (f^*u^* + v^*)^{-1}(e - f^*f)(uf + v)^{-1}, \tag{1.9}$$

where f is an element of $\mathcal{M}_4$ such that $e - f^*f$ is positive definite in $\mathcal{M}$. Furthermore, formula (1.9) gives a 1-1 correspondence between all such f and all positive extensions of k.

Note that the u and the v appearing in Theorem I.1.2 can be obtained from the elements x and y in Theorem I.1.1 as follows. Since $P_d x >_{\mathcal{M}_d} 0$, there is an invertible $r \in \mathcal{M}_d$ such that $P_d x = r^* r$. Also there is an invertible $s \in \mathcal{M}_d$ such that $P_d y = s^* s$. Put now $u = xr^{-1}$ and $v = ys^{-1}$.

Theorem I.1.2 is also true when, instead of assuming that $\mathcal{M}$ is a *-subalgebra of an B^*-algebra such that Axioms (A1) and (A2) hold, one requires that $\mathcal{M}$ is a Banach algebra satisfying the following three axioms:

(B1) If $e - g^*g$ is positive definite in $\mathcal{M}$ and $g \in \mathcal{M}_\pm$, then $e - g$ is invertible and $(e-g)^{-1} \in \mathcal{M}_\pm$;

(B2) If $\|g\|_{\mathcal{M}} < 1$, then $e - g^*g$ is positive definite in $\mathcal{M}$;

(B3) If a is positive definite in $\mathcal{M}$, then $e + a$ is also positive definite in $\mathcal{M}$.

Theorems I.1.1 is similar to an earlier result in [7] but now concerns positive extensions in the setting of an algebra with an involution. In [7] the algebra has no involution and the extensions are required to be invertible. Theorem I.1.2 is inspired by earlier concrete versions (see [3]). The proofs of Theorems I.1.1 and I.1.2 can be found in [11] and [12], respectively.

I.2. Strictly contractive extensions. Let $\mathcal{B}$ be a vector space, and suppose that $\mathcal{B}$ admits a direct sum decomposition

$$\mathcal{B} = \mathcal{B}_- \dot{+} \mathcal{B}_+,$$

where $\mathcal{B}_-$ and $\mathcal{B}_+$ are subspaces of $\mathcal{B}$. We are interested in the following problem: given $\phi \in \mathcal{B}_-$, when does there exists an element $\psi \in \mathcal{B}$ such that $\|\psi\| < 1$ (for some specified norm) and $\psi - \phi \in \mathcal{B}_+$? Such an element ψ is called a strictly contractive extension of ϕ. Furthermore, if a strictly contractive extension of ϕ exists, we want to describe all strictly contractive extensions of ϕ. We shall solve the problem using the results in the previous section. In order to be able to do this we need some more structure on $\mathcal{B}$. In what follows we shall assume that $\mathcal{B}$ can be embedded in an algebra of 2×2 matrices with a unit and an involution.

We shall assume that the space $\mathcal{B}$ appears as the space of (1,2)-elements of the following algebra of 2×2 block matrices:

$$\mathcal{N} = \left\{ f = \begin{pmatrix} a & b \\ c & d \end{pmatrix} : a \in \mathcal{A}, b \in \mathcal{B}, c \in \mathcal{C}, d \in \mathcal{D} \right\}. \tag{2.1}$$

Here $\mathcal{A}$ and $\mathcal{D}$ are algebras with identities $e_{\mathcal{A}}$ and $e_{\mathcal{D}}$, respectively, and involutions $*$, and $\mathcal{C}$ is a vector space which is isomorphic to $\mathcal{B}$ via an operator $*$ whose inverse is also denoted by $*$, such that for every choice of $a \in \mathcal{A}, b \in \mathcal{B}, c \in \mathcal{C}$ and $d \in \mathcal{D}$:

$$\begin{aligned} &bc \in \mathcal{A}, (bc)^* = c^*b^*; ab \in \mathcal{B}, (ab)^* = b^*a^*; \\ &bd \in \mathcal{B}, (bd)^* = d^*b^*; ca \in \mathcal{C}, (ca)^* = a^*c^*; \\ &dc \in \mathcal{C}, (dc)^* = c^*d^*; cb \in \mathcal{D}, (cb)^* = b^*c^*; \end{aligned} \tag{2.2}$$

It is easy to see that $\mathcal{N}$ is an algebra (with respect to the natural rules for matrix multiplication and addition) with unit

$$e := \begin{pmatrix} e_{\mathcal{A}} & 0 \\ 0 & e_{\mathcal{D}} \end{pmatrix}.$$

We define an involution $*$ on $\mathcal{N}$ by setting

$$\begin{pmatrix} a & b \\ c & d \end{pmatrix}^* := \begin{pmatrix} a^* & c^* \\ b^* & d^* \end{pmatrix}.$$

We will assume some additional structure within each of the four spaces $\mathcal{A} - \mathcal{D}$. The algebras $\mathcal{A}$ and $\mathcal{D}$ are assumed to admit direct sum decompositions

$$\mathcal{A} = \mathcal{A}_-^0 \dot{+} \mathcal{A}_d \dot{+} \mathcal{A}_+^0, \mathcal{D} = \mathcal{D}_-^0 \dot{+} \mathcal{D}_d \dot{+} \mathcal{D}_+^0 \tag{2.3}$$

in which all six of the newly indicated spaces are subalgebras and are such that

$$\begin{aligned} &e_{\mathcal{A}} \in \mathcal{A}_d, (\mathcal{A}_-^0)^* = \mathcal{A}_+^0, (\mathcal{A}_d)^* = \mathcal{A}_d, \\ &e_{\mathcal{D}} \in \mathcal{D}_d, (\mathcal{D}_-^0)^* = \mathcal{D}_+^0, (\mathcal{D}_d)^* = \mathcal{D}_d, \end{aligned} \tag{2.4}$$

and the inclusions

$$\mathcal{A}_d\mathcal{A}_\pm^0 \subset \mathcal{A}_\pm^0, \mathcal{A}_\pm^0\mathcal{A}_d \subset \mathcal{A}_\pm^0, \mathcal{D}_d\mathcal{D}_\pm^0 \subset \mathcal{D}_\pm^0, \mathcal{D}_\pm^0\mathcal{D}_d \subset \mathcal{D}_\pm^0 \tag{2.5}$$

are in force. It is then readily checked that

$$\mathcal{A}_\pm := \mathcal{A}_\pm^0 \dotplus \mathcal{A}_d, \mathcal{D}_\pm := \mathcal{D}_\pm^0 \dotplus \mathcal{D}_d$$

are algebras. Moreover, if $a \in \mathcal{A}_d$ (resp. $d \in \mathcal{D}_d$) and is invertible, then $a^{-1} \in \mathcal{A}_d$ (resp. $d^{-1} \in \mathcal{D}_d$). Finally, we suppose that $\mathcal{B}$ and $\mathcal{C}$ admit decompositions

$$\mathcal{B} = \mathcal{B}_- \dotplus \mathcal{B}_+, \mathcal{C} = \mathcal{C}_- \dotplus \mathcal{C}_+, \tag{2.6}$$

where $\mathcal{B}_\pm \subset \mathcal{B}$ and $\mathcal{C}_\pm \subset \mathcal{C}$ are subspaces satisfying

$$\begin{gathered} \mathcal{C}_- = \mathcal{B}_+^*, \mathcal{C}_+ = \mathcal{B}_-^*, \\ \mathcal{B}_\pm \mathcal{D}_\pm \subset \mathcal{B}_\pm, \mathcal{A}_\pm \mathcal{B}_\pm \subset \mathcal{B}_\pm, \\ \mathcal{C}_\pm \mathcal{B}_\pm \subset \mathcal{D}_\pm^0, \mathcal{B}_\pm \mathcal{C}_\pm \subset \mathcal{A}_\pm^0, \\ \mathcal{C}_\pm \mathcal{A}_\pm \subset \mathcal{C}_\pm, \mathcal{D}_\pm \mathcal{C}_\pm \subset \mathcal{C}_\pm. \end{gathered} \tag{2.7}$$

Now let us introduce the following subspaces of $\mathcal{N}$:

$$\begin{aligned} \mathcal{N}_1 &= \begin{pmatrix} 0 & \mathcal{B}_+ \\ 0 & 0 \end{pmatrix} = \left\{ \begin{pmatrix} 0 & b \\ 0 & 0 \end{pmatrix} \,|b \in \mathcal{B}_+ \right\}, \\ \mathcal{N}_2^0 &= \begin{pmatrix} \mathcal{A}_+^0 & \mathcal{B}_- \\ 0 & \mathcal{D}_+^0 \end{pmatrix} = \left\{ \begin{pmatrix} a & b \\ 0 & d \end{pmatrix} \,|a \in \mathcal{A}_+^0, b \in \mathcal{B}_-, d \in \mathcal{D}_+^0 \right\}, \\ \mathcal{N}_d &= \begin{pmatrix} \mathcal{A}_d & 0 \\ 0 & \mathcal{D}_d \end{pmatrix} = \left\{ \begin{pmatrix} a & 0 \\ 0 & d \end{pmatrix} \,|a \in \mathcal{A}_d, d \in \mathcal{D}_d \right\}, \\ N_3^0 &= \begin{pmatrix} \mathcal{A}_-^0 & 0 \\ \mathcal{C}_+ & \mathcal{D}_-^0 \end{pmatrix} = \left\{ \begin{pmatrix} a & 0 \\ c & d \end{pmatrix} \,|a \in \mathcal{A}_-^0, c \in \mathcal{C}_+, d \in \mathcal{D}_-^0 \right\}, \\ \mathcal{N}_4 &= \begin{pmatrix} 0 & 0 \\ \mathcal{C}_- & 0 \end{pmatrix} = \left\{ \begin{pmatrix} 0 & 0 \\ c & 0 \end{pmatrix} \,|c \in \mathcal{C}_- \right\}. \end{aligned}$$

Obviously,

$$\mathcal{N} = \mathcal{N}_1 \dotplus \mathcal{N}_2^0 \dotplus \mathcal{N}_d \dotplus \mathcal{N}_3^0 \dotplus \mathcal{N}_4, \tag{2.8}$$

and this decomposition satisfies the conditions (i) and (ii) in Section I.1. With respect to positive elements we assume that the algebra $\mathcal{N}$ satisfies the following axiom:

AXIOM (A0a). If $\begin{pmatrix} a & b \\ c & d \end{pmatrix}$ is positive definite in $\mathcal{N}$, then a is positive definite in $\mathcal{A}$ and d is positive definite in $\mathcal{D}$.

Note that the identities

$$\begin{aligned} k_h := \begin{pmatrix} e_\mathcal{A} & h \\ h^* & e_\mathcal{D} \end{pmatrix} &= \begin{pmatrix} e_\mathcal{A} & 0 \\ h^* & e_\mathcal{D} \end{pmatrix} \begin{pmatrix} e_\mathcal{A} & 0 \\ 0 & e_\mathcal{D} - h^*h \end{pmatrix} \begin{pmatrix} e_\mathcal{A} & h \\ 0 & e_\mathcal{D} \end{pmatrix} \\ &= \begin{pmatrix} e_\mathcal{A} & h \\ 0 & e_\mathcal{D} \end{pmatrix} \begin{pmatrix} e_\mathcal{A} - hh^* & 0 \\ 0 & e_\mathcal{D} \end{pmatrix} \begin{pmatrix} e_\mathcal{A} & 0 \\ h^* & e_\mathcal{D} \end{pmatrix} \end{aligned} \tag{2.9}$$

imply that for $h \in \mathcal{B}$ the element k_h is positive definite in $\mathcal{N}$ if and only if $e_{\mathcal{D}} - h^*h$ is positive definite in $\mathcal{D}$, or equivalently, $e_{\mathcal{A}} - hh^*$ is positive definite in $\mathcal{A}$. From now on we shall write e for both $e_{\mathcal{A}}$ and $e_{\mathcal{D}}$.

Let $\phi \in \mathcal{B}_-$ be given. An element $\psi \in \mathcal{B}$ is called a *strictly contractive extension of* ϕ if $\psi - \phi \in \mathcal{B}_+$ and $e - \psi^*\psi$ is positive definite in $\mathcal{D}$. Recall that an element $d \in \mathcal{D}$ is called positive definite in $\mathcal{D}$ if there exists an invertible element $c \in \mathcal{D}$ such that $d = cc^*$. The term "strictly contractive extension" is justified by the fact that in a B^*-algebra an element b has the norm less than one if and only if $e - b^*b$ is positive definite. We are interested to find all strictly contractive extensions of a given $\phi \in \mathcal{B}_-$. We call $g \in \mathcal{B}$ a *(strictly contractive) triangular extension of* ϕ if g is a strictly contractive extension of ϕ and $g(e - g^*g)^{-1}$ belongs to $\mathcal{B}_-$. In what follows we will omit the words strictly contractive and just talk about a triangular extension. As in Section I.1 we say that a positive element $d \in \mathcal{D}$ admits a *left (right) spectral factorization* (with respect to the decomposition of $\mathcal{D}$ in (2.3)) if there is an invertible $c \in \mathcal{D}_+(\mathcal{D}_-)$ such that $d = cc^*$ and $c^{-1} \in \mathcal{D}_+(\mathcal{D}_-)$. In $\mathcal{A}$ we have similar definitions. Note that k_ϕ admits a right spectral factorization in $\mathcal{N}$ if and only if $e - \phi^*\phi$ admits a right spectral factorization in $\mathcal{D}$ and that k_ϕ admits a left spectral factorization in $\mathcal{N}$ if and only if $e - \phi\phi^*$ admits a left spectral factorization in $\mathcal{A}$ (use (2.9)).

The following theorem, which follows from Theorem I.1.1, gives a way to construct a triangular extension of a given $\phi \in \mathcal{B}_-$. We need some additional notation. If $\mathcal{E}_0$ is a subspace of the space $\mathcal{E}$, we let $P_{\mathcal{E}_0}$ denote the projection in $\mathcal{E}$ on $\mathcal{E}_0$ along a natural complement. So, for instance, $P_{\mathcal{A}_+}$ is the projection on $\mathcal{A}_+$ along $\mathcal{A}_-^0$.

THEOREM I.2.1. *Let $\mathcal{N}$ be the algebra (2.1) with band structure (2.8), and assume that Axiom (A0a) is satisfied. Let $\phi \in \mathcal{B}_-$. The element ϕ has a triangular extension g such that $e - gg^*$ admits a left and $e - g^*g$ admits a right spectral factorization if and only if the equations*

$$e = a - P_{\mathcal{A}_-}(\phi(P_{\mathcal{C}_+}(\phi^*a)), e = d - P_{\mathcal{D}_+}(\phi^* P_{\mathcal{B}_-}(\phi d)). \tag{2.10}$$

have solutions a and d with the following properties:

(i) $a \in \mathcal{A}_-, d \in \mathcal{D}_+$,

(ii) *a and d are invertible,* $a^{-1} \in \mathcal{A}_-, d^{-1} \in \mathcal{D}_+$,

(iii) *$P_{\mathcal{A}_d}a$ and $P_{\mathcal{D}_d}d$ are positive definite in $\mathcal{A}_d$ and $\mathcal{D}_d$, respectively.*

In that case ϕ has a unique triangular extension g for which $e - gg^*$ admits a left and $e - g^*g$ admits a right spectral factorization, and this g is given by

$$g := bd^{-1} = a^{*-1}c^*,$$

where

$$b = P_{\mathcal{B}_-}(\phi d), c = P_{\mathcal{C}_+}(\phi^* a). \tag{2.11}$$

The spectral factorizations of $e - gg^*$ and $e - g^*g$ are given by

$$(2.12) \qquad e - gg^* = a^{*-1}(P_{\mathcal{A}_d}a)a^{-1}, e - g^*g = d^{*-1}(P_{\mathcal{D}_d}d)d^{-1}.$$

In many applications the algebra $\mathcal{N}$ has the additional property that every positive definite element admits a left and right spectral factorization. For such an algebra $\mathcal{N}$ the hypothesis of Theorem I.2.1 imply that there exists a unique triangular extension of ϕ.

If $\phi \in \mathcal{B}$, we let $\Xi := P_{\mathcal{A}_-}\phi : \mathcal{C}_+ \to \mathcal{A}_-$ denote the operator defined by the following action:

$$\Xi(c) = (P_{\mathcal{A}_-}\phi)(c) := P_{\mathcal{A}_-}(\phi c), c \in \mathcal{C}_+.$$

We shall employ this notation also for other subspaces.

THEOREM I.2.2. *Let $\mathcal{N}$ be the algebra (2.1) with band structure (2.8), and assume that Axiom (A0a) is satisfied. Further, assume that $\mathcal{N}$ is a Banach algebra. Let $\phi \in \mathcal{B}_-$ be given. Introduce the following operators*

$$\Xi := P_{\mathcal{A}_-}\phi : \mathcal{C}_+ \to \mathcal{A}_-; \Xi_* := P_{\mathcal{C}_+}\phi^* : \mathcal{A}_- \to \mathcal{C}_+;$$

$$\widetilde{\Xi} := P_{\mathcal{B}_-}\phi : \mathcal{D}_+ \to \mathcal{B}_-; \widetilde{\Xi}_* := P_{\mathcal{D}_+}\phi^* : \mathcal{B}_- \to \mathcal{D}_+.$$

Suppose that for each $0 \leq \epsilon \leq 1$ the operators $I - \epsilon^2\Xi\Xi_*$ and $I - \epsilon^2\widetilde{\Xi}_*\widetilde{\Xi}$ are invertible, and that the elements

$$P_{\mathcal{A}_d}[(I - \epsilon^2\Xi\Xi_*)^{-1}e], P_{\mathcal{D}_d}[(I - \epsilon^2\widetilde{\Xi}_*\widetilde{\Xi})^{-1}e]$$

are positive definite in $\mathcal{A}_d$ and $\mathcal{D}_d$, respectively. Let $r \in \mathcal{A}_d$ and $s \in \mathcal{D}_d$ be such that

$$P_{\mathcal{A}_d}[(I - \Xi\Xi_*)^{-1}e] = r^*r, P_{\mathcal{D}_d}[(I - \widetilde{\Xi}_*\widetilde{\Xi})^{-1}e] = s^*s,$$

and put

$$\alpha := ((I - \Xi\Xi_*)^{-1}e)r^{-1}, \gamma := P_{\mathcal{C}_+}(\phi^*\alpha),$$

$$\delta := ((I - \widetilde{\Xi}_*\widetilde{\Xi})^{-1}e)s^{-1}, \beta := P_{\mathcal{B}_-}(\phi\delta).$$

Then ϕ has a unique triangular extension g for which $e - g^*g$ admits a left and $e - gg^*$ admits a right spectral factorization, and this g is given by

$$g := \beta\delta^{-1} = \alpha^{*-1}\gamma^*.$$

Moreover, $\alpha^{\pm 1} \in \mathcal{A}_-, \delta^{\pm 1} \in \mathcal{D}_+$, and the spectral factorizations of $e - gg^*$ and $e - g^*g$ are given by

$$e - gg^* = \alpha^{*-1}\alpha^{-1}, e - g^*g = \delta^{*-1}\delta^{-1}.$$

The following theorem describes the set of all strictly contractive extensions of a given element in $\mathcal{B}_-$.

THEOREM I.2.3. *Let $\mathcal{N}$ be the algebra (2.1) with band structure (2.8), and assume that the algebra $\mathcal{N}$ is a *-subalgebra of a B^*-algebra $\mathcal{R}$ such that the unit e of $\mathcal{R}$ belongs to N and Axioms (A0a), (A1) and (A2) hold. Let $\phi \in \mathcal{B}_-$, and suppose that ϕ has a triangular extension g such that $e - gg^*$ admits a left and $e - g^*g$ admits a right spectral factorization. Let $\alpha \in \mathcal{A}_-$ and $\delta \in \mathcal{D}_+$ be invertible elements such that $\alpha^{-1} \in \mathcal{A}_-, \delta^{-1} \in \mathcal{D}_+$ and*

$$(e - gg^*)^{-1} = \alpha\alpha^* \ , (e - g^*g)^{-1} = \delta\delta^*,$$

and put

$$\beta = P_{\mathcal{B}_-}(\phi\delta), \gamma = P_{\mathcal{C}_+}(\phi^*\alpha).$$

Then each strictly contractive extension $\psi \in \mathcal{B}$ of ϕ is of the form

$$\psi = (\alpha h + \beta)(\gamma h + \delta)^{-1}, \tag{2.13}$$

where h is an element in $\mathcal{B}_+$ such that $e - h^*h$ is positive definite in $\mathcal{D}$. Furthermore, equation (2.13) gives a one-one correspondence between all such h and all strictly contractive extensions ψ of ϕ. Alternatively, each strictly contractive extension $\psi \in \mathcal{B}$ of ϕ is of the form

$$\psi = (\alpha^* + f^*\beta^*)^{-1}(\gamma^* + f^*\delta^*), \tag{2.14}$$

where f is an element in $\mathcal{B}_+$ such that $e - ff^*$ is positive definite in $\mathcal{A}$. Furthermore, equation (2.14) gives a one-one correspondence between all such f and all strictly contractive extensions ψ of ϕ.

Note that one can find α and δ in Theorem I.2.3 from the elements a and d in Theorem I.2.1 very easily by using (2.12). In order to prove Theorem I.2.3 one puts

$$u = \begin{pmatrix} -e & \beta \\ 0 & -\delta \end{pmatrix}, v = \begin{pmatrix} \alpha & 0 \\ -\gamma & e \end{pmatrix},$$

and applies Theorem I.1.2.

The approach in this section is the same as the abstract approach in [10], except that here the algebra is enriched with an involution. Both Theorems I.2.2 and I.2.3 are inspired by earlier concrete versions in [9], [10] (see also [3]). The proofs of Theorems I.2.1 and I.2.3 appear in [12] (it is not essential in those proofs that $\mathcal{A} = \mathcal{B} = \mathcal{C} = \mathcal{D}$). The proof of Theorem I.2.2 can be found in [11].

I.3. Maximum entropy principles. This section concerns maximum entropy principles in the general setting of the band method. These principles provide alternative ways to identify the band extension (in the positive extension problem) and the triangular extension (in the strictly contractive extension problem).

I.3.1. Positive extensions. Let $\mathcal{M}$ be an algebra with unit e and involution $*$ and with the band structure (1.1). We introduce the following notions. Let b be a positive definite element of $\mathcal{M}$ which admits a right spectral factorization $b = b_- b_-^*, b_-^{\pm 1} \in \mathcal{M}_-$. We define the *right multiplicative diagonal* $\Delta_r(b)$ of b to be the element

$$\Delta_r(b) := P_d(b_-)P_d(b_-)^*.$$

The right multiplicative diagonal of b is well-defined; it does not depend on the particular choice of the spectral factorization (see [13]). Note that $\Delta_r(b)$ is positive definite in $\mathcal{M}_d$. This follows from $P_d(b_-)^{-1} = P_d(b_-^{-1})$ (see Lemma I.1.2 in [11]). When b admits a left spectral factorization $b = b_+ b_+^*, b_+^{\pm 1} \in \mathcal{M}_+$, we define its *left multiplicative diagonal* $\Delta_l(b)$ to be the element

$$\Delta_l(b) := P_d(b_+)P_d(b_+)^*.$$

Again $\Delta_l(b)$ is well-defined and $\Delta_l(b)$ is positive definite in $\mathcal{M}_d$.

Recall that an element $a \in \mathcal{M}$ is called nonnegative definite in $\mathcal{M}$ if there is an element $c \in \mathcal{M}$ such that $a = c^*c$. With respect to nonnegative definite elements we introduce the following two axioms.

AXIOM (A3). The element $P_d(c^*c)$ is nonnegative definite in $\mathcal{M}$ for all $c \in \mathcal{M}$.

AXIOM (A4). If $P_d(c^*c) = 0$, then $c = 0$.

When $\mathcal{M}$ satisfies these two axioms we have the following general maximum entropy principle.

THEOREM I.3.1. *Let $\mathcal{M}$ be the algebra with band structure (1.1), and assume that $\mathcal{M}$ satisfies Axioms (A3) and (A4). Let $k \in \mathcal{M}_c$ have a band extension b which admits a right (left) spectral factorization. Then for any positive extension a of k which admits a right (left) spectral factorization*

$$\Delta_r(b) \geq \Delta_r(a) \quad (\Delta_l(b) \geq \Delta_l(a)). \tag{3.1}$$

Furthermore, equality holds in (3.1) if and only if $a = b$.

Theorem I.3.1 is inspired by earlier concrete versions (see [4], [6]). Its proof can be found in [13].

Let $\mathcal{M}_d^>$ denote the set of all elements in $\mathcal{M}_d$ that are positive definite:

$$\mathcal{M}_d^> := \{a \in \mathcal{M}_d \mid a > 0\}.$$

We call a function $F : \mathcal{M}_d^> \to \mathbf{R}$ *strictly monotone* if $d_1 \geq d_2$ and $d_1 \neq d_2$ implies $F(d_1) > F(d_2)$.

COROLLARY I.3.2. *Let $\mathcal{M}$ be the algebra with band structure (1.1), and assume that $\mathcal{M}$ satisfies Axioms (A3) and (A4), and let $F : \mathcal{M}_d^{>} \to \mathbf{R}$ be strictly monotone. Let $k \in \mathcal{M}_c$. Suppose that k has a band extension b which admits a right (left) spectral factorization. Then for any positive extension a of k which admits a right (left) spectral factorization*

$$F(\Delta_r(b)) \geq F(\Delta_r(a)) \quad (F(\Delta_l(b)) \geq F(\Delta_l(a))). \tag{3.2}$$

Furthermore, equality holds in (3.2) if and only if $a = b$.

I.3.2. Strictly contractive extensions. In this subsection $\mathcal{N}$ is the algebra (2.1) with band structure (2.8). We shall use the following axiom.

AXIOM (A0b). If $\begin{pmatrix} a & b \\ c & d \end{pmatrix}$ is nonnegative definite in $\mathcal{N}$, then a is nonnegative definite in $\mathcal{A}$ and d is nonnegative definite in $\mathcal{D}$.

The notions of right and left multiplicative diagonals of elements in $\mathcal{A}$ and $\mathcal{D}$ are introduced in the same way as is done in Subsection I.3.1 for elements in $\mathcal{M}$. We now have the following theorems

THEOREM I.3.3. *Let $\mathcal{N}$ be the algebra (2.1) with band structure (2.8), and assume that $\mathcal{N}$ satisfies Axioms (A0a), (A0b), (A3) and (A4). Let $\phi \in \mathcal{B}_-$, and suppose that ϕ has a triangular extension g such that $e - g^*g$ admits a right spectral factorization. Then for any strictly contractive extension ψ of ϕ such that $e - \psi^*\psi$ admits a right spectral factorization*

$$\Delta_r(e - g^*g) \geq \Delta_r(e - \psi^*\psi),$$

and equality holds if and only if $\psi = g$.

THEOREM I.3.4. *Let $\mathcal{N}$ be the algebra (2.1) with band structure (2.8), and assume that N satisfies Axioms (A0a), (A0b), (A3) and (A4). Let $\phi \in \mathcal{B}_-$, and suppose that ϕ has a triangular extension g such that $e - gg^*$ admits a left spectral factorization. Then for any strictly contractive extension ψ of ϕ such that $e - \psi\psi^*$ admits a left spectral factorization*

$$\Delta_l(e - gg^*) \geq \Delta_l(e - \psi\psi^*),$$

and equality holds if and only if $\psi = g$.

Both theorems can be proved by using Theorem I.3.1 (see [13]). As in Subsection I.3.1 any strictly monotone function on $\mathcal{D}_d^{>}$ (or on $\mathcal{A}_d^{>}$) can be used to pick out the triangular extension.

CHAPTER II. AN APPLICATION.

In this chapter we deal with one particular application of the general theory, namely the nonstationary version of the positive and strictly contractive extension problems. This application is a very general one; it also includes the solution of the classical Carathéodory-Toeplitz and Nehari extension problems.

II.1. A nonstationary positive extension problem. Let $\mathcal{L}$ denote the linear space of all semi-infinite operator matrices $V = \left(V_{jk}\right)_{j,k=0}^{\infty}$ such that

$$\|V\|_{\mathcal{L}} = \sum_{v=-\infty}^{\infty} \sup_{k-j=v} \|V_{jk}\| < \infty. \tag{1.1}$$

The entry V_{jk} of V is assumed to be an operator from the Hilbert space H_k into the Hilbert space H_j. The space $\mathcal{L}$ is an algebra under the usual operations of addition and multiplication for infinite matrices, and $\mathcal{L}$ endowed with the norm (1.1) is a Banach algebra. For $V = \left(V_{jk}\right)_{j,k=0}^{\infty}$ we define

$$V^* = \left(V_{kj}^*\right)_{j,k=0}^{\infty},$$

and this operation $*$ is an involution on $\mathcal{L}$. The element $E = \text{diag}(I_{H_j})_{j=0}^{\infty}$ is the unit in $\mathcal{L}$.

We write Z for the Hilbert space $\bigoplus_{j=0}^{\infty} H_j$. Thus Z consists of all square summable sequences $(\eta_j)_{j=0}^{\infty}$ with $\eta_j \in H_j$. Note that each element $V \in \mathcal{L}$ induces a bounded linear operator on Z. We write $\|V\|_O$ for the norm of V as an operator on Z. Writing an element $V \in \mathcal{L}$ as a sum of diagonals one easily checks that $\|V\|_O \leq \|V\|_{\mathcal{L}}$. It is known (see [12], Lemma II.3.1) that $V \in \mathcal{L}$ is invertible as an operator if and only if V is invertible in $\mathcal{L}$.

We will call an element $V \in \mathcal{L}$ *positive definite* if

$$V = UU^* \tag{1.2}$$

for some invertible $U \in \mathcal{L}$. In [12] the following lemma is proved.

LEMMA II.1.1. *Let $V \in \mathcal{L}$. Then V is positive definite as an element of the algebra $\mathcal{L}$ if and only if V is positive definite as an operator on Z. In that case one may choose the semi-infinite operator matrix U in (1.2) to be lower or upper triangular.*

We are interested in the following extension problem. Let $A_{ij} = A_{ji}^*, i, j = 0, 1, \ldots, |j - i| \leq p$, be given operators. Determine all $V \in \mathcal{L}$ such that

(a) V is positive definite,

(b) $V_{ij} = A_{ij}, |j - i| \leq p$.

Such an infinite matrix $V \in \mathcal{L}$ will be called a *positive extension* of the band $\{A_{ij} \mid |j - i| \leq p\}$. Let $A^{(n,n+p)}$ be defined by (1.3) below. Note that if there exists a positive extension V of the band $\{A_{ij} \mid |j - i| \leq p\}$, then the sequences $A^{(n,n+p)}$, $n = 0, 1, \ldots$, and $[A^{(n,n+p)}]^{-1}, n = 0, 1 \ldots$, are uniformly bounded (by $\|V\|$ and $\|V^{-1}\|$, respectively) sequences of positive definite matrices. This necessary condition for the existence of a positive extension is also sufficient as the next theorem shows.

THEOREM II.1.2. *Let* $A_{ij} = A_{ji}^*, i,j = 0,1,\ldots,\ |j-i| \leq p$, *be an operator from the Hilbert space* H_j *into the Hilbert space* H_i, *and suppose that*

$$A^{(n,n+p)} := \left(A_{ij}\right)_{i,j=n}^{n+p}, n = 0,1,2,\ldots\ , \tag{1.3}$$

are positive definite operators such that $A^{(n,n+p)}$ *and* $[A^{(n,n+p)}]^{-1}$ *are uniformly bounded in norm. For* $n = 0,1,\ldots$ *let*

$$\begin{pmatrix} y_{nn} \\ y_{n+1,n} \\ \vdots \\ y_{n+p,p} \end{pmatrix} = [A^{(n,n+p)}]^{-1} \begin{pmatrix} I \\ 0 \\ \vdots \\ 0 \end{pmatrix}, \tag{1.4}$$

and

$$\begin{pmatrix} x_{\gamma(n),n} \\ \vdots \\ x_{n-1,n} \\ x_{n,n} \end{pmatrix} = [A^{(\gamma(n),n)}]^{-1} \begin{pmatrix} 0 \\ \vdots \\ 0 \\ I \end{pmatrix}, \tag{1.5}$$

where $\gamma(q) = \max\{0, q-p\}$. *Let the infinite matrices* U *and* V *be defined by*

$$V_{ij} = \begin{cases} y_{ij}y_{jj}^{-1/2} & , j \leq i \leq j+p; \\ 0 & , \text{elsewhere}; \end{cases} \tag{1.6}$$

$$U_{ij} = \begin{cases} x_{ij}x_{jj}^{-1/2} & , \gamma(j) \leq i \leq j; \\ 0 & , \text{elsewhere}. \end{cases} \tag{1.7}$$

Then the semi-infinite operator matrix F *given by the following factorizations*

$$F := U^{*-1}U^{-1} = V^{*-1}V^{-1} \tag{1.8}$$

is the unique positive definite extension of the given band with $(F^{-1})_{ij} = 0,\ |j-i| > p$.

Let us explain how the above theorem may be derived from Theorem I.1.1. The first step is to define a band structure on $\mathcal{L}$. This is done as follows:

$$\mathcal{L}_1 = \mathcal{L}_4^* = \{\left(F_{ij}\right)_{i,j=0}^{\infty} \in \mathcal{L} \mid F_{ij} = 0, j-i \leq p\},$$

$$\mathcal{L}_2^0 = \mathcal{L}_3^{0*} = \{\left(F_{ij}\right)_{i,j=0}^{\infty} \in \mathcal{L} \mid F_{ij} = 0, j-i > p \text{ and } j-i \leq 0\},$$

$$\mathcal{L}_d = \{\left(F_{ij}\right)_{i,j=0}^{\infty} \in \mathcal{L} \mid F_{ij} = 0, j \neq i\}.$$

It is easy to see that

$$\mathcal{L} = \mathcal{L}_1 \dotplus \mathcal{L}_2^0 \dotplus \mathcal{L}_d \dotplus \mathcal{L}_3^0 \dotplus \mathcal{L}_4, \tag{1.9}$$

and that the conditions (i) and (ii) in Section I.1 are fulfilled. In other words $\mathcal{L}$ is an algebra with band structure (1.9).

Next, we may apply Theorem I.1.1 to $K = \left(K_{ij}\right)_{i,j=0}^{\infty}$, where $K_{ij} = A_{ij}$ for $|j-i| \leq p$ and $K_{ij} = 0$ otherwise. Note that $K \in \mathcal{L}_c := \mathcal{L}_2^0 \dotplus \mathcal{L}_d \dotplus \mathcal{L}_3^0$. The two equations in Theorem I.1.1 take the following form

$$P_2(KX) = E\ , P_3(KY) = E\ , \tag{1.10}$$

where P_2 (resp. P_3) is the projection in $\mathcal{L}$ on $\mathcal{L}_2 := \mathcal{L}_2^0 \dotplus \mathcal{L}_d$ (resp. $\mathcal{L}_3 := \mathcal{L}_d \dotplus \mathcal{L}_3^0$) along $\mathcal{L}_1 \dotplus \mathcal{L}_3^0 \dotplus \mathcal{L}_4$ (resp. $\mathcal{L}_1 \dotplus \mathcal{L}_2^0 \dotplus \mathcal{L}_4$). Now, let $X = \left(x_{ij}\right)_{i,j=0}^{\infty}$ and $Y = \left(y_{ij}\right)_{i,j=0}^{\infty}$ be the upper and lower triangular band matrices of which the entries in the band $|j-i| \leq p$ are given by (1.5) and (1.4), respectively, and which have zero entries outside this band. One checks easily that X and Y are solutions of (1.10). In order to apply Theorem I.1.1 it remains to prove that the main diagonals of X and Y (which we denote by X_0 and Y_0, respectively) are positive definite operators on H, and that X and Y are invertible in $\mathcal{L}_+ := \mathcal{L}_1 \dotplus \mathcal{L}_2$ and $\mathcal{L}_- := \mathcal{L}_3 \dotplus \mathcal{L}_4$, respectively. The proofs of these two statements are rather involved and can be found in [12]. It follows that $U := XY_0^{-1/2}$ and $V := YX_0^{-1/2}$ are invertible elements of $\mathcal{L}$, and $U^{-1} \in \mathcal{L}_+$ and $V^{-1} \in \mathcal{L}_-$. Now Theorem I.1.1 guarantees that the infinite matrix F defined by (1.8) is precisely the unique band extension of K.

The algebra $\mathcal{L}$ (with band structure (1.9)) satisfies Axioms (A1) and (A2) with $\mathfrak{R}$ equal to the B^*-algebra of all bounded linear operators on $Z := \bigoplus_{j=0}^{\infty} H_j$. Indeed, Lemma II.3.1 in [12] shows that Axiom (A1) is fulfilled. In order to check Axiom (A2), let $V^{(n)} \in \mathcal{L}_+, V \in \mathcal{L}$ and $\|V^{(n)} - V\|_O \to 0$ as $n \to \infty$. Then

$$\|V_{ij}\| = \|(V^{(n)} - V)_{ij}\| \leq \|V^{(n)} - V\|_O, j < i,$$

so it follows that $V_{ij} = 0, j < i$. Hence $V \in \mathcal{L}_+$ and Axiom (A2) is fullfilled. Thus we may apply Theorem I.1.2 to the algebra $\mathcal{L}$, which yields the following description for the set of all positive extensions.

THEOREM II.1.3. *Let $A_{ij} = A_{ji}^*, i,j = 0,1,\ldots,$ $|j-i| \leq p$, be an operator from the Hilbert space H_j into the Hilbert space H_i. In order that there exists a positive extension of the band $\{A_{ij} |\ |j-i| \leq p\}$ it is necessary and sufficient that $A^{(n,n+p)}, n = 0,1,\ldots$ and $[A^{(n,n+p)}]^{-1}$, $n = 0,1,\ldots$ are uniformly bounded (in norm) sequences of positive definite operators. Assume that the latter conditions hold. Let U and V be the infinite matrices defined by (1.4)-(1.7). Then each positive extension F of the given band is of the form*

$$F = (G^*V^* + U^*)^{-1}(I - G^*G)(VG + U)^{-1}, \tag{1.11}$$

where G is an element of $\mathcal{L}$ with $\|G\|_O < 1$ and $G_{ij} = 0, j-i \leq p$. Furthermore, formula (1.11) gives a 1-1 correspondence between all such G and all positive extensions F. Alternatively, each positive extension F of the given band is of the form

$$F = (V^* + U^*H^*)^{-1}(I - H^*H)(V + UH)^{-1}, \tag{1.12}$$

where H is an element of $\mathcal{L}$ with $\|H\|_O < 1$ and $H_{ij} = 0,\ j-i \geq -p$. Furthermore, formula (1.12) gives a 1-1 correspondence between all such H and all positive extensions F.

We know from Lemma II.1.1 that $V \in \mathcal{L}$ is positive definite if and only if V is positive definite as an operator on Z. From the same lemma we also deduce that each positive definite V in $\mathcal{L}$ can be written uniquely as

$$V = (E+U)^*\Delta_r(V)(E+U), \tag{1.13}$$

where $U = \left(U_{jk}\right)_{j,k=0}^{\infty}$ is such that $E+U$ is invertible in $\mathcal{L}$, $U_{jk} = 0$ and $[(E+U)^{-1} - E]_{jk} = 0, j \leq k$, and $\Delta_r(V) = \operatorname{diag}(\Delta_i^r(V))_{i=0}^{\infty}$, where $\Delta_i^r(V) : H_i \to H_i$. The entries of the diagonal matrix $\Delta_r(V)$ can be calculated as follows:

$$\Delta_0^r(V) = V_{00}, \tag{1.14}$$

$$\Delta_i^r(V)^{-1} = V_{ii} - \begin{pmatrix} V_{i1} \dots V_{i,i-1} \end{pmatrix} \begin{pmatrix} V_{11} & \dots & V_{1i-1} \\ \vdots & \vdots & \vdots \\ V_{i-1,i} & \dots & V_{i-1,i-1} \end{pmatrix}^{-1} \begin{pmatrix} V_{1i} \\ \vdots \\ V_{i-1,i} \end{pmatrix}, i = 1, 2, \dots$$

(see [1]). From (1.13) it follows that $\Delta_r(V)$ is precisely the right multiplicative diagonal of V as an element of the algebra $\mathcal{L}$. Since Axioms (A3) and (A4) are fulfilled for $\mathcal{L}$, Theorem I.3.1 yields the following maximum entropy principle.

THEOREM II.1.4. *Let $A_{ij} = A_{ji}^*, i,j \geq 0,\ |j-i| \leq p$, be a given operator acting from a Hilbert space H_j into a Hilbert space H_i, and suppose that*

$$A^{(n,n+p)} = \left(A_{ij}\right)_{i,j=n}^{n+p}, n = 0, 1, \dots,$$

are positive definite operators on $H_n \oplus \dots \oplus H_{n+p}$ such that $A^{(n,n+p)}$ and $[A^{(n,n+p)}]^{-1}$ are uniformly bounded. Put

$$M_i = \begin{pmatrix} 0 & \dots & 0 & I_{H_i} \end{pmatrix} \begin{pmatrix} A_{\gamma(i),\gamma(i)} & \dots & A_{\gamma(i),i} \\ \vdots & & \vdots \\ A_{i,\gamma(i)} & \dots & A_{ii} \end{pmatrix}^{-1} \begin{pmatrix} 0 \\ \vdots \\ 0 \\ I_{H_i} \end{pmatrix}, i = 0, 1, \dots, \tag{1.15}$$

with $\gamma(i) = \max\{1, i-p\}$. Then for any positive extension V of the given band

$$\Delta_i(V) \leq M_i^{-1}, i = 0, 1, \dots. \tag{1.16}$$

Moreover, equality holds for all i in (1.16) if and only if V is the unique band extension of the given band.

For the case that $H = \mathbb{C}$ Theorem II.2.1 was established earlier in [2].

II.2. A nonstationary version of the Nehari problem. As in Section II.1 the symbol $\mathcal{L}$ denotes the algebra of all semi-infinite operator matrices $V = \left(V_{jk}\right)_{j,k=0}^{\infty}$ such that

$$\|V\|_{\mathcal{L}} = \sum_{v=-\infty}^{\infty} \sup_{k-j=v} \|V_{jk}\| < \infty.$$

We consider the following extension problem in the space $\mathcal{L}$. Let $A_{ij}, 0 \leq j \leq i < \infty$, be an operator from a Hilbert space H_j into the Hilbert space H_i. Determine all $V \in \mathcal{L}$ such that

(1) $\|V\|_O < 1$,

(2) $V_{ij} = A_{ij}, j \leq i$.

Such a semi-infinite matrix V will be called a *strictly contractive extension* of the lower triangular part $\{A_{ij} \mid j \leq i\}$. Recall that $\|V\|_O$ stands for the norm of V as a bounded linear operator on $Z = \overset{\infty}{\underset{j=0}{\oplus}} H_j$. Note that if there exists such an extension V it is necessary that the operators

$$\Lambda_k := \left(A_{ij}\right)_{i=k,j=0}^{\infty\quad k}, \tag{2.1}$$

Λ_k viewed as an operator acting from $H_0 \oplus \cdots \oplus H_k \to Z$, have the property that

$$\sup_{k=0,1,\ldots} \|\Lambda_k\| < 1. \tag{2.2}$$

Since $V \in \mathcal{L}$, also

$$\sum_{v=-\infty}^{0} \sup_{j-i=v} \|A_{ij}\| < \infty. \tag{2.3}$$

It turns out that the two conditions above are also sufficient for the existence of a strictly contractive extension of the lower triangular part $\{A_{ij} \mid j \leq i\}$.

THEOREM II.2.1. *For $0 \leq j \leq i < \infty$ let A_{ij} be an operator from a Hilbert space H_j into the Hilbert space H_i, and let Λ_k, $k = 0, 1, \ldots$, be defined by (2.1).*

Suppose that (2.2) and (2.3) hold. Put

$$\begin{pmatrix} \hat{\alpha}_{ii} \\ \hat{\alpha}_{i+1,i} \\ \vdots \end{pmatrix} = (I - \mathbf{\Lambda}_i \mathbf{\Lambda}_i^*)^{-1} \begin{pmatrix} I \\ 0 \\ \vdots \end{pmatrix} \quad , i = 0, 1, \ldots \tag{2.4}$$

$$\begin{pmatrix} \hat{\beta}_{ii} \\ \hat{\beta}_{i+1,i} \\ \vdots \end{pmatrix} = \mathbf{\Lambda}_i (I - \mathbf{\Lambda}_i^* \mathbf{\Lambda}_i)^{-1} \begin{pmatrix} 0 \\ \vdots \\ 0 \\ I \end{pmatrix} \quad , i = 0, 1, \ldots \tag{2.5}$$

$$\begin{pmatrix} \hat{\gamma}_{0i} \\ \hat{\gamma}_{1i} \\ \vdots \\ \hat{\gamma}_{i,i} \end{pmatrix} = \mathbf{\Lambda}_i^* (I - \mathbf{\Lambda}_i \mathbf{\Lambda}_i^*)^{-1} \begin{pmatrix} I \\ 0 \\ \vdots \end{pmatrix} \quad , i = 0, 1, \ldots \tag{2.6}$$

$$\begin{pmatrix} \hat{\delta}_{0i} \\ \vdots \\ \hat{\delta}_{i-1,i} \\ \hat{\delta}_{ii} \end{pmatrix} = (I - \mathbf{\Lambda}_i^* \mathbf{\Lambda}_i)^{-1} \begin{pmatrix} 0 \\ \vdots \\ 0 \\ I \end{pmatrix}, \quad i = 0, 1, \ldots, \tag{2.7}$$

and let

$$\alpha := \left(\alpha_{ij}\right)_{i,j=0}^{\infty}, \alpha_{ij} = \begin{cases} \hat{\alpha}_{ij}\hat{\alpha}_{jj}^{-1/2} & , i \geq j; \\ 0 & , i < j; \end{cases} \tag{2.8}$$

$$\beta := \left(\beta_{ij}\right)_{i,j=0}^{\infty}, \beta_{ij} = \begin{cases} \hat{\beta}_{ij}\hat{\delta}_{jj}^{-1/2} & , i \geq j; \\ 0 & , i < j; \end{cases} \tag{2.9}$$

$$\gamma := \left(\gamma_{ij}\right)_{i,j=0}^{\infty}, \gamma_{ij} = \begin{cases} \hat{\gamma}_{ij}\hat{a}_{jj}^{-1/2} & , i \leq j; \\ 0 & , i > j; \end{cases} \tag{2.10}$$

$$\delta := \left(\delta_{ij}\right)_{i,j=0}^{\infty}, \delta_{ij} = \begin{cases} \hat{\delta}_{ij}\hat{\delta}_{jj}^{-1/2} & , i \leq j; \\ 0 & , i > j; \end{cases} \tag{2.11}$$

then the semi-infinite matrix G defined by

$$G := \beta\delta^{-1} = \alpha^{*-1}\gamma^*$$

is the unique strictly contractive extension of the given lower triangular part $\{A_{ij} \mid i \geq j \geq 0\}$ with $(G(I - G^*G)^{-1})_{ij} = 0$ for $j > i$.

To derive the above theorem we apply Theorem I.2.1 with $\mathcal{N} = \mathcal{L}^{2\times 2}$, the algebra of 2×2 matrices with entries in $\mathcal{L}$. To give the desired band structure we take

$$\mathcal{A}_+^0 = \mathcal{A}_-^{0*} = \{\left(F_{ij}\right)_{i,j=0}^{\infty} \mid F_{ij} : H_j \to H_i, F_{ij} = 0, j - i \leq 0\}(= \mathcal{L}_+^0),$$

$$\mathcal{A}_d = \{\left(F_{ij}\right)_{i,j=0}^{\infty} \mid F_{ij} : H_j \to H_i, F_{ij} = 0, j - i \neq 0\}(= \mathcal{L}_d),$$

and define $\mathcal{D}_\pm^0$ and $\mathcal{D}_d$ in the same way. Furthermore, let

$$\mathcal{B}_+ = \mathcal{C}_-^* = \{\left(F_{ij}\right)_{i,j=0}^{\infty} \mid F_{ij} : H_j \to H_i\ ,\ F_{ij} = 0, j - i \leq p\},$$

$$\mathcal{B}_- = \mathcal{C}_+^* = \{\left(F_{ij}\right)_{i,j=0}^{\infty} \mid F_{ij} : H_j \to H_i\ ,\ F_{ij} = 0, j - i > p\}.$$

It is easy to see that the conditions (2.2)–(2.7) in Section 1.2 are satisfied. Also the required Axiom (A0a) is fulfilled.

Let $\phi = \left(A_{ij}\right)_{i,j=0}^{\infty}$, where $A_{ij} = 0$ for $j - i > p$, and introduce

$$a := \left(a_{ij}\right)_{i,j=0}^{\infty}, a_{ij} = \begin{cases} \hat{\alpha}_{ij} & , i \geq j; \\ 0 & , i < j; \end{cases}$$

$$d := \left(d_{ij}\right)_{i,j=0}^{\infty}, d_{ij} = \begin{cases} \hat{\delta}_{ij} & , i \geq j; \\ 0 & , i < j. \end{cases}$$

One easily checks that a and d are solutions of equations (2.10) in Section I.2. Now one needs to show that a and d also satisfy the condition (i)–(iii) in Theorem I.2.1. Condition (i) is trivially satisfied. The proofs that conditions (ii) and (iii) are satisfied are rather elaborate and can be found in [12], Section II.4. Now we can apply Theorem I.2.1 and obtain that the semi-infinite matrix G is the unique triangular extension of ϕ. Thus the theorem follows.

The algebra $\mathcal{N} = \mathcal{L}^{2\times 2}$ (endowed with the band structure defined above) satisfies Axioms (A1) and (A2) with $\mathcal{R}$ equal to the B*-algebra of all bounded linear operators on $Z \dotplus Z$ (where $Z = \overset{\infty}{\underset{j=0}{\oplus}} H_j$). Thus for the description of the set of all strictly contractive extensions one may simply apply Theorem I.2.3.

THEOREM II.2.2. *For $0 \leq j \leq i < \infty$ let A_{ij} be an operator from a Hilbert space H_j into the Hilbert space H_i, and let $\Lambda_k, k = 0, 1, \ldots,$ be defined by (2.1). Suppose that (2.3) holds. Then the lower triangular part $\{A_{ij} | j - i \leq 0\}$ has a strictly contractive extension if and only if (2.2) holds. Suppose that (2.2) holds, and let α, β, γ, and δ be defined by (2.4)–(2.11). Then each strictly contractive extension F of the given lower triangular part is of the form*

$$F = (\alpha E + \beta)(\gamma E + \delta)^{-1}, \tag{2.12}$$

where $E = \left(E_{ij}\right)_{i,j=0}^{\infty}$ *is an element of* $\mathcal{L}$ *with* $\|E\|_O < 1$ *and* $E_{ij} = 0,\ j < i$. *Furthermore, (2.12) gives a one-one correspondence between all such* E *and all strictly contractive extensions* F. *Alternatively, each strictly contractive extension* F *of the given lower triangular part is of the form*

$$F = (\alpha^* + K\beta^*)^{-1}(\gamma^* + K\delta^*) \tag{2.13}$$

where $K = \left(K_{ij}\right)_{i,j=0}^{\infty}$ *is an element of* $\mathcal{L}$ *with* $\|K\|_O < 1$ *and* $K_{ij} = 0,\ j < i$. *Furthermore, (2.13) gives a one-one correspondence between all such* K *and all strictly contractive extensions* F.

Since Axioms (A3) and (A4) are fulfilled for $N = \mathcal{L}^{2\times 2}$, Theorem I.3.3 yields the following maximum entropy principle. Recall that for an element $V \in \mathcal{L}$ the entries of its right multiplicative diagonal are given in (1.14) in Section II.1.

THEOREM II.2.3. *For* $0 \le j \le i < \infty$ *let* A_{ij} *be an operator from a Hilbert space* H_j *into the Hilbert space* H_i, *and let* Λ_k, $k = 0, 1, \dots$, *be defined by (2.1). Suppose that (2.2) and (2.3) hold. Put*

$$M_i := \begin{pmatrix} I & 0 & \cdots \end{pmatrix}(I - \Lambda_i^* \Lambda_i)^{-1}\begin{pmatrix} I \\ 0 \\ \vdots \end{pmatrix}, i = 0, 1, \dots . \tag{2.14}$$

Then for the right multiplicative diagonal $\operatorname{diag}\left(\Delta_i^r(I - \psi^*\psi)\right)_{i=0}^{\infty}$ *of* $I - \psi^*\psi$, *where* ψ *is a strictly contractive extension of the given lower triangular part, the following inequalities hold:*

$$\Delta_i^r(I - \psi^*\psi) \le M_i^{-1}, i = 0, 1, \dots . \tag{2.15}$$

Moreover, equality holds for all i *in (2.15) if and only if* ψ *is the unique triangular extension of the given lower triangular part.*

REFERENCES

[1] M.A. BARKAR AND I.C. GOHBERG, *On factorizations of operators relative to a discrete chain of projectors in Banach space*, Mat. Issled. 1 (1966), 32–54. [Eng. Transl. Amer. Math. Soc. Transl. (2) 90 (1970), 81–103].

[2] A. BEN-ARTZI AND I. GOHBERG, *Nonstationary Szegö theorem, band sequences and maximum entropy*, Integral Equations Operator Theory 11 (1988), 10–27.

[3] H. DYM, *J-contractive matrix functions, reproducing kernel Hilbert spaces and interpolation*, CBMS Regional Conference Series in Mathematics, Providence (1989).

[4] H. DYM AND I. GOHBERG, *Extensions of matrix valued functions with rational polynomial inverse*, Integral Equations Operator Theory 2 (1979), 503–528.

[5] H. DYM AND I. GOHBERG, *On an extension problem, generalized Fourier analysis and an entropy formula*, Integral Equations Operator Theory 3 (1980), 143–215.

[6] H. DYM AND I. GOHBERG, *Extensions of band matrices with band inverses*, Linear Algebra Appl. 36 (1981), 1–24.

[7] H. DYM AND I. GOHBERG, *Extensions of kernels of Fredholm operators*, J. Analyse Math. 42 (1982/83), 51–97.

[8] H. DYM AND I. GOHBERG, *Unitary interpolants, factorization indices and infinite Hankel block matrices*, J. Funct. Anal. 54 (1983), 229–289.

[9] H. DYM AND I. GOHBERG, *A maximum entropy principle for contractive interpolants*, J. Funct. Anal. 65 (1986), 83–125.

[10] H. DYM AND I. GOHBERG, *A new class of contractive interpolants and maximum entropy principles, in: Topics in operator theory and interpolation (Ed. I. Gohberg)*, Operator Theory: Adv. Appl. OT 29, Birkhäuser Verlag, Basel (1988), 117–150.

[11] I. GOHBERG, M.A. KAASHOEK AND H.J. WOEDERMAN, *The band method for positive and contractive extension problems*, (to appear in J. Operator Theory).

[12] I. GOHBERG, M.A. KAASHOEK AND H.J. WOERDEMAN, *The band method for positive and contractive extension problems: an alternative version and new applications*, Integral Equations Operator Theory 12 (1989), 343–382.

[13] I. GOHBERG, M.A. KAASHOEK AND H.J. WOERDEMAN, *A maximum entropy principle in the general framework of the band method*, preprint, Department of Mathematics and Computer Science, Vrije Universiteit, Amsterdam, 1gdg.

ON THE COMPLEXITY OF PATTERN RECOGNITION ALGORITHMS ON A TREE-STRUCTURED PARALLEL COMPUTER*

A.L. GORIN, D.B. ROE and A.G. GREENBERG†

1. Introduction. Pattern Recognition is the problem of automatically classifying an unknown pattern as one of a known set of reference patterns. Applications include speech recognition, speaker recognition, and shape recognition of imaged objects. As the number and variability of the reference patterns increases, so does the computational burden, leading one to consider parallel processing in order to speed up the computation. Several researchers [B 79], [S 84], [S 85], [G 87] have proposed trees of processors for accelerating pattern recognition computations in a variety of applications, such as databases, expert systems, speech and vision. One reason is that the important interprocessor communication operations that arise in pattern recognition can be executed quickly and simply on a tree interconnection. We call these operations *communication kernels.*

This paper describes the theory and practice of efficiently implementing the communication kernels for pattern recognition on a tree machine. We first define algorithms for these kernels, then derive complexity models that express execution time as a function of problem and device parameters. In particular, rather than a traditional asymptotic complexity analysis, we explicitly model and evaluate the interaction between communications and processing parameters in several application scenarios. In the worst cases, one observes logarithmic communications overhead in the parallel algorithms. We demonstrate, however, that in some important situations this overhead can be made essentially zero by pipelining the communications kernels. Furthermore, we show that these tree kernels are minimal in comparison to all other architectures whose interconnection topology is of bounded-degree.

Examples of the communications kernels that we describe arose in application studies of real-time large-vocabulary speech recognition on the AT&T BT-100 parallel processor, which is a tree machine developed by AT&T's ASPEN Project with the Defense Advanced Research Projects Agency (DARPA). Speech recognition is an important example of pattern recognition, which we will draw upon for motivation and illustration throughout this paper. Our experience has shown that while the details of the pattern recognition computation vary greatly amongst applications, the pattern recognition communication kernels remain remarkably similar. Benchmark analyses of several communications kernels on the BT-100 will be presented to support and quantify the theoretical complexity models.

Complexity models of the communications kernels are of practical use for at least two reasons. The first is to predict improvements in performance as device parameters of an architecture evolve. The second is to provide guidelines within

*This work was partially supported by DARPA Contract No. N00039-86-C-0173.
†AT&T Bell Laboratories, Murray Hill, New Jersey 07974

which to optimize the hardware for the kernel operations that occur often in practice. Interestingly, this is an application of the *reduced instruction set* concept to the optimization of interprocessor communications networks.

Independent of the application, trees have two attractive and distinguishing characteristics within the set of parallel computer architectures. First, the sparseness of a tree interconnection allows for compact high-density implementations. For example, as described by Gorin and Shively [G 87], the ASPEN Project's BT-100 has a current computing density of 2 GigaFLOPS per cubic foot, with an eventual goal of 40 GigaFLOPS per cubic foot [S 88]. Second, the tree is universal in the sense that it can be embedded within any lattice of processors while maintaining communications diameter. This is important from the perspective of providing reconfigurability and fault-tolerance in parallel computers, as described by Snyder [S 82] and Pradhan [P 85].

The paper has four sections. Section 2 provides a brief introduction to tree machines, emphasizing those well-known characteristics that make them scalable and well-suited to VLSI and advanced packaging. In particular, the processing element and board-level modules do not change as the machine size is scaled, maintaining a small and constant pinout via Leiserson's expansion scheme [L 81]. Section 3 describes how to partition and map pattern recognition algorithms onto a tree, building upon the pattern matching paradigm of Bentley and Kung [B 79] to encompass structural pattern recognition via dynamic programming search. We describe a host-orchestrated software environment based on Single Program Multiple Data (SPMD) processing which suffices to support these parallel algorithms, and allows us to treat separately their computational and communications complexity. In Section 4, we present algorithms for the communication kernels, derive complexity models, and show minimality. Finally, Section 5 reports on experimental benchmarks on the BT-100 tree machine that quantify and support these models, in particular demonstrating the advantages of pipelined communications.

2. Tree Machines. This section briefly reviews the definition and important properties of a tree machine. The goal is to motivate our focusing on a tree interconnection by appeal to its scalability and suitability for VLSI. These properties lead to compact, high-density implementations.

For our purposes in this paper, a *tree machine* is a parallel computer wherein each processing element (PE) is connected to its parent, left child and right child. This is in particular a binary tree machine. Each PE comprises a processor, local memory and communications logic. Large machines are constructed by connecting identical boards via Leiserson's recursive expansion scheme [L 81] as shown in Figure 1.

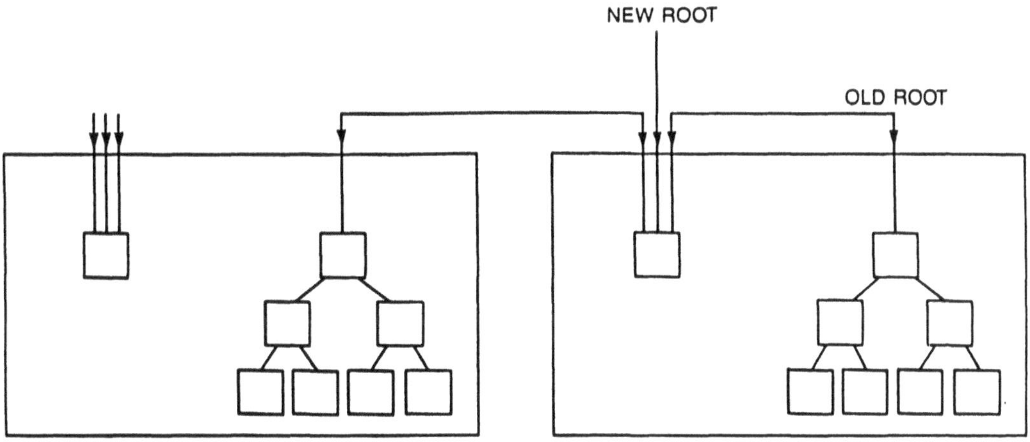

Figure 1

Leiserson Expansion Scheme for Tree Machines

The recursive scheme applies to interconnecting larger modues as well, such as cabinets. We view the tree machine as a peripheral to a host computer, with external communications via the root PE. Programmable tree machines targeted at a broad class of applications were first investigated by Browning [B 80]. Several prototypes have been implemented since then, differing in PE granularity, software architecture, and application focus. These include NON_VON [S 84], CORAL [T 85], DADO [S 85] and ASPEN (BT-100) [G 87].

The binary tree architecture is scalable in two senses. First the PEs do not change as the size of the tree increases, remaining 3-port devices communicating with only a parent and two children. Second, the board-level modules do not change as the packaging density increases, remaining 4-port devices regardless of the number of PEs on a board (df. Figure 1). This 4-port property extends recursively to larger submodules such as cabinets. The small and constant pinouts of the PEs and submodules make the tree well-suited for miniaturization via VLSI and advanced packaging. The BT-100 illustrates the high-density computing that can be achieved via this approach. Each PE comprises an 8 MFLOP programmable 32-bit processor and 64 kbytes of memory. The board-level module comprises 8 PEs on an 8″ × 13″ board, with a 127 PE GigaFLOP processor housed in 0.5 cubic feet. This computing density, 2 GigaFLOPs per cubic foot, is the highest known to the authors for a programmable parallel computer.

3. Pattern Recognition on a Tree Machine. *Pattern Recognition* is the problem of automatically classifying un unknown input pattern as one of a known set of reference patterns. In some cases, the references can be effectively enumerated and stored in a library. Then, one can proceed, as for example described by Duda and Hart [D 73], by comparing the unknown pattern to each reference and then classifying it as that reference to which it best matches. The comparison can be via a distance function, wherein the best match is a minimum distance, or via a probability scoring, wherein the best match is a maximum likelihood. This is the basic pattern matching problem, with extensions to be described below. As the number of patterns in the reference library increases, so does the computational burden of the system. The goal is therefore to exploit parallelism to maintain execution speed as the problem size increases.

In this paper, we restrict ourselves to viewing an individual pattern match function as an atomic computation, and focus on the nature and complexity of the operations necessary to distribute the matches over many processors. Thus, our results are independent of the details of the match function, depending only on surface characteristics such as the memory required to describe the patterns and the execution time of the match function.

The computational core of a pattern recognizer is the matching of the unknown pattern against the library of reference pattern. Expanding on the paradigm introduced by Bentley and Kung [B 79], one partitions and maps the computation onto a tree as follows. The machine is initialized by first distributing the library of reference patterns amongst the PEs, storing a subset in each PE's memory. The unknown pattern is then broadcast from the tree's root to all PEs. Each PE then concurrently matches the input patterns with its subset of the reference library. The best score is then resolved by bubbling it up the tree. Associated information on the model with the best score is then reported up the tree. In this scheme, the matching is performed concurrently, at a communications cost proportional to the tree depth. This cost is small, since the tree depth grows only as the logarithm of the tree size. I.e., a complete binary tree with 1023 PEs has a depth of merely 10.

We now describe a software architecture that both supports this pattern-matching control strategy and also provides program scalability. Both processing and global communications are orchestrated via function calls from a serial host program, so that data flow within the tree is controlled by the host, independently of the tree size. Therefore, the application source code is unchanged with the addition of more reference patterns and PEs. Furthermore, as long as the number of PEs is scaled proportional to the problem size, the computation time remains the same, and one can separately analyze the logarithmic growth of the communications functions, as we do in section 4.

The major parallel construct is a *sliced procedure*, in which the identical program is executived simultaneously in each PE, but on different data sets. The multiple executions of this single procedure can follow different instruction streams (still within the procedure, of course) depending upon the data. These potentially different instruction streams are forced to converge and synchronize at the completion of the sliced procedure. This synchronization scheme is known as barrier synchronization, and the control strategy known as Single Program Multiple DATA (SPMD) processing. For adherents of Flynn's [F 66] SIMD/MIMD categories, SPMD may be alternatively viewed as coarse grain SIMD or as data-driven MIMD.

In support of the SPMD processing mode, we use several global burst communications functions. These communications are invoked by function calls from the host, and are examples of traffic-specific interprocessor communications. ***Broadcast*** transmits data from the host to the PEs. ***Resolve*** identifies the PE in the tree with the maximum value of some variable, such as a probability score. The ***Report*** command, typically following a *Resolve*, transmits data from the identified PE to the host, such as the name of the reference pattern that has the maximum probability. Such a software architecture has been implemented on the BT-100, where the host program and sliced procedures are programmed in standard C, and the invocation of sliced procedures and communication functions are via C function calls from the host program.

Structural Pattern Recognition. When a reference set becomes too large to effectively list, one can instead describe it as a set of structural patterns formed by gluing together simpler patterns according to some set of rules. For example, in speech recognition [L 85], one usefully views a sentence as a structural pattern comprising concatenated words, rather than as a gestalt. Similarly, in large-vocabulary recognition, one views words as concatenations of subword units such as phonemes. One does not try to match the unknown pattern with all such structural patterns, since that is a combinatorially explosive search problem. The basic approach used in speech recognition to prune the search while still guaranteeing an optimal match is dynamic programming, as described by Rabiner and Levinson [R 81].

Bentley and Kung's pattern matching paradigm has been extended to accelerating dynamic programming search for structural pattern recognition, and then applied to real-time speech recognition by Gorin, Shively and Roe [G 87], [G 88]. The concept underlying this extension is as follows. Dynamic programming is a graph search, whose outer loop in this context comprises an iteration of the following.

1. starting at some state(s) in the search space, evaluate subsequent states via matching the next input epoch against the library of reference pattern primitives;
2. select the best or several best matching references;
3. make the transition(s) to subsequent states in the search space.

As shown in Figure 2, step 1 is performed in parallel via a sliced procedure, step 2 via a resolve/report, and step 3 via a broadcast. In many searches, one wishes to maintain multiple hypotheses. This motivates a K-best resolve/report function, which will be described in Section 4.

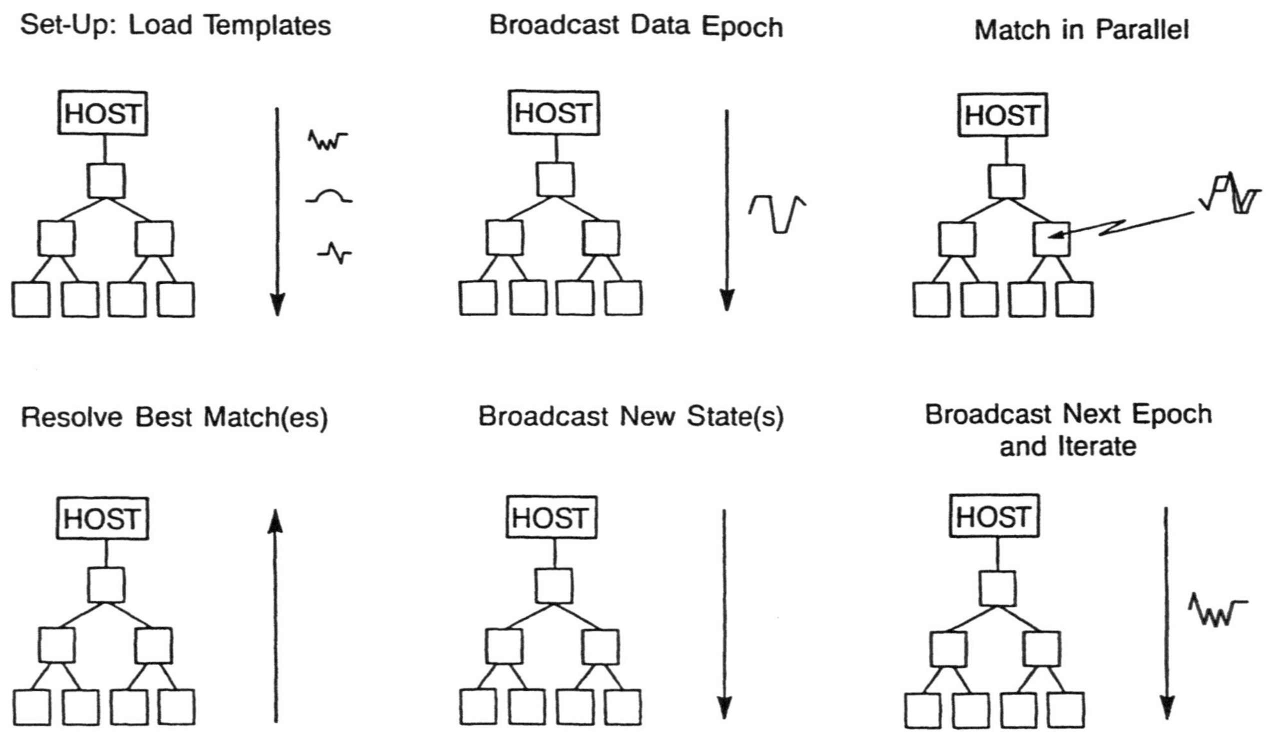

Figure 2

Parallel Dynamic Programming

For illustration, we briefly review the connected-word recognition benchmark described by Gorin and Roe [G 88]. That algorithm is a dynamic programming search that computes a Viterbi scoring of the probability that an unknown spoken utterance was generated by a reference Hidden Markov Model [L 85]. For real-time execution, the search must complete the above three-step cycle for each 15 milisecond input frame of speech. I.e., all processing and communications for an input frame needs be completed before the next frame arrives. A set of 25 short-term acoustic spectral features are measured from the frame, then broadcast to all PEs. The local pattern recognition computation involves evaluation of a continuous probability density function defined via a 9-component Gaussian mixture in the 25-dimensional feature space. For each of several syntactic hypotheses, the K-best scores are resolved and reported, then broadcast to all PE's along with the next frame of spectral features. Both the number of hypotheses and K depend greatly upon the vocabulary and grammar. For a digit-recognition example, there were 7 hypotheses maintained with the 3 best scores each. These parameters can increase a hundred-fold as vocabulary and language complexity increase.

4. Complexity Models. In this section, we derive formulae for execution time of the pattern recognition communication and processing kernels on a tree, expressed in terms of problem and device parameters. We also discuss application scenarios and show minimality. In particular, we will show that in several important application scenarios, the parallel pattern recognition algorithm achieves perfect linear acceleration of the match computation, while incurring a communications cost proportional to the number, K, of scores resolved at each cycle of the search plus a logarithmic pipeline fillup term.

We will show that this is the best that can be achieved on any architecture whose interconnection topology is of bounded degree. We say that an interconnection is of bounded degree b if each PE is connected to at most b other PEs. This will be true, for some value of b, for all physically realizable machines. For example, a binary tree of any size has degree 3, a hypercube with 1024 PEs has degree 10, and a rectangular array has degree 4.

4.1 Broadcast The first stage in the basic parallel pattern recognition algorithm is to *broadcast* the set of features characterizing the unkown pattern. This is achieved by each PE receiving a value from its parent, storing it locally, and then re-transmitting it to its children. For a single data element, the delay between entering the root and arriving at the leaves is proportional to the logarithm of the number of PEs, i.e the tree depth. We will show, however, that for many situations of interest, pipelining can effectively eliminate the apparent logarithmic communications cost.

Let B denote the number of bits in the feature vector. (We choose the units of bits for exposition's sake. It is straightforward to modify these arguments to, for example, 16-bit wide pathways.) Let c denote the communications time to propagate a bit one level down the tree. Let N denote the number of PEs, and $\log_2 N$ the depth of the tree. Finally, let S denote the host startup/shutdown costs.

One method to measure elapsed time for a broadcast is from initiation until the last bit arrives in the leaves, denoted by T_{B_L} and given by

$$T_{B_L} = S + c(B + \log_2 N), \tag{1}$$

costing cB cycles for the transfer of the data to the root, and then $c\log_2 N$ cycles for a pipeline shutdown time as the data propagates down the tree.

However, in practice, a broadcast function call is embedded within a host C program. The actual time of interest is *not* T_{B_L}, but *rather* when the host program proceeds on to its next instruction. Since the broadcast function returns no value to the host, the host instruction counter is incremented as soon as the last bit enters the root of the tree. Thus, in *all* situations of interest, the pipeline shutdown occurs concurrently with the next program step, and can be ignored. Denote this effective execution time for a broadcast by T_B, given by

$$T_B = S + cB. \tag{2}$$

Minimality. Consider the problem of broadcasting B bits of data to all PEs in some parallel architecture. I.e., at completion, the data should be resident in local memory at each PE. Assuming that writing to a PE's local memory is a serial process, then it requires time at least cB to accomplish this write to any individual PE. Thus, a lower bound for completing the global broadcast is cB, which is achieved modulo a constant factor by the tree broadcast kernel.

Finally, observe that when the block size is large, the host startup cost $S \ll cB$, and T_B is approximated by

$$T_B \approx cB. \tag{3}$$

In the AT&T BT-100, experimental data bear out formulae (2) and (3), as will be seen in Section 5.

Example. In a speech recognition application, one broadcasts measures of the short-term acoustic spectra, which in the example of [G 88] comprise a batch of 25 32-bit numbers at 15 millisecond intervals. On the BT-100, communications is achieved via software protocols between PEs, and as will be shown in Section 5, the communications parameter c is 1.4 microseconds/bit/level, and the startup cost, S, on the host PC is 120 microseconds. Thus, summing, the broadcast time T_B is 920 microseconds, well within the 15 millisecond real-time budget. A significantly faster value for c is projected for the machine described by Shively and Gorin in [S 88], based on 16-bit wide data path clocked at 20 Mhz, yielding 320 Mbits per level per second. In that case, the broadcast time for 25 32-bit numbers is 2.5 microseconds. While this is overkill for a single real-time speech channel, it extends the applicability to pattern recognition of higher-bandwidth signals.

4.2 Match. The second phase in parallel pattern recognition is for each PE to concurrently *match* the unknown against the subset of reference patterns in that PE. The communication kernel associated with invocation of a sliced procedure is twofold: first, initiation of the process, then global synchronization at completion. We will show that for many situations of interest, the logarithmic initiation/synchronization overhead is negligible, and that one can achieve essentially perfect linear acceleration of the match computation.

Let L denote the number of reference patterns and N the number of PEs. Since in this paper we treat a match function as an atomic computation, we assume that there is at least one reference per PE. I.e. the number of PEs is assumed less than or equal to the number of references. For simplicity of exposition, we assume that L is an integral multiple of N, so that each PE has the same number of reference patterns. Thus, there are L/N reference patterns per PE.

Let m denote the number of machine cycles necessary to perform a single match, and let p denote the time necessary for a PE to execute one of those cycles. For simplicity of exposition we assume that m is the same for all reference patterns. Thus, each PE will spend time pmL/N computing its matches. For initiation and resynchronization, the communications overhead is $2c \log_2 N$, where c is the communications time per level as before. Denote by $T_M(L, N)$ the total execution time for computing L matches distributed over N PEs, which is given by

$$T_M(L, N) = pmL/N + 2c \log_2 N. \tag{4}$$

Observe that the time to execute these matches on a single PE is given by

$$T_M(L, 1) = pmL. \tag{5}$$

The best that one can expect in accelerating the match computation is to achieve *perfect linear acceleration*, i.e., $T_M(L, N) = T_M(L, 1)/N$. Let us examine the deviation from perfect linear acceleration in equation (4), rewriting it as

$$T_M(L, N) = [T_M(L, 1)/N][1 + (2cN \log_2 N)/(pLm)]. \tag{6}$$

Thus, in order to approach perfect linear acceleration in the match phase, the deviation term $(cN \log_2 N)/(pLm)$ must be much smaller than unity. We consider three different sufficient cases, by no means exhaustive, for which this is true. Each is based on a situation in which the amount of computation that the PE performs is much greater than the communications costs. To clarify the exposition, we first rewrite equation 6 as

$$T_M(L, N) = [T_M(L, 1)/N][1 + (\log_2 N/m)(2c/p)(N/L)]. \tag{7}$$

The first case is if the match function is complex, i.e. the number of machine cycles per march, m, is large. For sake of example, assume that communications and processing are equal, i.e. $c = p$. Upon recalling that $(N/L) \leq 1$, then if the ratio $\log_2 N/m$ is very small then so is the deviation. I.e., the deviation from perfect linear acceleration is small when the number of machine cycles per match is much greater that the tree depth. This occurs in many situations of interest. For example, the computation of a Gaussian likelihood is quadratic in the dimension of the feature vector. In speech recognition, a typical acoustic feature vector comprises 25 dimensions, implying on the order of 625 floating point computations to performs the match. This is much larger than any conceivable value of $\log_2 N$, and thus leads to essentially perfect linear acceleration of the match phase. The efficiency of the speech benchmarks on the BT-100 in [G 88] are due to the complex match functions found in signal pattern recognition.

The second case is if there are many references per PE. Again, for sake of example, assume that $c = p$. Then many references per PE implies that N/L is very small, again leading to a small deviation and thus essentially perfect linear acceleration. This situation occurs, for example, when there are large numbers of simple matches to perform. E.g., if there are millions of references, and we distribute them over thousands of PEs, then each PE will have plenty of work to do regardless of how simple the match might be.

The third case, although not as interesting, occurs if one links together relatively slow processors with a relatively fast communications network, i.e. $p \gg c$. This will lead to perfect linear acceleration, but would be the result of an unbalanced design yielding an efficient albeit slow implementation. For example, consider a situation in which $p > 10c$, i.e. a processor instruction cycle requires more than ten times the communications time per level. Then, in a 1023 PE tree of depth 10, an initiation or synchronization communication would appear to be instantaneous, since $c \log_2 N$ is less than one instruction cycle.

4.3 Resolve. The resolve kernel has two components: first, initiation of the process, then bubbling up of the best score via local three-way comparisons. Both require time proportional to the logarithm of the number of PEs, which we will show is minimal over all parallel architectures of bounded degree.

As before, let L be the number of reference patterns and N be the number of PEs. Let c be the communication time per word between levels, and p the processing time for a two-way comparison at each PE. Denote by $T_R(L, N)$ the time for resolving a single global best score from L values evenly distributed over an N-PE tree, which is given by

$$T_R(L, N) = pL/N + 2(c + p)\log_2 N, \tag{8}$$

since pL/N is the time for each PE concurrently to select the best of its local L/N references, $c \log_2 N$ the time to initiate the process, and $(c + 2p)\log_2 N$ the time to bubble up the global best score to the root.

Although Resolve is a communications function, it involves computation as well. Thus, it is reasonable to compare the time required to select the best of L scores on 1 PE to the time on N PEs. Perfect linear acceleration of the resolve phase would occur when $T_R(L, N) = T_R(L, 1)/N$, where

$$T_R(L, 1) = pL. \tag{9}$$

Minimality. The deviation from perfect linear acceleration of the resolve operator is the term $2(c + p)\log_2 N$ in equation (8). We will show that, when $L = N$, i.e. for one reference pattern per PE, this deviation is minimal over all architectures whose interconnections are of bounded degree. Consider an interconnection whose branching factor is at most $b+1$, and pose the problem of resolving the minimum of a set that is distributed over the ensemble with one value per PE. The worst-case time for gathering that value to a specific PE is the diameter of the interconnection, which is at least $\log_b N$. Thus, for any interconnection, the resolve operation requires at least logarithmic time.

4.4 K-Best Resolve. Thus far, we have shown that the basic pattern matching problem of finding the single best match can be computed minimally on a tree of processors . In some applications, however, one requires not merely the single best score, but the K-best scores. Furthermore, in many applications it is desirable to maintain several alternative contextual hypotheses that will weight the decision via modification of prior probabilities, and one is interested in the K-best scores for each of these several alternate hypotheses. In these situations, it is possible to pipeline the multiple resolves up the tree, absorbing the logarithmic overhead in a pipeline. We now consider the problem of finding the K-Best scores, given a set of L scores distributed evenly among the N nodes of the tree.

We simplify the problem for sake of exposition: Consider a collection of L values, $z_1, z_2, \ldots, z_L$, distinct real numbers, distributed among the N nodes. Assume that L is an integral multiple of N, and each node initially contains L/N values. The problem is to produce at the root a decreasing list of the K largest values, $z_{i(1)}, \ldots, z_{i(K)}$ such that $z_{i(1)} > z_{i(2)} > \cdots > z_{i(K)} > z_j$, for all $j \notin \{i(1), i(2), \ldots, i(K)\}$. All of what follows is easily adapted to the case where N need not divide L and the z_i need not be distinct, and a generalization to an arbitrary tree of processors is also discussed. In a typical instance of the problem for machines that should be soon be available, one might have $K \approx 100, N \approx 1000$, and $L \approx 100,000$.

We will consider and compare two solutions to finding the K-best scores. The first is to merely iterate the Resolve function. The second is to pipeline the Resolve operations. We denote these times by T_{nopipe} and T_{pipe} respectively.

Non-Pipelined K-Best Resolve. The simplest solution is to repeat the following steps K times, where $M = \min(K, L/N)$.

1. *Local select and sort*: In parallel each node computes the list of the M largest local values in decreasing order.
2. *Merge sort via iterated resolves*: Do the following K times. Select the maximum of the values at the heads of the N local lists via a resolve, broadcast this global maximum, remove it from the head of the local list where it originated, and promote the next largest local value to the head of that list.

Let $T_s(L/N, M)$ denote the time for the first phase: the serial computation, concurrently in each PE, of the list of M largest values from its set of L/N values. The second phase involves a sequence of K resolves, whose individual times were given in equation (8). Thus, the worst case running time $T_{nopipe}(L, N, K)$ of the algorithm is

$$T_{nopipe}(L, N, K) = T_s(L/N, M) + K[S + 2(p + c)] \log_2 N, \tag{10}$$

where p and c are processing and communication times respectively, and S is a startup cost. We will discuss the complexity of the local serial sort, T_s, in the next subsection.

For example, consider the case of one reference pattern per PE, where $L = N$. Then, $T_s = 0$ and

$$T_{nopipe}(N, N, K) = K(a_1 + a_2 \log_2 N), \tag{11}$$

where a_1 and a_2 are constants derived from S, p and c in formula 10. Observe that the time is linear in K, but the slope depends on $\log_2 N$. This behavior is illustrated quantitatively in the benchmarks reported on in the Section 5.

Pipeplined K-Best Resolve. We now define and analyze a pipelined version of the above algorithm. We show that, in the analog of equation (11), execution time will still be linear in K, but the slope will be independent of $\log_2 N$. Rather, the $\log_2 N$ term appears as a very small additive bias representing a pipeline fillup and shutdown time. This makes the communications time essentially independent of tree size, and thus scalable. This behavior is illustrated quantitatively in the kernel benchmarks of Section 5. The pipelined algorithm is as follows, where as before $M = \min(K, L/N)$.

1. *Local select and sort*: In parallel each node computes the list of its M largest local values in decreasing order. (This is the same as in the non-pipelined algorithm.)
2. *Merge sort via pipelined resolves*: The nodes merge (as described below) the lists computed in phase one to produce a list of the K largest values at the root.

First consider the local select and sort function of phase 1, which is performed concurrently in all PEs and is the same as in the non-pipelined algorithm. The pipelining impacts only the merge sort phase of the algorithm, but it will be useful to understand the relative magnitudes of the two phases. It is difficult to give a single sorting method that would be best in practice for all problem sizes, however, the following is asymptotically minimal. Blum's algorithm [A 74] can be utilized to select the M-th largest element in time $O(L/N)$ and heapsort [A 74] used to sort the M largest values in time $O(M \log_2 M)$, to achieve, where A is some constant,

$$T_s(L/N, K) \leq A(L/N + M \log M) \leq A(L/N + K \log_2 K). \tag{12}$$

The Merge Phase. After the first phase of the algorithm, each node holds a list of its M largest local values in decreasing order. We now describe how to carry out the second phase where the N lists are merged, using three auxiliary buffers per node, each with the capacity to store up to K values. At the end of this subsection we adapt the algorithm to use less memory.

In the course of the algorithm, a node manipulates three FIFO queues:

- q_0 : used to hold values received from the left child;
- q_1 : used to hold values received from the right child;
- q_2 : used to hold local values.

Initially, q_2 holds the list of the K largest local values obtained via the local sort and select phase, padded with $-\infty$ if $K > L/N$. Queues q_0 and q_1 are initially empty. The algorithm is such that all three queues will always be in decreasing order; the largest values will be first and the smallest values last.

The task of a leaf is to write its local list q_2 to its parent one value at a time. An internal node first waits until each child has a value to send upwards. Then, for K steps, the PE accepts two values from its two children and enqueues the two in q_0 and q_1. The PE then selects a maximum value from among the three at the heads of q_0, q_1, and q_2, removes that value from its queue, and outputs the value to its parent. What follows is the program (a sliced procedure involving communication) that each PE concurrently executes within the K-Best kernel.

The program is data driven, in that when one PE wants to communicate with another it waits until the other PE is ready and then carries out the communication. In this way, the outputs of the leaves trigger the data bubbling up to the root.

```
do K times {
    if not a leaf then {
      wait until left child supplies a value; insert that value in q0
      wait until right child supplies a value; insert that value in q1
      v0= value at the head of q0
      v1= value at the head of q1
      v2= value at the head of q2
      j = index in {0,1,2} such that vj = max{v0,v1,v2}
    }
    else j = 2
    remove vj from qj
    if a root then append vj to the solution list
    else wait until parent is ready and then supply vj to parent
}
```

Denote the running time of the merge phase of the pipelined algorithm by $T_{merge}(L, N, K)$, which is given by

$$T_{merge}(L, N, K) = S + 2(c+p)(K + \log_2 N), \tag{13}$$

where c and p are communications and processing parameters respectively, and S is the host startup cost. The root PE is inactive for $\log_2 N$ cycles while waiting for the initial values to bubble up from the leaves.

In the case of one reference pattern per PE, where $L = N$, then $T_s = 0$ and we can model the total execution time for the pipelined algorithm, T_{pipe}, as

$$T_{pipe}(N, N, K) \doteq b_0 + b_1 K + b_2 \log_2 N, \tag{14}$$

where b_0, b_1 and b_2 are constants derived from S, c and p in formula 13. Comparing this time to that for the non-pipelined algorithm in equation (11), we observe that execution time is still linear in K, but the $\log_2 N$ term appears additively rather than multiplicatively. Thus, the execution time is essentially independent of the tree size if K is large. This behavior will be supported quantitatively by a benchmark in Section 5.

We combine equations (12) and (13) to bound the total run time of the pipelined K-Best Resolve,

$$T_{pipe}(L, N, K) \leq A\ (L/N + M \log M + \log_2 N), \tag{15}$$

where A represents some positive constant and $M = \min(K, L/N)$.

Minimality. We do not know, for all parameter values, if T_{pipe} is within a constant factor of minimal. However, for particular scenarios, it is indeed minimal to within a constant factor. This occurs when the parameter values are such that the cost of the local sort phase is proportional to L/N, (the number of data elements per PE, rather than $L/N + M \log_2 M$. One example of such a scenario is for a 2-Best resolve, i.e. $K = 2$. A second example is when there is only one reference pattern per PE, i.e. $L = N$. In those cases, the net running time is bounded by $T_{pipe}(L, K, N) \leq A(L/N + K + \log_2 N)$. To show minimality, a simple fan-in argument yields that for A_1 some positive constant, at least time $A_1(L/N + K + \log_b N)$ time is needed on any network of processors where the K largest must be collected at one processor and each processor communicates directly with at most $b+1$ other processors. Thus, in the stated scenarios, the K-Best Resolve kernel is minimal, within a constant factor, over all parallel architectures whose interconnections are of bounded degree.

It is straightforward to show the merge algorithm is correct:

THEOREM. *For each node x and each $i = 1, 2, \ldots, K$, the i-th output of node x to its parent is the i-th largest value stored among all nodes in the subtree rooted at x.*

Proof. We say that a leaf has level 0, and any other PE has level one greater than the maximum level of its children. We proceed by induction on the level index. By construction, the Theorem holds for the i-th output of each leaf, for each $i = 1, 2, \ldots, K$. Suppose that, for some $j \geq 1$, the Theorem holds for the i-th output of each node at level j, for each $i = 1, 2, \ldots, K$. It then follows that during the K steps at which each node x at level $j + 1$ is active, the node computes and then outputs a three way merge of

- the sorted list consisting of the K largest values in the subtree rooted at the left child of x,
- the sorted list consisting of the K largest values in the subtree rooted at the right child of x,
- the sorted list of values local to node x,

which establishes the result at level $j + 1$, and by induction at all levels. □

In practice it would be best to use two buffers to implement q_0 and q_1, where each consists of K consecutive slots, the unit of memory needed to represent a value. In that case, the algorithm requires at least $2K$ slots of auxiliary, or scratch, memory to support the two queues.

The algorithm is easily modified to use just $2s$ auxiliary slots, for any $1 \leq s \leq K$, at the cost of some extra communication. More precisely, q_0 and q_1 would each be implemented as a FIFO queue with capacity s using a circular buffer [A 74] of s slots. If, for example, at a given node x, queue q_0 is at capacity, then node x blocks the output from its left child. (That blocked data will not figure into the calculation of the current output of node x anyway.) A blocked output is simply left at the head of the queue from which it arose and retried at each step until it is unblocked and accepted. This modification hardly changes the processing time. The worst case running time approximately agrees with (16), except that the constant c is amplified to account for the blocking and unblocking signals.

To adapt the algorithm for an arbitrary tree of processors, a node having b children maintains $b+1$ queues: one for local values and one for values received from each child. A $b+1$ way maximization replaces the three way one. This changes the running time of the merge phase to $(c_b + p_b(K+H)$, where H is the depth of the tree, and c_b and p_b represent communication and processing parameters dependent on the maximum branching factor.

5. Experimental Kernel Benchmarks. These benchmark experiments were performed on an AT&T BT-100 processor running at a 20 MHz clock rate attached to an AT&T PC6300+ host. The processor was reconfigured to provide (sometimes skewed) trees of depths ranging from 3 to 9, corresponding to balanced trees of sizes 7 through 511. The times were measured by embedding the kernel function call within a loop in the host C program, and then averaging. The first set of benchmarks show that the broadcast kernel is indeed invariant to tree depth, regardless of block size. The second set of benchmarks demonstrate the speedup and insensitivity to tree depth obtained via the pipelined K-Best Resolve as compared to the nonpipedlined algorithm.

5.1 Broadcast. We recall that equation (2) predicted that, for a broadcast instruction, elapsed time until the next function is executed would be independent of tree depth and linearly dependent on the broadcast block size. Figure 3 shows execution time for integer broadcasts on blocks of size 1, 10 and 100, for trees of depth 3, 5 and 9. Observe that, for a fixed block size, the broadcast times are essentially constant with respect to tree depth. Figure 4 shows the same execution times plotted as a function of block size. Observe the linear relationship. There is only one plot in Figure 4 because the values for the three tree-depths overlap.

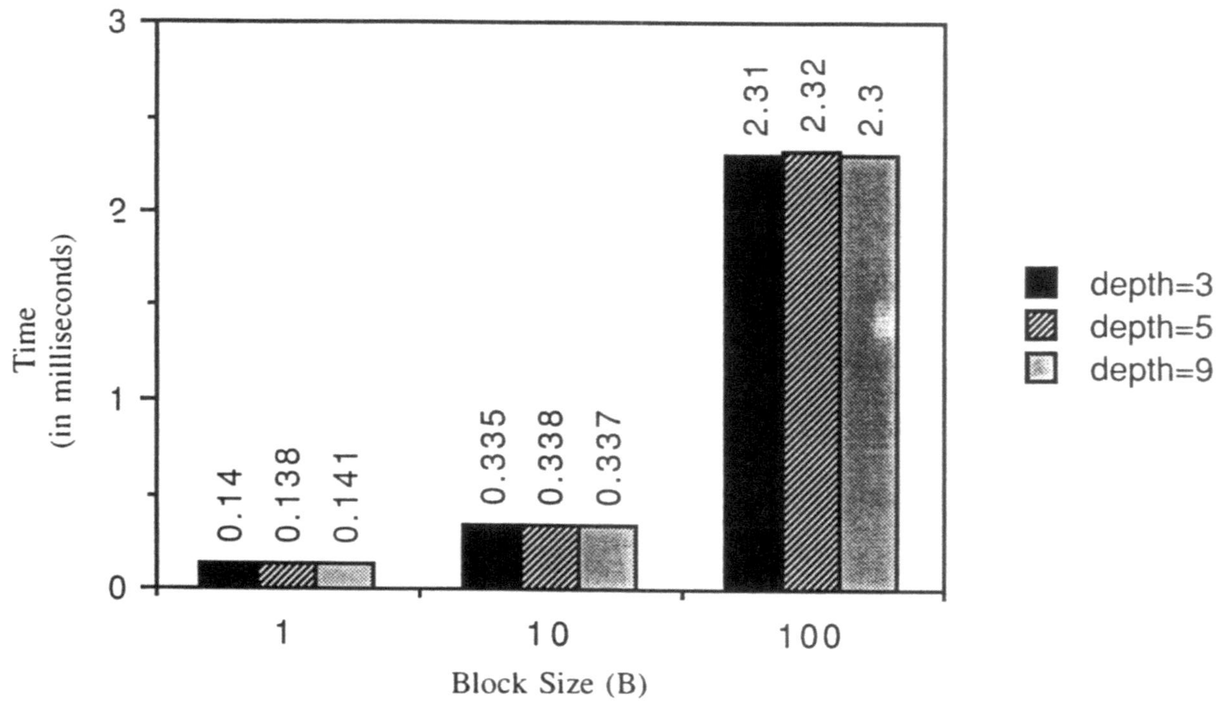

Figure 3

Execution Times for an Integer Block Broadcast

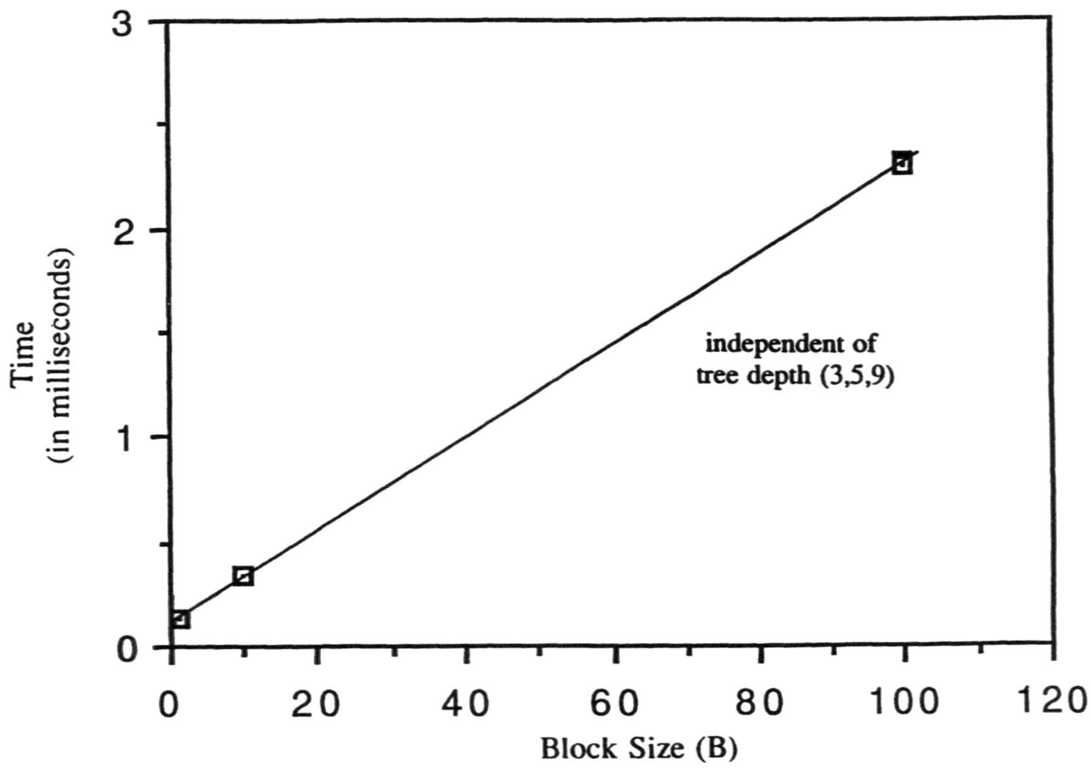

Figure 4

Execution Times for an Integer Block Broadcast

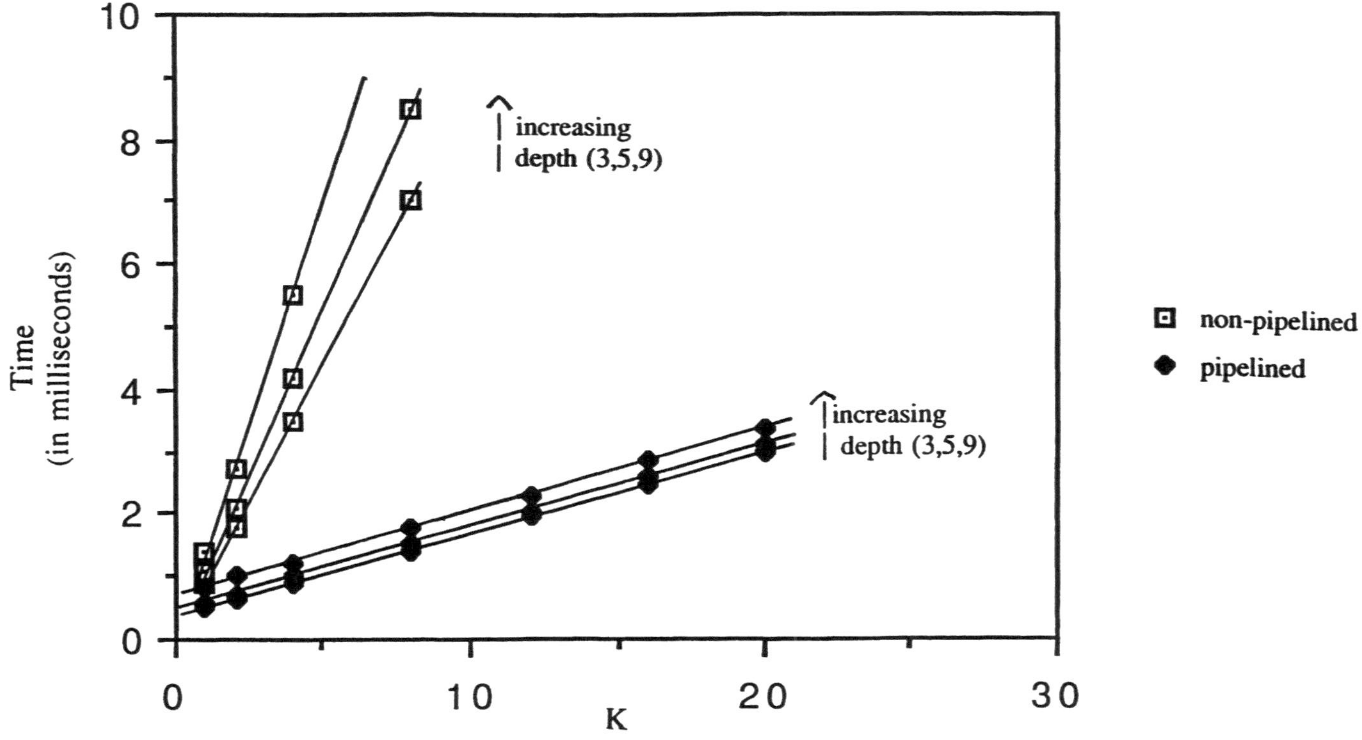

Figure 5

Benchmark Comparision of K-Best Resolve/Report

A data analysis yields a broadcast communications parameter of $c = 22$ microseconds/word/level, or equivalently 1.4 microseconds/bit/level. As observed in Section 4.1, this unremarkable value arises since communications is accomplished via software through a bit-serial I/O port. A more reasonable value is projected for the next generation machine [S 88], which executes the kernel in a semi-custom router chip, yielding a 16-bit wide data path at 20 Mhz. Finally, one can determine the startup overhead incurred by the current PC host, which is $S = 120$ microseconds.

5.2 K-Best Resolve Benchmarks. In this section, we report on several experiments that support the performance models for the K-best resolve algorithms of section 4.4. We consider just the second phase (merging) of the algorithms, since the first phases (local sorting) were identical and only involve processing. Recall that the execution times for the non-pipelined and pipelined K-best resolve algorithms were modeled by equations (11) and (14) as

$$T_{nopipe} = K(a_1 + a_2 \log_2 N), \tag{16}$$

$$T_{pipe} = b_0 + b_1 K + b_2 \log_2 N. \tag{17}$$

Each score was a 32-bit floating point quantity, with three associated 16-bit integers to be reported. Thus, the measured times are for a K-Best Resolve/Report. The times were measured to an accuracy of 0.5 % by looping several times through C code on the host that invoked the communication functions. Results for the pipelined and non-pipelined algorithms are shown in Figure 3 for values of K between 1 and 20 and for tree depth ranging from 3 to 9. Observe that the pipelined approach is much faster even when K is small. Observe also that the slopes of the linear regressions for the pipelined approach are constant for all N, confirming that the time is independent of tree depth, and thus of the number of PEs. We can fit these timing measurements to the predicted behavior via a regression analysis. The coefficients in formulae 16 and 17 are given in milliseconds with a correlation coefficient of 0.99 as:

$$T_{nopipe} = K(0.70 + 0.061 \log_2 N) \tag{18}$$

$$T_{pipe} = 0.204 + 0.132K + 0.050 \log_2 N. \tag{19}$$

REFERENCES

[A 74] AHO, A., HOPCROFT J., AND ULLMAN, J., *The Design and Analysis of Computer Algorithms*, Addison Wesley, New York (1974).

[B 79] J.L. BENTLEY AND H.T. KUNG, *A Tree Machine for Searching Problems*, Proc. International Conference on Parallel Processing IEEE New York (1979), pp. 257–266.

[B 80] S.A. BROWNING, *The Tree Machine: A Highly Concurrent Computing Environment*, Ph.D. Thesis, CalTech (Jan 1980).

[D 73] R.O. Duda and P.E. Hart, *Pattern Classification and Scene Analysis*, Wiley (1973).

[F 66] M.J. Flynn, *Very High-Speed Computing Systems*, Proc. of the IEEE 54, 12 (Dec. 1966), pp. 1901–1909.

[G 87] A.L. Gorin and R.R. Shively, *The ASPEN Parallel Computer, Speech Recognition, and Parallel Dynamic Programming*, Proc. 1987 International Conference on Acoustics, Speech and Signal Processing (ICASSP) (April 1987), pp. 976–979.

[G 88] A.L. Gorin and D.B. Roe, *Parallel Level-Building on a Tree Machine*, Proceedings of ICASSP 88 (April 1988), pp. 295–298.

[L 81] C.E. Leiserson, *Area-Efficient VLSI Computation*, Ph.D. Thesis, Carnegie Mellon University (1981).

[L 85] S.E. Levinson, *Structural Methods in Automatic Speech Recognition*, Proceedings of the IEEE (Nov. 1985).

[L 87] M. B. Lowrie and K. Fuchs, *Reconfiguration Tree Architectures using Subtree Oriented Fault Tolerance*, IEEE Trans. on Comp., c.-36, 10, pp. 1172–1182 (Oct. 1987).

[P 85] Pradhan, D.K., *Dynamically Restructurable Fault-Tolerant Processor Network Architectures*, IEEE Trans. on Comp., c-34, 5 (May 1985).

[R 81] L.R. Rabiner and S.E. Levinson, *Isolated and Connected-Word Recognition-Theory and Selected Applications*, IEEE Trans. on Communications, 29, 5 (May 1981).

[S 82] L. Snyder, *Introduction to the Configurable, Highly Parallel Computer*, Computer (January 1982).

[S 84] D. Shaw, *SIMD and MSIMD variants of the Non-Von Supercomputer*, Proc. of ComCon, San Francisco, CA (Feb. 1984).

[S 85] S.J. Stolfo and D. Miranker, *DADO: A Parallel Processor for Expert Systems*, Advanced Computer Architecture (Agrawal, ed.) IEEE Comp. Soc. Press (1985).

[S 88] R.R. Shively and A.L. Gorin, *A Fault-Tolerant Reconfigurable Systolic Signal Processor*, Proc. I (ASSP 89 (May 1989), pp. 2397–2340.

[T 85] Y. Takahashi, et al, *Efficiency of Parallel Computation on the Binary Tree Machine CORAL '83*, Journal of Information Processing, 8, 4 (1985), pp. 288–299.

SOLITON MATHEMATICS IN SIGNAL PROCESSING

F. ALBERTO GRÜNBAUM†

TABLE OF CONTENTS

Introduction. The purpose of this lecture is to illustrate the use of tools initially developed in the field of "soliton mathematics" as a way to handle a specific problem in the apparently unrelated area of Nuclear Magnetic Resonance Imaging.

It will turn out that this application suggests new problems in the area of "inverse scattering".

We restrict ourselves to one problem in an area of great medical interest: the Selective Excitation problem, but in fact the entire area of NMR and coherent optics requires the design of "special purpose" pulse sequences designed to achieve a specified magnetization vector distribution.

Indeed some applications require a "broad band" population inversion", almost the opposite goal of the one considered here. The mathematical tools given here are however just as useful in both cases.

The reader interested in a good and very clear introduction to the use of NMR in the field of medical imaging should consult the important article by W. Hinshaw and A. Lent [1]. The problem of "selective excitation" considered here arises with the desire to excite only those spins lying in a specified plane. This is a step that precedes the application of special pulse sequences and the eventual processing involved in obtaining an image. Although this is such a specialized problem the quality of the resulting "plane selection" is crucial to the overall image resolution, and no completely satisfactory solution has yet been achieved.

We will give an ab-initio description of the physical problem in question, and at the appropriate point, we will give an introduction to the tools from inverse scattering and soliton theory that have been found relevant in the analysis of the problem.

The physics of the problem is adequately described by the Bloch equations without decay terms, i.e. just the Euler equations describing the dynamic of a top or spin.

†Mathematics Department, University of California, Berkeley, CA 94720

Some of the early analysis of this problem, see D. Hoult [2], is predicated on a "small tip–angle" approximation to the Bloch equations: this makes available the well understood Fourier Transform and all the standard toolkit of linear signal processing.

In a well defined sense one can conceive of "soliton mathematics" as the natural step after Fourier analysis. Within this picture it should not be too surprising that this beautiful piece of mathematics should rear its head in a host of physical problems where the Fourier approximations does not suffice.

The appropriate extension of the Fourier transform is the so called Bloch transform, see F.A. Grünbaum [3,4]. This nonlinear transform, whose linear approximation at zero input is the usual Fourier transform, is introduced below.

Some of this work leads to a number of open questions; we expect to return to them in the near future.

Other workers have made a connection between Coherent–Optical–Pulse propagation and inverse scattering, most importantly G. Lamb [5,6]. Intriguing analogies between these two fields can be seen in the pioneering work on self–induced transparency by McCall and Hahn [7]. It is also illuminating to take a look at the papers [8, 9, 10] listed in the references.

An NMR analog of this result has been discovered in important papers by M. Silver, R. Joseph and D. Hoult [11]. A good part of my own work in this area has been joint work with Andy Hasenfeld, [12, 13]. I have also benefited from conversations with M. Buonocore, some of whose work is reported in [14]. In conclusion, we make it clear that there are other ways of attacking this problem. See for instance the work of J. Baum, R. Tycko, A. Pines [15], as well as the optimization work of S. Conolly, D. Nishimura and A. Macovski [16], and J. Murdoch, A. Lent, M. Kritzer [17].

The Bloch Equations. Consider a magnetization distribution

$$\overline{M}(x, y, z, t)$$

defined in a region of space and subject to the effect of an imposed magnetic field distribution

$$\overline{B}(x, y, z, t)$$

If one ignores decay effects, the time evolution of $\overline{M}$ is given by the Euler equations that result from simplifying the Bloch equations, namely

$$\frac{d}{dt}\,\overline{M} = \gamma\,\overline{B} \times \overline{M} \tag{1}$$

Here γ is a constant, the gyromagnetic ratio of the nucleus in question.

For the problem of "slice selection" the times involved are so short that it is entirely appropriate to neglect the decay terms in the full Bloch equation.

For our purposes we take B to consists of a linear gradient in the direction of the main field (with the purpose of encoding spatial information) and an arbitrary "radio frequency" amplitude and phase modulation components in the transversal directions

$$\overline{B}(x,y,z,t) = B_1(t)\cos w_1(t)\overline{i} + B_1(t)\sin w_1(t)\overline{j} + (B_0 + Gz)\overline{k}$$

It is customary to follow the evolution of $\overline{M}$ on a rotating frame. Set, for a fixed ω,

$$N(t) = \begin{pmatrix} \cos\omega t & -\sin wt & 0 \\ \sin\omega t & \cos\omega t & 0 \\ 0 & 0 & 1 \end{pmatrix} M(t)$$

and then convert (1) into

(2)

$$\dot{N}(t) = \begin{pmatrix} 0 & \gamma(B_0 + Gz) - \omega & -\gamma B_1(t)\sin(w_1(t) - \omega t) \\ -\gamma(B_0 + G_z) + \omega & 0 & \gamma B_1(t)\cos(w_1(t) - \omega t) \\ \gamma B_1(t)\sin(w_1(t) - \omega t) & -\gamma B_1(t)\cos(w_1(t) - \omega t) & 0 \end{pmatrix}$$

The natural choice for ω is $\omega = \gamma B_0$, and then one defines the "resonance offset" $\Delta\omega$ by the expression

$$\Delta\omega \equiv \gamma Gz$$

The identification between $\Delta\omega$ and the spatial coordinate z should be kept in mind throughout.

Define also the new amplitude and phase modulation functions

$$\omega_1(t) = \gamma B_1(t)$$

and

$$\phi(t) = w_1(t) - \gamma B_0 t$$

respectively. If we denote by $Q(t)$ the solution to the equation

$$(3) \quad \dot{Q}(t) = \begin{pmatrix} 0 & \Delta\omega & -\omega_1(t)\sin\phi(t) \\ -\Delta\omega & 0 & \omega_1(t)\cos\phi(t) \\ \omega_1(t)\sin\phi(t) & -\omega_1(t)\cos\phi(t) & 0 \end{pmatrix} Q(t); Q(0) = I$$

it is clear that the solution to (2) is given by

$$N(t) = Q(t)\ N(0)$$

At this point we can formulate the MAIN PROBLEM, namely, given an initial distribution of magnetization

$$\overline{M}(x,y,z,0)$$

and a final desired distribution

$$\overline{M}(x,y,z,T)$$

find appropriate functions

$$\omega_1(t) \quad , \quad \phi(t) \qquad 0 \le t \le T$$

such that $\overline{M}(T)$ is the evolution of $\overline{M}(0)$ under the dynamics given above.

In general this may turn out to be impossible, and in fact we will be concerned with special cases of this problem.

A typical initial condition will be

$$\overline{M}(x,y,z,0) = \begin{pmatrix} 0 \\ 0 \\ 1 \end{pmatrix} \quad \text{or} \quad \begin{pmatrix} 0 \\ 0 \\ -1 \end{pmatrix}$$

i.e. " all spins up" or "all spins down".

Given such an initial distribution, the final vector $\overline{M}(x,y,z,T)$ will only a function of z. *Recall that z is identified with the "resonance offset $\Delta\omega$".*

We are thus faced with a "control" type problem of driving an initial state into a desired final state.

A natural approach to this problem is to visualize the output $\overline{M}(T)$ as a transform of the input $\overline{M}(0)$ and to study the inversion properties of the transformation thus defined. We undertake this task along three different paths.

In the remainder of this section we introduce what I have termed the "Bloch transform", see [3,4]. We do not make any attempts at studying either the range of this nonlinear map, or its invertibility properties.

In section 2 we convert the equations given above into a differential equation for a complex valued function, and then make a connection with a well studied inversion problem.

In section 3 we will give an equivalent formulation of the problem in terms of parameters introduced by Cayley and Klein, i.e. the spinor formalism. This permits us to connect our problem with another well known inversion problem.

The "Bloch transform" comes about by expressing the solution to equation (3) as the sum of the series

$$Q(t) = T(t) + \int_0^t T(t-s_1)B(s_1)T(s_1)\, ds_1 +$$

$$+ \int_0^t \int_0^{s_1} T(t-s_1)B(s_1)T(s_1-s_2)B(s_2)T(s_2)\, ds_2 ds_1 + \ldots$$

Here $T(t)$ denotes the matrix

$$T(t) = \begin{pmatrix} \cos\Delta\omega t & \sin\Delta\omega t & 0 \\ -\sin\Delta\omega t & \cos\Delta\omega t & 0 \\ 0 & 0 & 1 \end{pmatrix}$$

and $B(t)$ denotes the matrix

$$B(t) = \begin{pmatrix} 0 & 0 & -\omega_1(t)\sin\phi(t) \\ 0 & 0 & \omega_1(t)\cos\phi(t) \\ \omega_1(t)\sin\phi(t) & -\omega_1(t)\cos\phi(t) & 0 \end{pmatrix}$$

Suppose now that initial distribution is

$$\overline{M}(x,y,z,0) = \begin{pmatrix} 0 \\ 0 \\ 1 \end{pmatrix}$$

In this case we have $\overline{N}(0) = \overline{M}(0)$ and it is a matter of straightforward computation to observe that if one looks at the complex valued quantity whose real and imaginary parts are the x and y components of $\overline{N}(T)$ respectively one obtains a convergent infinite series depending on the complex valued function

$$\omega_1(t)e^{i\phi(t)} \qquad 0 \le t \le T$$

If one linearizes this expression at zero one obtains the Fourier transform of this complex valued function as a function of the "resonance offset": the higher order terms give the discrepancy between the Bloch transform thus defined and its linearized version.

It is clear that for applied radio frequency pulses that are not too weak, or for times that are not too short, specially any time that the "small tip–angle" approximation is invalid, an analysis that replaces the Bloch transform by the Fourier transform becomes rather suspect.

The Ricatti and Schroedinger Equations. We will consider the solution to (1) under the initial condition

$$\overline{M}(x,y,z,0) = \begin{pmatrix} 0 \\ 0 \\ -1 \end{pmatrix}$$

i.e. "all spins pointing down".

Since the length of the vector $\overline{M}$ is clearly preserved by (1), and thus the vector $\overline{M}(x,y,z,t)$ is restricted to the unit sphere for each (x,y,z) and all times, it is natural to consider the stereographic projection of $\overline{M}$ onto a plane. As we will see two different such projections can be used to some advantage: either from the north pole $(0,0,1)$ or from $(1,0,0)$.

It is worth pointing out the similarity of equation (1) with the Frenet–Serret equations describing the infinitesimal variation of a moving frame along a curve in space in terms of its curvature and torsion. In this geometric setup Sophus Lie seems to be the first one to have used this trick of stereographic projection, see

[18]. In terms of optics, Lamb [5,6] found an equation that looks exactly like the Frenet–Serret equations. He was the first one to point out the relative advantages of using two different "poles" to perform the projection, and he also noticed the advantages of going over to a "spinor formulation".

We consider first stereographic projection from the North Pole.

If we define

$$\eta = \frac{Nx + i\, N_y}{N_z - 1}$$

and recall that equation (1) for $M(t)$ has become

$$\dot{N}(t) = \begin{pmatrix} 0 & \Delta\omega & -\omega_1(t)\sin\phi(t) \\ -\Delta\omega & 0 & \omega_1(t)\cos\phi(t) \\ \omega_1(t)\sin\phi(t) & -\omega_1(t)\cos\phi(t) & 0 \end{pmatrix} N(t) \tag{2}$$

we get the Riccati equation

$$\begin{aligned}\dot{\eta} &= -\left(\frac{\omega_1 \sin\phi + i\omega, \cos\phi}{2}\right)\eta^2 - i\Delta\omega\eta + \frac{-\omega_1\sin\phi + i\omega_1\cos\phi}{2} \\ &= a\eta^2 + b\eta + c\end{aligned}$$

with $a = -\frac{i}{2}\,\omega_1 e^{-i\phi}$, $b = -i\Delta\omega$, $c = \frac{i}{2}\,\omega_1 e^{i\phi} = a^*$

Setting

$$\eta = -\frac{u'}{au}\,, \quad \text{i.e.} \quad u = e^{-\int a\eta}$$

and eventually

$$u = e^{\frac{1}{2}\int b}\sqrt{a}\; s$$

we get for $s(t)$ the equation

$$s'' + s\left[\frac{1}{2}\,(\log a)'' + \frac{\omega_1^2}{4} + \frac{1}{4}\,(i(\log a)' + \Delta\omega)^2\right] = 0 \tag{4}$$

The presence of the term

$$(\log a)' = \frac{\omega_1'}{\omega_1} - i\dot{\phi}$$

prevents this last equation (4) from taking the standard Schroedinger form

$$-S'' + V(t)S = \frac{(\Delta\omega)^2}{4}\,S \tag{5}$$

with $\frac{(\Delta\omega)^2}{4}$ playing the role of energy, and

$$V(t) = \frac{1}{2}\,\log a'' - \frac{\omega_1^2}{4}$$

If we had obtained equation (5) in this transcription of the original Bloch equation (1') one could try to use the well known fact that sufficient "scattering information" allows one to reconstruct the potential $V(t)$, see [19,20,21,22].

In fact the problem of recovering a pair of functions $A(x)$ and $B(x)$ from scattering information for the problem

$$S'' + S\left(\frac{(\Delta\omega)^2}{4} + i\frac{\Delta\omega}{2}A(t) + B(t)\right) = 0 \tag{4'}$$

has been solved by P. Sabatier [23]. It was later observed by Jaulent and Miodek [24] that this is equivalent to the consideration of a Zakharov–Shabat two component problem. In the context of optics this had already been pointed out by Lamb [5,6].

We will consider a two component formulation of our problem in the next section.

We consider equation (2) once again, but this time we use a different stereographic projection, namely

$$\eta = \frac{N_z + i\,N_y}{N_x - 1}$$

We find that the new equation is

$$\dot{\eta} = \frac{\omega_1 \sin\phi + i\Delta\omega}{2}\eta^2 + i\omega_1 \cos\phi\eta + \frac{\omega_1 \sin\phi - i\Delta\omega}{2}$$

For appropriate a, b, c this is again of the Riccati form $\dot{\eta} = a\eta^2 + b\eta + c$ and setting, as before

$$\eta = -\frac{u'}{au}$$

and

$$u = e^{\frac{1}{2}\int b}\sqrt{a}\ s$$

we get the awful looking expression

$$\begin{aligned} & s'' + s\left[\frac{\omega_1^2}{4} + \frac{\Delta\omega^2}{4}\right] + \\ & s\,\frac{-2i\Delta\omega^2(\cos\phi\omega_1)' + 2i\Delta\omega\left((\sin\phi\omega_1)'' + i\sin\phi\cos\phi\omega_1\omega_1' - i(1+\sin^2\phi)\phi'\omega_1^2\right)}{4(\omega_1\sin\phi + i\Delta\omega)^2} + \\ & + \frac{2\sin^2\phi\omega_1\omega_1'' - 3\sin^2\phi w_1' - 2\cos\phi\sin\phi\phi'\omega_1\omega_1' - 2i\sin\phi\phi'\omega_1^3}{4(\omega_1\sin\phi + i\Delta\omega)^2} + \\ & \frac{2\cos\phi\sin\phi\phi'' w_1^2 - 3\phi'^2\omega_1^2 + \sin^2\omega\phi\phi'^2 u}{4(\omega_1\sin\phi + i\Delta\omega)^2} \\ & = 0 \end{aligned} \tag{6}$$

The only redeeming value of this way of doing the projection appears in the case of "amplitude modulation" when

$$\phi(t) \equiv 0$$

in this case we get

$$s'' + s\left[\frac{\omega_1^2}{4} + \frac{\Delta\omega^2}{4} + \frac{i}{2}\,\omega_1'\right] = 0 \tag{7}$$

as found in [12, 13].

It appears very unlikely that any other simplifications of (6) into a nice form like (7) are possible.

The advantage of (7) is that it is written in the standard Schroedinger form, and thus all the machinery of inverse scattering becomes available. Part of this program was carried out in [12, 13].

If we are interested in the case of "amplitude and phase modulation" it appears clear that a different way of handling the Bloch equation is called for. This is done in the next section.

The Two Component Formulation. We give now an alternative description of the spin dynamics embodied in equations (1) or (2).

The only reason for going through this is that we can take advantage of a ready made setup for which the "inverse scattering" situation is well understood.

One replaces the three dimensional unit vector

$$\overline{N}(x, y, z, t)$$

by a two by two complex matrix

$$P = \frac{i}{2}\begin{pmatrix} -N_z & N_x + iN_y \\ N_x - iN_y & N_z \end{pmatrix}$$

and notices that the time evolution of (2) is given by the rule

$$P(t) = U(t)P(0)\;U^{+}(t) \tag{8}$$

with $U(t)$ a unitary matrix. This follows from the fact that the eigenvalues of $P(t)$ are independent of t.

Differentiation of (8), and repeated use of the relation

$$U(t)U^{+}(t) = U(t)U^{-1}(t) \equiv I$$

gives

$$P'(t) = \left[U'(t)U^{-1}(t), P(t)\right] \tag{9}$$

We write the matrix $U(t)$, an element of $SU(2)$, in the form

$$U(t) = \begin{pmatrix} \alpha & \beta \\ -\beta^{*} & \alpha^{*} \end{pmatrix}$$

with

$$|\alpha|^2 + |\beta|^2 = 1$$

Since we get

$$U'U^{-1} = \begin{pmatrix} \alpha'\alpha^* + \beta'\beta^* & -\alpha'\beta + \beta'\alpha \\ -\alpha^*\beta^{*\prime} + \alpha^{*\prime}\beta^* & \alpha\alpha^{*\prime} + \beta\beta^{*\prime} \end{pmatrix}$$

we see that (9) leads us to impose on the quantities $\alpha(t, \Delta\omega)$ and $\beta(t, \Delta\omega)$ the evolution equation

$$\frac{d}{dt}\begin{pmatrix} \alpha \\ \beta^* \end{pmatrix} = \begin{pmatrix} -\frac{i}{2}\,\Delta\omega & -\frac{i}{2}\,\omega_1(t)e^{+i\phi(t)} \\ -\frac{i}{2}\,\omega_1(t)e^{-i\phi(t)} & \frac{+i}{2}\,\Delta\omega \end{pmatrix}\begin{pmatrix} \alpha \\ \beta^* \end{pmatrix}$$

We remark that this is an evolution of the Zakharov–Shabat type, considered by Lamb [5,6] in the case of coherent optics.

In this case the "inverse scattering" situation is well understood. Appropriate information on the solutions allows one to recover the off diagonal elements of the two by two matrix, i.e. the individual functions

$$\omega_1(t) \quad \text{and} \quad \phi(t)$$

This will be briefly reviewed at the end of this section.

Working out the evolution (8) we can read off the components of

$$\overline{N}(x, y, z, t)$$

in terms of its initial values and the values of

$$\alpha(t, \Delta\omega) \quad \text{and} \quad \beta(t, \Delta\omega)$$

We get

$$N_z(t) = (|\alpha|^2 - |\beta|^2)N_z(0) - 2Re\ \alpha^*\beta(N_x(0) - iN_y(0))$$
$$N_x(t) + iN_y(t) = 2\alpha\beta\ N_z(0) + \alpha^2(N_x(0) + iN_y(0)) - \beta^2(N_x(0) - iN_y(0))$$

Notice that the initial condition

$$\overline{N}(x, y, z, 0) \equiv \begin{pmatrix} 0 \\ 0 \\ -1 \end{pmatrix}$$

allows for the choice

$$\left|\alpha(0, \Delta\omega)\right| \equiv 1 \quad \text{and} \quad \beta(0, \Delta\omega) \equiv 0$$

If we are interested in "selective excitation", say

$$\overline{N}(x,y,z,T) \cong \begin{pmatrix} 0 \\ 0 \\ 1 \end{pmatrix} \qquad \text{for } |z| \text{ very small}$$

$$\overline{N}(x,y,z,T) \cong \begin{pmatrix} 0 \\ 0 \\ -1 \end{pmatrix} \qquad \text{for other values of } z$$

we should aim at getting $\alpha(T,\Delta\omega)$ and $\beta(T,\Delta\omega)$ that satisfy

$$|\alpha(T,\Delta\omega)| \cong \begin{cases} 0 & \text{for } |z| \text{ very small} \\ 1 & \text{for other values of } z \end{cases}$$

$$|\beta(T,\Delta\omega)| \cong \begin{cases} 1 & \text{for } |z| \text{ very small} \\ 0 & \text{for other values of } z \end{cases}$$

We close this section with a *minimal* presentation of the relevant "inverse scattering" for a two component system, included here to show that at least in principle the "pulse design" problem corresponding to selecting excitation can be handled with these tools.

Given the system

$$\begin{pmatrix} v_1 \\ v_2 \end{pmatrix}_t = \begin{pmatrix} -i\eta & q(t) \\ r(t) & i\eta \end{pmatrix} \begin{pmatrix} v_1 \\ v_2 \end{pmatrix} \quad \begin{matrix} -\infty < t < \infty \\ \eta \text{ complex} \end{matrix}$$

with complex valued $q(t)$ and $r(t)$ vanishing fast enough $\pm\infty$, we can find, for real values of η, a pair of linearly independent solutions ϕ and $\overline{\phi}$, specified by their behaviours at $-\infty$

$$\phi \sim \begin{pmatrix} 1 \\ 0 \end{pmatrix} \bar{e}^{i\eta t}$$

$$\overline{\phi} \sim \begin{pmatrix} 0 \\ -1 \end{pmatrix} e^{i\eta t}$$

One can introduce, still for real η, the functions

$$a(\eta),\ b(\eta),\ \overline{a}(\eta),\ \overline{(}\eta)$$

by looking at the behaviour of these same solutions at $+\infty$. We have, for $t \sim +\infty$

$$\phi \sim \begin{pmatrix} a(\eta) & e^{-i\eta t} \\ b(\eta) & e^{i\eta t} \end{pmatrix} \qquad \overline{\phi} \sim \begin{pmatrix} \overline{b}(\eta) & e^{-i\eta t} \\ -\overline{a}(\eta) & e^{i\eta t} \end{pmatrix}$$

One can then show that the functions $a(\eta), \overline{a}(\eta)$ can be extended analytically into the upper and lower half planes respectively. Their zeros in these half planes give the "eigenvalues" corresponding to square summable solutions of the system of equations.

Under enough decay assumptions on $q(t)$ and $r(t)$ the functions $b(\eta)$ and $\overline{b}(\eta)$ can also be extended analytically away from the real axis. For instance if q and r have compact supports then all four functions are entire functions of η.

It turns out that the knowledge for real η - of the ratios

$$\frac{b(\eta)}{a(\eta)} \quad , \quad \frac{\overline{b}(\eta)}{\overline{a}(\eta)}$$

along with the complex zeros of $a(\eta)$ and $\overline{a}(\eta)$ and the residues at these zeros for the ratios just mentioned, allows one *at least in principle* to reconstruct $q(t)$ and $r(t)$. Slightly more complicated cases arise for nonsimple zeroes.

In our case we have

$$\begin{pmatrix} \alpha \\ \\ \beta^* \end{pmatrix}_t = \begin{pmatrix} -i\eta & q(t) \\ \\ -q^*(t) & i\eta \end{pmatrix} \begin{pmatrix} \alpha \\ \\ \beta^* \end{pmatrix}$$

(the condition $r = -q^*$ brings along all sort of simplifications in the general picture) and we are looking for a driving force $q(t)$ such that a solution ϕ should exist that looks at $t = -\infty$ like

$$\phi = \begin{pmatrix} \alpha \\ \\ \beta^* \end{pmatrix} \sim \begin{pmatrix} \overline{e}^{i\eta t} \\ \\ 0 \end{pmatrix}$$

i.e.

$$|\alpha(-\infty, \eta)| \equiv 1 \qquad \beta(-\infty, \eta) \equiv 0$$

and such that at $t = +\infty$ should look like

$$\begin{pmatrix} a(\eta)e^{-i\eta t} \\ \\ b(\eta)e^{i\eta t} \end{pmatrix}$$

This calls for a choice of the functions

$$a(\eta) \quad \text{and} \quad b(\eta)$$

such that their modulus are (as close as possible) to

$$|b(\eta)| = \begin{cases} 1 & \text{for } |\eta| \text{ very small, real} \\ 0 & \text{for other values of real } \eta \end{cases}$$

$$|a(\eta)|^2 = 1 - |b(\eta)|^2$$

Return to the Fourier Transform. We noticed in section 1 that if

$$\overline{N}(x, y, z, 0) \equiv \begin{pmatrix} 0 \\ 0 \\ 1 \end{pmatrix}$$

then the complex valued function of the "resonance offset" $\Delta\omega$

$$N_x(\Delta\omega, T) + iN_y(\Delta\omega, T)$$

is given as a "nonlinear Fourier Transform" of the complex valued function

$$\omega_1(t)e^{i\phi(t)} \qquad 0 \le t \le T$$

Loosely speaking, this means that in the limit of vanishing applied magnetic field, the linearization of this nonlinear map is just the usual Fourier Transform.

Since we do not have an effective way to invert this nonlinear transformation we went on in sections 2 and 3 to different formulations. These approaches have the advantage that one can "stand on the shoulders" of Gelfand, Levitan, Marchenko, Krein, Zakharov, Shabat and others, see [19, 20, 21, 22] if one views the problem as one of going from "scattering data" back to the potential.

In a gathering like this one, devoted to signal processing, it may be fitting to conclude by recalling the well known fact that this "inverse scattering transform" is yet another nonlinear map whose linear part, in the limit of vanishing potential, is given by the Fourier Transform, see [19, 20, 21, 22].

REFERENCES

[1] W. Hinshaw, A. Lent, Proc. IEEE 71, 3, 1983 pp 175.
[2] D. Hoult, J. of Magnetic Resonance 35 1979 pp 69–86.
[3] F.A. Grünbaum, Proc. 3rd Annual Meeting Soc. Magnetic Resonance in Medicine, NY, 1984.
[4] F.A. Grünbaum, *Inverse Problems 1*, L25 1985.
[5] G. Lamb, 1, 1974 pp 422–430.
[6] G. Lamb, Physical Rev. Letters 31, 4 1973 pp 196–199.
[7] S. Mc Call, E. Hahn, Phys. Rev. 183, 1969 pp 457.
[8] D. Kaup, Physical Review A 16, 2 1977 pp 704–719.
[9] R. Michalska-Trautman, Physical Review A 23, 1 1981 pp 352–359.
[10] M. Ablowitz, D. Kaup, A. Newell, J. Math. Physics 15, 1974 pp 1852.
[11] M. Silver, R. Joseph, D. Hoult, *Physical Review A 31.4 1985 pp 2753–2755.*
[12] F.A. Grünbaum, A. Hasenfeld, Inverse Problems 2 1986 pp 75–81.
[13] F.A. Grünbaum, A. Hasenfeld, Inverse Problem 4 1988.
[14] M. Buonocore, Report at the Center for Pure and Applied Math, UC Berkeley, 1987 #375.
[15] J. Baum, R. Tycko, A. Pines, J. Chem. Physics 79, 9, 1983 pp 4643–4644.
[16] S. Conolly, D. Nishimura, A. Macovski, IEEE Trans. on Medical Imaging Vol MI–5, 2, 1986 pp 106–115.
[17] J. Murdoch, A. Lent, M. Kritzer, J. Magnetic Resonance 74, 1987 pp 226–263.

[18] S. Lie, *Complete Works # 37*, pp 531–535.

[19] S. Novikov, S. Manakov, L. Pitaevski, V. Zakharov, *Theory of Solitons*, Plenum, NY 1984.

[20] A. Newell, *Solitons in Mathematics and Physics*, Siam, Philadelphia 1985.

[21] M. Ablowitz, H. Segur, *Solitons and the Inverse Scattering Transform*, Siam, Philadelphia 1984.

[22] G. Lamb, *Elements of Soliton Theory*, John Wiley, New York, 1980.

[23] P. Sabatier, *J. Math. Physics 9, 1969 pp 1241.*

[24] M. Jaulent and I. Miodek, Lettere al Nuovo Cimento 20, 18 1977 pp–655–660.

SELECTIVE 'COMPLEX' REFLECTIONLESS POTENTIALS

ANDREW HASENFELD†

Abstract. Previous results obtained in attempts to invert the Bloch transform for the reduced case of amplitude modulation are extended to the general case of simultaneous amplitude and phase modulation. Numerical results are presented which illuminate this general case, and these results are interpreted in terms of one-dimensional non-relativistic quantum mechanical scattering. We find that the full (complex) case of simultaneous amplitude and phase modulation is more robust than the reduced (real) case of amplitude modulation.

1. Preamble. This lecture should be viewed as a companion to the one entitled "Soliton Mathematics in Signal Processing" by F. Alberto Grünbaum [1]. Indeed, the numerical results presented here, along with the experimental verification contained in [5], stimulated much of the new perspective provided by [1]. Before elaborating on the title of this paper, a few remarks should be addressed to the current situation.

A nice feature of the inverse scattering reformulation of the NMR "selective excitation" issue is in its physical motivation. The concerns fundamental to this NMR problem are nonlinear, and so borrowing techniques from the nonlinear wave community should be seen as natural, and moreover physically intuitive. In particular, the general prescription of viewing the desired response (namely the amplitude and phase modulations of the imposed radiofrequency field that achieve a given excitation profile) as comprising scattering data, and the unknown modulations as sitting in the potential that produces that scattering data, is an appealing metaphor for getting a handle on the relevant questions and their solutions.

We close this brief sermon with a few remarks concerning the many unexplained mysteries run into along the tortuous path to this point (perhaps the words "conspiracy of miracles" describe it best), and some deeply felt thanks for the pioneering work of G.L. Lamb, Jr. (referenced in [1]). One would like to have a clear physical picture of the many exotic phenomena exhibited by simple spin systems, not only encompassing the miraculous response of self-induced transparency [6], but also the rectangular inversions described here. The inverse scattering perspective unifies the explanations for these things under one umbrella, so that the generalization from pure amplitude modulation to simultaneous amplitude and phase modulation comes with a nice physical meaning (corresponding to the transition from a real-valued to a complex-valued potential in the associated scattering picture). As we will show in section 7, this approach is equivalent to the general framework constructed by physicists to describe statistical ensembles of spins [10], and so in principle should describe all of the relevant phenomena. Finally, it should be stated that there exist other approaches to the solution of this problem, most notably the one arising from ideas in optimal control [7].

---

†Department of Chemistry, Princeton University, Princeton, NJ 08544. This work was supported by the National Institutes of Heath under grant GM35253 BBCB, and by the National Science Foundation under grant CHE-8502199.
Present address: Center for Pure and Applied Mathematics, University of California, Berkeley, CA 94720.

2. Formal introduction. We have written at length about attempts to invert a complicated nonlinear map called the Bloch transform ([2-4] and references therein) in the reduced case of amplitude modulation. This transformation

$$\mathcal{B}: \gamma \mathrm{H}_1(t) = \omega_1(t)e^{i\varphi(t)} \rightarrow \mathbf{M}(\Delta\omega, T) \tag{2.1}$$

represents integration of the Bloch equations, where the pulse of radiation lasts in time t from 0 to T, and $\Delta\omega$ is a parameter called the resonance offset (linearly proportional to distance). The underlying physical motivation in all of this work has been to comprehend the quantum mechanical interaction of electromagnetic radiation with spin, and the use of the language of inverse scattering has provided a backdrop upon which to realize an appropriate description. Most recently, we have had success extending the formalism to the general case of simultaneous amplitude and phase modulation [5], and it is now clear that viewing $\mathcal{B}^{-1}$ as an analogue of the inverse scattering transform is both fruitful and remarkably accurate. One sees that the general case follows as a simple extension of the scattering formalism to scattering in the presence of a complex potential.

We will begin with an analytical treatment of the simple quantum mechanical equations of motion for an isolated two-level system in an external field. We then move to a consideration of numerical findings (in an attempt to identify the subset of reflectionless potentials that produce 'rectangular' inversion profiles), using some remarkably simple arguments from one-dimensional non-relativistic quantum mechanical Schrödinger scattering for an interpretation of the dynamics of uncoupled two-level systems. Finally, we connect the two component formulation of Zakharov and Shabat to the physicists' language of the density matrix. Some mysteries still remain, but we are now confident that our elementary scattering viewpoint can ultimately handle them.

3. Bloch → Riccati → Schrödinger. We now describe a certain mathematical pathway which, in the reduced case of amplitude modulation, led to a successful description (see [2-4]). In trying to extend this method to the general situation embodied in (3.1), however, one runs into problems that cannot be ignored. Nevertheless, it is instructive to see how it all breaks down.

We begin with the Bloch equations for amplitude and phase modulation

$$\frac{d}{dt}\begin{pmatrix} M_x \\ M_y \\ M_z \end{pmatrix} = \begin{pmatrix} 0 & \Delta\omega & -\omega_1(t)\sin\varphi(t) \\ -\Delta\omega & 0 & \omega_1(t)\cos\varphi(t) \\ \omega_1(t)\sin\varphi(t) & -\omega_1(t)\cos\varphi(t) & 0 \end{pmatrix}\begin{pmatrix} M_x \\ M_y \\ M_z \end{pmatrix} \tag{3.1}$$

with the initial condition $\mathbf{M}(\Delta\omega,0)=(0,0,-1)^T$, and where relaxation terms are neglected. By stereographically projecting $\mathbf{M}$ onto the y-z plane

$$\eta(\Delta\omega, t) = \frac{M_z + iM_y}{M_x - 1}, \tag{3.2}$$

we find that η satisfies a Riccati equation

$$\begin{aligned} \dot{\eta} &= \left(\frac{\omega_1 \sin\varphi + i\Delta\omega}{2}\right)\eta^2 + (i\omega_1\cos\varphi)\eta + \left(\frac{\omega_1\sin\varphi - i\Delta\omega}{2}\right) \\ &= a\eta^2 + b\eta + c. \end{aligned} \tag{3.3}$$

(Here is one of the mysteries: why does stereographic projection yield a Riccati equation?) The general approach is then to construct a

$$g \equiv e^{-\int_{-\infty}^{t}(a\eta + \frac{\dot{a}}{2a} + \frac{b}{2})} \tag{3.4}$$

to linearize (3.3) into the form

$$\begin{aligned} -\ddot{g} &= -\frac{1}{4}[3(\frac{\dot{a}}{a})^2 - \frac{2\ddot{a}}{a} + \frac{2\dot{a}b}{a} + b^2 - 4ac - 2\dot{b}]g \\ &= [E - V_1 - V_2]g \end{aligned} \tag{3.5}$$

with

$$\begin{aligned} V_1 &= \tfrac{1}{4}[3(\tfrac{\dot{a}}{a})^2 - \tfrac{2\ddot{a}}{a} + \tfrac{2\dot{a}b}{a}] \\ V_2 - E &= \tfrac{1}{4}[b^2 - 4ac - 2\dot{b}] \\ a &= \tfrac{\omega_1 \sin\varphi + i\Delta\omega}{2} \\ b &= i\omega_1 \cos\varphi \\ c = a^* &= \tfrac{\omega_1 \sin\varphi - i\Delta\omega}{2} \end{aligned} \tag{3.6}$$

In this form, by replacing t by a spatial parameter $\overline{x}$, we arrive at the "time-independent" Schrödinger equation

$$-\frac{d^2 g}{d\overline{x}^2} + (V - E)g = 0, \tag{3.7}$$

where one identifies the energy E and the potential V as

$$\begin{aligned} V - E = V_1 + V_2 - E &= \tfrac{1}{4}[3(\tfrac{\dot{a}}{a})^2 - \tfrac{2\ddot{a}}{a} + \tfrac{2\dot{a}b}{a} + b^2 - 4ac - 2\dot{b}] \\ E \equiv \lambda^2 &= (\tfrac{\Delta\omega}{2})^2 \end{aligned} \tag{3.8}$$

At this point, we run into difficulties. When $\varphi \equiv 0$ (amplitude modulation), we see from (3.6) that V_1 vanishes identically, and that V_2-E becomes

$$V_2 - E = -\frac{1}{4}[\omega_1^2 + 2i\dot{\omega}_1 + \Delta\omega^2]. \tag{3.9}$$

¿From the argument in [3], the term -$\frac{i\dot{\omega}_1}{2}$ is neglected, and so the ansatz

$$\omega_1 \sim \sqrt{-V(\overline{x})} \tag{3.10}$$

results. In the general case (3.1), however, V_1 no longer vanishes, and (as noted in [1]) our simple inverse scattering picture based on the Schrödinger equation begins to unravel.

However, proceeding ahead blindly, we evaluate

$$V_2 - E = -(\frac{\omega_1}{2})^2 - i\frac{\dot{\omega}_1 \cos\varphi}{2} + i\frac{\omega_1 \dot{\varphi} \sin\varphi}{2} - (\frac{\Delta\omega}{2})^2. \tag{3.11}$$

Recall we wish to reconstruct ω_1 and φ from the real and imaginary parts of V. But if

$$\dot{\omega_1} \sim \omega_1 \dot{\varphi}, \tag{3.12}$$

from (3.6) and (3.8) we have

$$Re(V_2) \sim -(\frac{\omega_1}{2})^2\, ,\, Im(V_2) \sim -\frac{\dot{\omega}_1}{2}(\cos\varphi - \sin\varphi). \tag{3.13}$$

Now (3.12) is easy to integrate

$$\varphi = \mu\, ln(\omega_1), \tag{3.14}$$

for constant μ, which yields the ansatz

$$\gamma H_1(t) = \omega_1 e^{i\mu\, ln(\omega_1)} = \omega_1^{(1+\mu i)} \sim (\sqrt{-V})^{(1+\mu i)}. \tag{3.15}$$

Call it divine inspiration, but the rest of this paper will present some very surprising numerical results using this ansatz (3.15), and we will also interpret the results in the language of one-dimensional Schrödinger scattering. The mystery here concerns the neglect of the terms in V_1, together with the simple explanation of these results in terms of a Schrödinger picture. Yet as we shall see in section 7, a two component formulation is able to account for this omission nicely.

4. Asymptotics. By examining the large positive and negative $\overline{x}$ (equivalently t) behavior of g in (3.4), we see that

$$g \sim \begin{cases} e^{-i\lambda\overline{x}}, & \text{as } \overline{x} \to -\infty \\ pe^{-i\lambda\overline{x}} + se^{i\lambda\overline{x}}, & \text{as } \overline{x} \to \infty \end{cases} \tag{4.1}$$

since V→0 as $\overline{x} \to \pm\infty$, and because of the initial condition $\mathbf{M}(\Delta\omega,0)=(0,0,-1)^T$. By also imposing boundary conditions on the bound states ($E=-\lambda_i^2, \lambda_i > 0$, i=1,...,N)

$$g \sim \begin{cases} e^{\lambda\overline{x}}, & \text{as } \overline{x} \to -\infty \\ s_i(\lambda_i)e^{-\lambda\overline{x}}, & \text{as } \overline{x} \to \infty \end{cases} \tag{4.2}$$

for s_i the ith normalization constant, we have translated the Bloch evolution in t into a Schrödinger scattering problem in $\overline{x}$ (note that the interval [0,T] is replaced by $[-\infty, \infty]$) in which a plane wave is incident with amplitude p from the right, and gets transmitted with unit amplitude and reflected with amplitude s.

We can say more, by examining the large t behavior of $\frac{\dot{g}}{g}$. ¿From the definition of g in (3.4), we see that

$$\frac{\dot{g}}{g} \sim -i\lambda\eta = -i\lambda\frac{M_z + iM_y}{M_x - 1} \tag{4.3}$$

and on the other hand, from (4.1)

$$\frac{\dot{g}}{g} \sim -i\lambda \frac{pe^{-i\lambda t} - se^{i\lambda t}}{pe^{-i\lambda t} + se^{i\lambda t}}. \tag{4.4}$$

Moreover, at large t, from (3.1) it is easy to see that

$$\begin{aligned} M_x(2\lambda, t) &= \tfrac{m_x}{2}(e^{2i\lambda t} + e^{-2i\lambda t}) + \tfrac{m_y}{2i}(e^{2i\lambda t} - e^{-2i\lambda t}) \\ M_y(2\lambda, t) &= \tfrac{m_y}{2}(e^{2i\lambda t} + e^{-2i\lambda t}) - \tfrac{m_x}{2i}(e^{2i\lambda t} - e^{-2i\lambda t}). \\ M_z(2\lambda, t) &= m_z \end{aligned} \tag{4.5}$$

Substituting (4.5) into (4.3) and cross-multiplying, when the dust settles one finds the eight conditions

$$\begin{array}{cc} -\frac{m_x s}{2} = -\frac{m_x s}{2} & \frac{m_x p}{2} = \frac{m_x p}{2} \\ \frac{m_y s}{2} = \frac{m_y s}{2} & \frac{m_y p}{2} = \frac{m_y p}{2} \\ \frac{m_y s}{2} = -\frac{m_y s}{2} & \frac{m_y p}{2} = -\frac{m_y p}{2} \\ m_z s - \frac{m_x p}{2} = s + \frac{m_x p}{2} & m_z p + \frac{m_x s}{2} = -p - \frac{m_x s}{2} \end{array}, \tag{4.6}$$

with only two possible solutions:
(i) p=0 and $M_z = m_z$=1, and
(ii) s=0 and $M_z = m_z$=-1.
We discuss these two cases in turn.

Examining (4.1) when p vanishes, the potential V becomes 'super-radiant', in the sense that no wave impinges on V, but there exist outgoing waves in both directions. While this is unusual physical behavior, it is allowed when V is complex, and so the inverted state M_z=1 corresponds to a situation in which waves emanate from V in both directions, even though no wave is incident.

Conversely, when s vanishes, from (4.1) we see that the potential V is "reflectionless", since this time the plane wave propagates unimpeded through V, and no part of the wave is reflected. This case is identical to the previous results for amplitude modulation [4], that V is a reflectionless potential [8].

We close this section with a brief summary of how one computes such reflectionless V. The determinant of the matrix

$$A_{ij} = \delta_{ij} + \frac{s_i}{\lambda_i + \lambda_j} e^{-(\lambda_i + \lambda_j)\overline{x}} \tag{4.7}$$

is differentiated to obtain

$$V(\overline{x}) = -2\frac{d^2}{d\overline{x}^2} \ln det(A(\overline{x})). \tag{4.8}$$

5. Numerics. We show numerical simulations [9] of the Bloch equations (3.1) using the ansatz (3.15) with reflectionless V. While the analytic approach is exact in the limit as t $(=\overline{x}) \to \pm\infty$, in the finite digital simulations, the behavior is not perfect (i.e., there is a transition region from M_z=-1 to M_z=1). Nevertheless, as shown in figures 1 and 2, the behavior is very close to the limiting theoretical results.

In figures 3 and 4, we exhibit results for reflectionless potentials that do not achieve rectangular inversions. We find increased robustness in the parameter space ($\{\lambda_i, s_i\}$, i=1,...,N) compared to the real-valued case of amplitude modulation, so that now the result (equation (20) of [3])

$$|\lambda_{i+1} - \lambda_i| > O(1) \tag{5.1}$$

becomes

$$|\lambda_{i+1} - \lambda_i| > O(10^{-1}). \tag{5.2}$$

The general question raised in [3] remains: namely, what values of the 2N parameters $\{\lambda_i, s_i\}$ that determine a particular reflectionless V produce "good" NMR pulses? (The phase ϕ_i determines the position of the (i-1)th soliton).

One comment about our graphical display: in previous papers, we have usually shown 50 experiments in a single figure. However, in [3] we mention that this technique is misleading (it led to a misplaced correspondence with the Korteweg-de Vries equation), and so we now display a single reflectionless potential in each figure, to stress that the important object here is a set of 2N numbers, and not an isospectral flow of some nonlinear PDE.

6. Quantum mechanics. We will now give some arguments from non-relativistic quantum mechanics [11] to provide explanations of the observed numerical findings. The first is heuristic, to give a qualitative indication that we are on the right track. The other arguments are based on solid principles of non-relativistic quantum theory, and offer a perspective that we hope can explain all of the simple spin physics. There are also recent indications that some well-known approximation techniques, such as the Born approximation and the WKB method, can provide additional insight, but those approaches would take us too far afield here, and so will be discussed elsewhere.

One can understand the response purely in terms of scattering, with no need to appeal to the concept of adiabatic rapid passage [10]. Loosely speaking, think of the potential well V instead as a potential barrier of height $\sim \omega_1^2$, and the incident wave as having energy $\sim \Delta\omega^2$. The wave will be exponentially damped if the energy in the wave is too small to get over the barrier (corresponding to M_z=1). Likewise, when the energy of the wave surpasses the height of the barrier, there is no reflection, and the wave propagates unimpeded (corresponding to M_z=-1). The transition occurs when the energy in the wave is close to the height of the barrier.

There is a second interpretation that reinforces our belief in the relevance of quantum scattering theory. One of the first computations done in learning about quantum theory is the derivation of the continuity equation for the probability amplitude $|\psi^*\psi|$ from the Schrödinger equation

$$i\hbar\frac{\partial\psi}{\partial t} = [-\frac{\hbar^2}{2m}\frac{\partial^2}{\partial x^2} + V(x)]\psi. \tag{6.1}$$

The usual result is

$$\frac{\partial \rho}{\partial t} + \frac{\partial}{\partial x}\mathrm{S} = 0, \tag{6.2}$$

where

$$\rho = \int \psi^* \psi \, dx \, , \, \mathrm{S} = \frac{\hbar}{2im} \int (\psi^* \frac{\partial \psi}{\partial x} - (\frac{\partial \psi^*}{\partial x})\psi) \, dx, \tag{6.3}$$

which is obtained when the potential V is real-valued. In the more general case of complex V, the zero on the r.h.s. of (6.2) is replaced by the source term $\frac{2}{\hbar}\mathrm{Im(V)}$. Now, from (3.12-14), it is easy to see that our Im(V) scales linearly with μ, so that as μ increases, so does the production of probability in the state M_z=1. Therefore, the size of the inverted region increases (linearly with μ!), as seen in figure 1.

The final argument says that our analogy to quantum mechanics is really one-dimensional. It is a consequence of the general properties of the motion in one dimension that the spectrum of discrete bound states, or negative energy levels, is nondegenerate. (An equivalent statement is that the Wronskian of two different eigenfunctions corresponding to the same negative value of energy vanishes identically in one dimension). But this is *exactly* the phenomenon observed in figure 3, and formalized in (5.2).

Another mystery occurs at this point, as illustrated in figure 5. In the amplitude modulation case, one has an additional constraint (N denotes the number of bound states)

$$\int_{-\infty}^{\infty} \omega_1(t) \, dt = 2\pi N. \tag{6.4}$$

Therefore, in this reduced (real-valued) case, there is extreme "ω_1 sensitivity", namely that (6.4) must hold. In the general case (3.1), the constraint (6.4) disappears, and in its place one finds the opposite behavior, that the profile is insensitive to the strength of the irradiating radiofrequency field (equivalently, extreme "ω_1 insensitivity").

7. Density matrix. The two component formalism described in [1] is identical to that of a statistical operator ρ (called the density matrix [10]). Modulo factors of ±i, and setting $\hbar$ and the gyromagnetic ratio γ everywhere to 1, this equivalence is given by

$$\rho \simeq P \equiv \sigma_1 M_x + \sigma_2 M_y + \sigma_3 M_z = \sigma \cdot \mathbf{M}. \tag{7.1}$$

Moreover, the time evolution for P (equation (9) in section 3 of [1]) is the same as the variation of ρ in time

$$\dot{\rho} = -[\mathcal{H}, \rho]. \tag{7.2}$$

We therefore see that the Hamiltonian $\mathcal{H}$ is identified with

$$\mathcal{H} \simeq -U'U^{-1} = -\sigma \cdot \mathbf{H}, \tag{7.3}$$

where **H** is the imposed magnetic field

$$\mathbf{H} = \begin{pmatrix} \omega_1(t)\cos\varphi(t) \\ \omega_1(t)\sin\varphi(t) \\ \Delta\omega \end{pmatrix}. \tag{7.4}$$

Thus the evolution of Zahkarov-Shabat type is the same as the density matrix motion (very similar to Heisenberg's equations of motion) for a statistical ensemble of free two-level spins.

If we try to write this evolution

$$\frac{d}{dt}\begin{pmatrix} v_1 \\ v_2 \end{pmatrix} = \begin{pmatrix} -i\zeta & q(t) \\ -q^*(t) & i\zeta \end{pmatrix}\begin{pmatrix} v_1 \\ v_2 \end{pmatrix} \tag{7.5}$$

in second order form, we have

$$-\ddot{v_1} + \frac{\dot{q}}{q}\dot{v_1} + (i\zeta\frac{\dot{q}}{q} - \zeta^2 + |q|^2)v_1 = 0. \tag{7.6}$$

Now, when N=1, our potential q is

$$q(t) = \frac{iA}{2}(\operatorname{sech}\beta t)^{1+\mu i}, \tag{7.7}$$

so that

$$\frac{\dot{q}}{q} = -\beta(1+\mu i)\tanh\beta t. \tag{7.8}$$

Using a Liouville substitution

$$v_1 = ue^{-\frac{\beta}{2}(1+\mu i)\int^t \tanh\beta s\, ds}, \tag{7.9}$$

we finally obtain a Schrödinger equation

$$-\ddot{u}+[\frac{\beta^2}{4}(1+\mu i)^2\tanh^2\beta t - i\beta\zeta(1+\mu i)\tanh\beta t - \frac{A^2+2\beta^2(1+\mu i)}{4\cosh^2\beta t} - \zeta^2]u = 0 \tag{7.10}$$

with a potential hole of "modified" Pöschl-Teller type [12].

Finally, we consider the general problem (3.1) in a two component formalism, so that (7.5) becomes

$$\frac{d}{dt}\begin{pmatrix} v_1 \\ v_2 \end{pmatrix} = -\frac{i}{2}\begin{pmatrix} \Delta\omega & \omega_1 e^{i\varphi} \\ \omega_1 e^{-i\varphi} & -\Delta\omega \end{pmatrix}\begin{pmatrix} v_1 \\ v_2 \end{pmatrix} \tag{7.11}$$

with the potential q and energy ζ^2 given as

$$q(t) = -\frac{i\omega_1(t)}{2}e^{i\varphi(t)}, \quad \zeta = \lambda = \frac{\Delta\omega}{2}. \tag{7.12}$$

Defining g=$\frac{v_1+v_2}{2}$ in (3.5), we recover (3.11) exactly in the form

$$\ddot{g} = \frac{\ddot{v_1} + \ddot{v_2}}{2} = [-(\frac{\omega_1}{2})^2 - i\frac{\dot{\omega}_1 \cos\varphi}{2} + i\frac{\omega_1 \dot{\varphi} \sin\varphi}{2} - (\frac{\Delta\omega}{2})^2]g. \tag{7.13}$$

In addition, for the reduced case of pure amplitude modulation (i.e., $\varphi(t) \equiv 0$), this expression reduces to the Schrödinger form given in (3.9), namely

$$\ddot{g} = -\frac{1}{4}[\omega_1^2 + 2i\dot{\omega}_1 + \Delta\omega^2]g. \tag{7.14}$$

8. Closing remarks. We have seen that non-relativistic scattering theory and uncoupled two-level system dynamics are intimately related. Indeed, we can see, by identifying the Zakharov-Shabat and density matrix formulations, that they are complementary views of a single phenomenon. Nevertheless, there are still some unresolved mysteries, and the complete picture remains unfinished.

The special initial and final states of the evolving magnetization we have considered has led to a large family of useful NMR pulses, namely those arising from the celebrated reflectionless potentials [8]. However, one would like to treat more general initial and final data for the magnetization **M**. While the Schrödinger formulation does appear tied to this reflectionless case, as we have seen in the previous section, a two component formulation of the inverse scattering problem arises as an equivalent representation to the density matrix. So one can look at two component versions of the inverse problem as a hopeful way to generalize this NMR reformulation to arbitrary configurations.

I gratefully acknowledge enlightening conversations with F.A. Grünbaum and W.S. Warren, and thank the Princeton Local Allocation Committee for five hours of supercomputer time at the John von Neumann Center, some of which was used in the development of the numerical evidence. Finally, I thank the officers and staff at the IMA for providing an extremely pleasant atmosphere in Vincent Hall.

Figure Captions

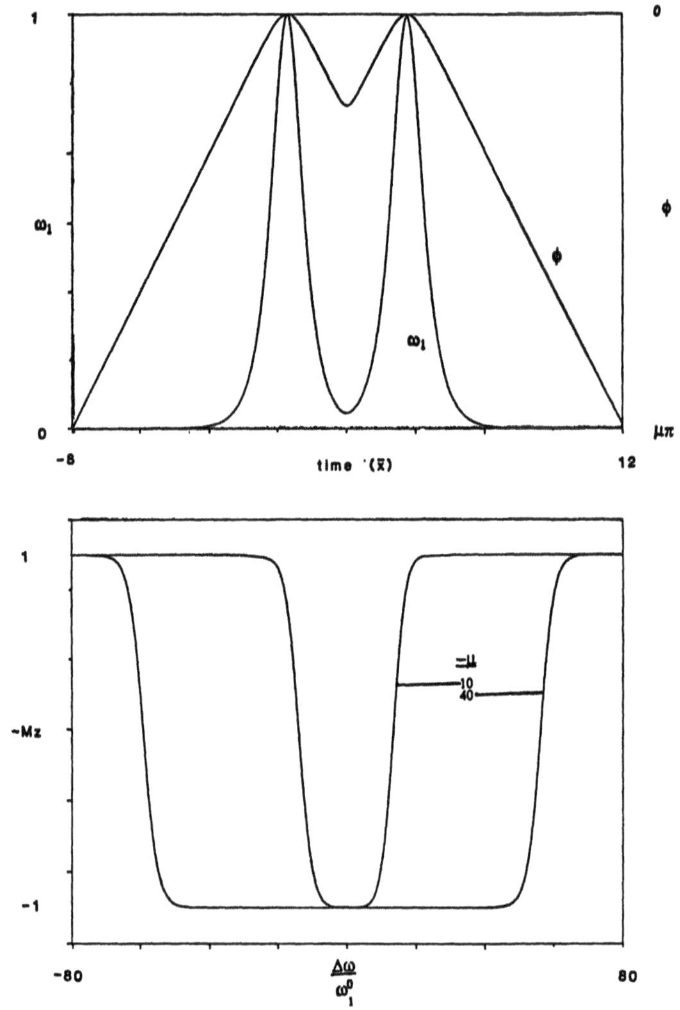

FIGURE 1: The N=2 reflectionless potential determined from the data $(\lambda_1, \lambda_2, \phi_1, \phi_2)$=(2 4.1,2.7) with -$\mu$=10,40 at the top is used in the ansatz (3.15) as an inverting pulse, where the response -M_z is displayed as a function of resonance offset $\frac{\Delta\omega}{\omega_1^o}$ at bottom.

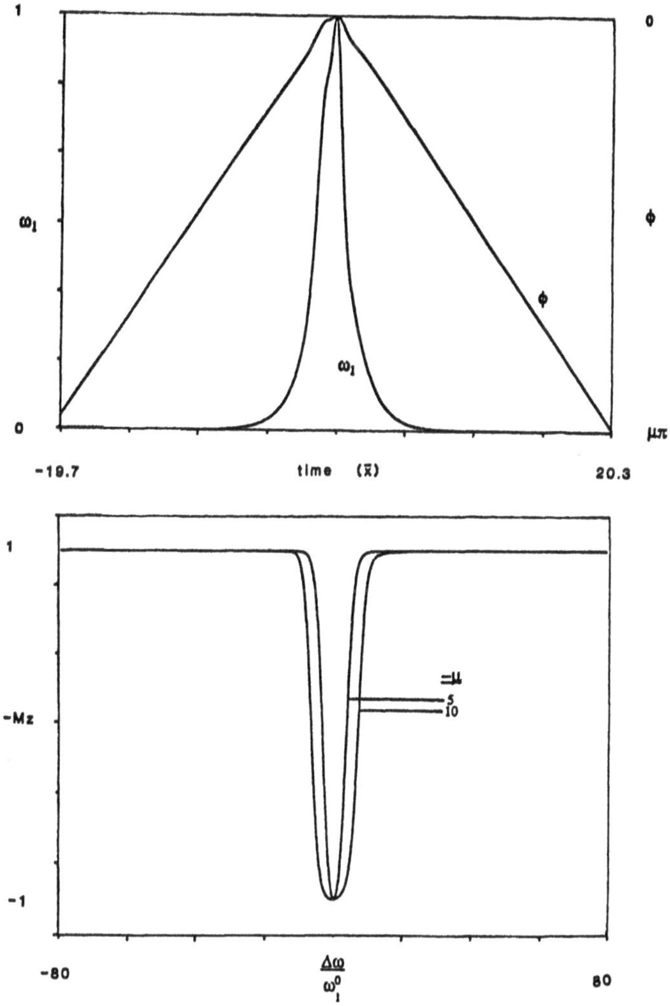

FIGURE 2: The N=3 reflectionless potential determined from the data $(\lambda_1, \lambda_2, \lambda_3, \phi_1, \phi_2, \phi_3, \phi_4)=$ (3.3,2.0,0.8,0.2,-0.3,0.25,-0.39) with -μ=5,10 at the top is used in the ansatz (3.15) as an inverting pulse, where the response -M_z is displayed as a function of resonance offset $\frac{\Delta\omega}{\omega_1^0}$ at bottom.

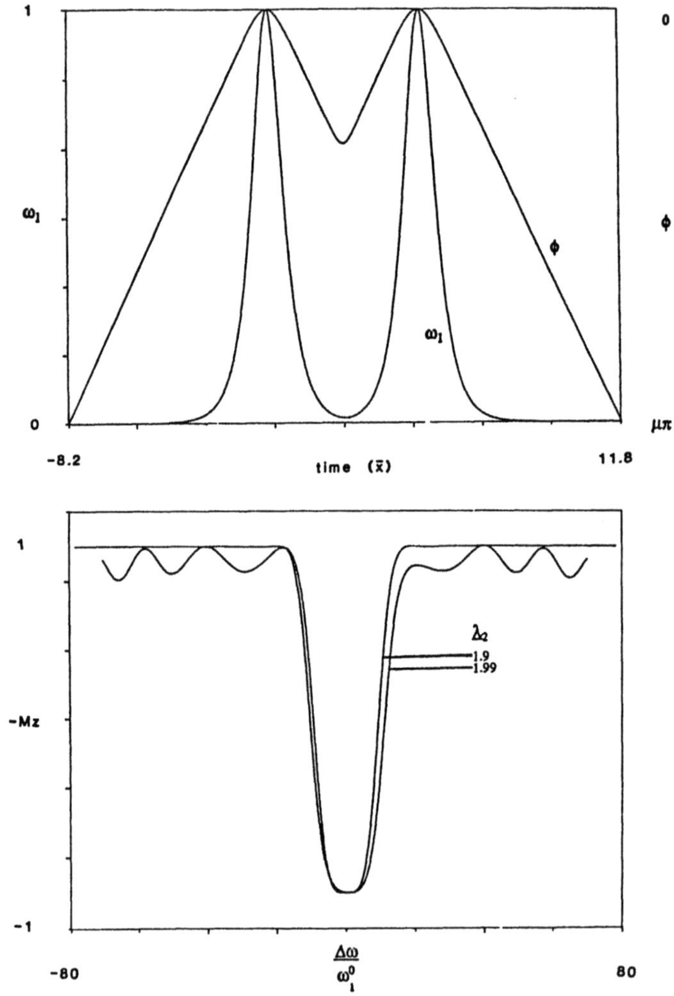

FIGURE 3: Everything is the same as in Figure 1, except that the small eigenvalue moves from $\lambda_2 = 1.9 \rightarrow 1.99$, and -$\mu$=6.

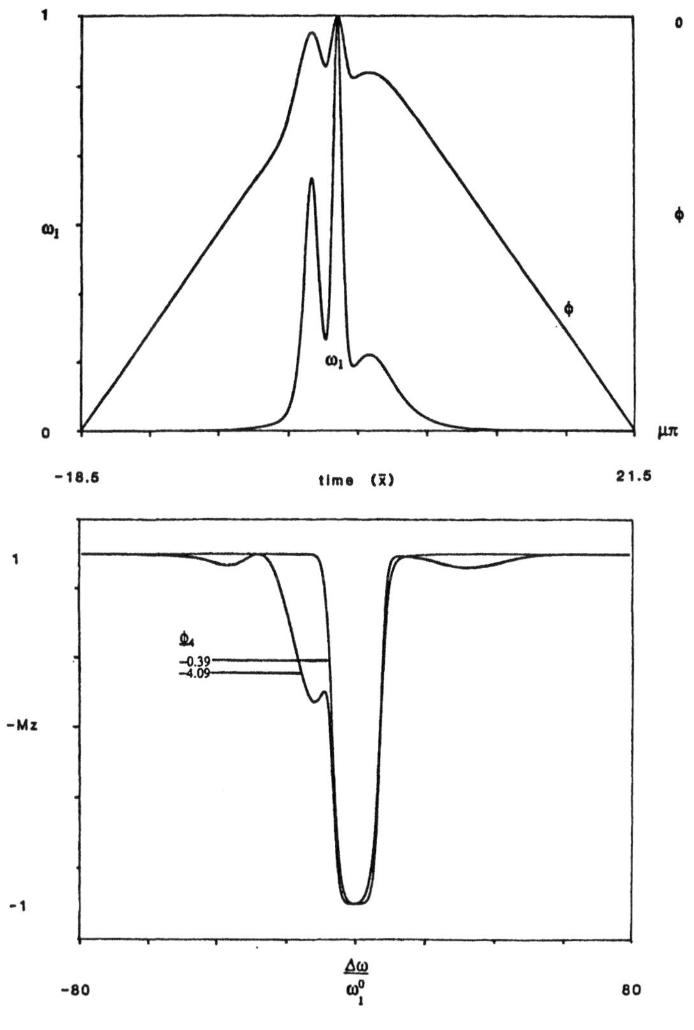

FIGURE 4: The eigenvalue data is the same as in Figure 2, but now the phases have moved $(\phi_1, \phi_2, \phi_3, \phi_4) = (0.2, -0.3, 0.25, -0.39) \rightarrow$ (2.3,-1.6,3.1,-4.09) to better separate the solitons, with -μ=10.

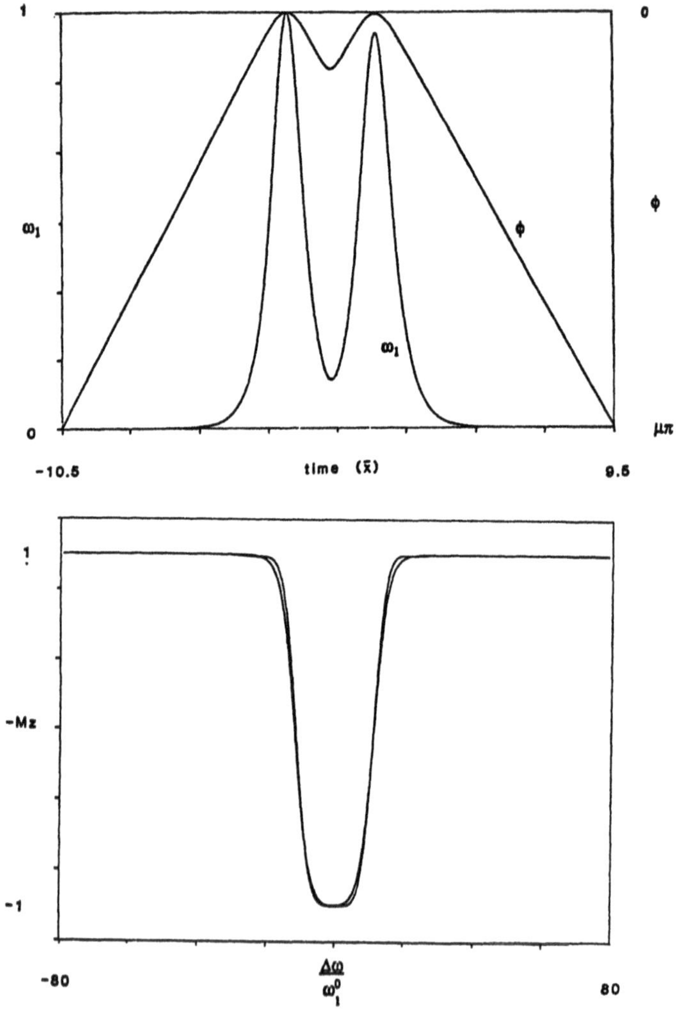

FIGURE 5: The N=2 reflectionless potential determined from the data $(\lambda_1, \lambda_2, \phi_1, \phi_2)$=(2.0,1.9,2.7,2.7) with -μ=8 at the top is used in the ansatz (3.15) as an inverting pulse for $|\, \omega_1 \,|$=25.14,101.1, where the response -M_z is displayed as a function of resonance offset $\frac{\Delta\omega}{\omega_1^0}$ at bottom.

(More details on notation are available for the N=2 and N=3 cases in [4] and [3], respectively, and $\frac{\Delta\omega}{\omega_1^0}$ denotes the average pulse height).

REFERENCES

[1] F.A. Grünbaum, *Soliton Mathematics in Signal Processing (this volume).*
[2] F.A. Grünbaum and A. Hasenfeld, *An exploration of the invertibility of the Bloch transform*, Inverse Problems, 2 (1986), pp. 75–81.
[3] ——————, *An exploration of the invertibility of the Bloch transform: II*, Inverse Problems, 4 (1988), pp. 485-493.
[4] A. Hasenfeld, *A connection between the Bloch equations and the Korteweg-de Vries equation*, J. Magn. Reson., 72 (1987), pp. 509–521.
[5] A. Hasenfeld, S.L. Hammes, and W.S. Warren, *Understanding phase modulation in two-level systems through inverse scattering*, Phys. Rev. A (to appear).
[6] S.L. McCall and E.L. Hahn, *Self-induced transparency*, Phys. Rev., 183 (1969), pp. 457–485.
[7] J.B. Murdoch, A.H. Lent, and M.R. Kritzer, *Computer-optimized narrowband pulses for multislice imaging*, J. Magn. Reson., 74 (1987), pp. 226–263.
[8] V. Bargmann, *On the connection between phase shifts and scattering potential*, Rev. Mod. Phys., 21 (1949), pp. 488–493.
[9] A. Hasenfeld, *SHARP biomedical NMR spatial localization*, Magn. Reson. Med., 2 (1985), pp. 505–511.
[10] A. Abragam, *The Principles of Nuclear Magnetism*, Oxford University Press, London, 1961.
[11] L.D. Landau and E.M. Lifshitz, *Quantum Mechanics (Non-relativistic theory), Third edition*, Pergamon Press, New York, 1977, p. 60.
[12] S. Flügge, *Practical Quantum Mechanics*, Springer-Verlag, Berlin, 1971, p. 94.

WAVELETS AND FRAMES

CHRISTOPHER HEIL*

Abstract. This paper presents basic results about frames in Hilbert spaces, and gives examples of two types of frames for $L^2(\mathbf{R})$. These are the Weyl-Heisenberg frames, which are translations and modulations of a single "mother wavelet", and the affine frames, which are translations and dilations of a mother wavelet.

0. Introduction. It is a well-known fact that every separable Hilbert space, and in particular $L^2(\mathbf{R})$, possesses an orthonormal basis. One of the major properties of such sequences is that they provide a "decomposition" of the space. That is, if $\{e_n\}$ is an orthonormal basis for a Hilbert space H then every $x \in H$ can be written $x = \sum_n \langle x, e_n \rangle e_n$. Unfortunately, orthonormal bases can often be difficult to find or inconvenient to work with. For example, one orthonormal basis for $L^2(\mathbf{R})$ is the sequence $\{\varphi_{mn}\}$, where

$$\varphi_{mn}(x) \;=\; e^{2\pi i(x-n)m}\, \chi_{[n,n+1)}(x).$$

However, these functions are discontinuous, which can make the representation of even smooth functions in $L^2(\mathbf{R})$ unpleasant.

Frames are an alternative to orthonormal bases. By giving up the requirements of orthogonality and uniqueness of decomposition we allow much more freedom in our choice of "basic vectors", while still retaining the ability to decompose the space. In Section 2 we define the notion of a frame and show that if $\{x_n\}$ is a frame then every $x \in H$ can be written $x = \sum_n c_n x_n$ in a good way, i.e., the scalars are computable, the series converges unconditionally, etc. The remainder of the paper is devoted to finding frames for $L^2(\mathbf{R})$. These fall into two general categories, which we call Weyl-Heisenberg and affine frames.

W-H frames, studied in Section 3, are similar in structure to the orthonormal basis $\{\varphi_{mn}\}$ given above. In particular, note that φ_{mn} can be written $\varphi_{mn} = T_n E_m \varphi$, where $\varphi = \chi_{[0,1)}$, E_m denotes multiplication by $e^{2\pi imx}$, and T_n denotes translation by n. In the same way, W-H frames are composed of discrete modulates and translates of a single function, called the "mother wavelet". Unlike orthonormal bases, we show that it is possible to find W-H frames whose mother wavelet is smooth. Moreover, in Section 4 we indicate that if a W-H frame should happen to form a basis then the mother wavelet is either not smooth or does not decay quickly. Section 4 is devoted to the Zak transform, also known as the Weil-Brezin map [AT], which is an important tool in the analysis of W-H frames.

In Section 5 we turn to a fundamentally different method of constructing frames for $L^2(\mathbf{R})$. These "affine frames" are obtained by taking discrete translates and dilates of the mother wavelet. A well-known example is the standard Haar system, whose mother wavelet is not smooth. After stating some of the basic results about

*University of Maryland, College Park, MD 20742 and Mitre Corp., McLean, VA 22102.

affine frames, we briefly mention the Meyer wavelet. This is a C^∞ function with compactly supported Fourier transform, and the affine frame it generates forms an orthonormal basis for $L^2(\mathbf{R})$. Comparing this with our above remarks on W-H frames and bases, we see that there is a great deal of difference between the two types of frames.

Although we omit or only sketch most of the proofs in this article, they can all be found in [HW], along with a great deal of additional material, such as the relationship of W-H and affine frames to continuous representations of $L^2(\mathbf{R})$. I would like to thank David Walnut, my colleague at the University of Maryland and Mitre Corp., for permission to use some of the material from [HW] in this article. Another superb reference on the subject of W-H and affine frames is [D1], from which many of the results quoted in this paper were taken.

1. Notation. We write $\mathbf{R}$ for the real line thought of as the time axis, and $\hat{\mathbf{R}}$ for its dual group, the real line thought of as the frequency axis. All sequences and series with undefined limits are to be taken over $\mathbf{Z}$, the set of integers. $L^2(\mathbf{R})$ is the Hilbert space of all complex-valued, square-integrable functions f on $\mathbf{R}$, normed by $\|f\|_2 = \left(\int_{\mathbf{R}} |f(x)|^2\, dx\right)^{1/2}$. The inner product of $f, g \in L^2(\mathbf{R})$ is $\langle f, g\rangle = \int_{\mathbf{R}} f(x)\,\overline{g(x)}\, dx$. The Fourier transform of an integrable f is

$$\hat{f}(\gamma) = \int_{\mathbf{R}} f(x)\, e^{-2\pi i \gamma x}\, dx, \qquad \text{for } \gamma \in \hat{\mathbf{R}}.$$

Given a function f we define

Translation:	$T_a f(x) = f(x-a),$	for $a \in \mathbf{R}$;
Modulation:	$E_a f(x) = e^{2\pi i a x} f(x),$	for $a \in \mathbf{R}$;
Dilation:	$D_a f(x) = \lvert a\rvert^{-1/2} f(x/a),$	for $a \in \mathbf{R}\backslash\{0\}$.

Each of these is a unitary operator of $L^2(\mathbf{R})$ onto itself, i.e., a linear bijective isometry. We also use the symbol E_a by itself to refer to the **exponential function**, i.e., $E_a(x) = e^{2\pi i a x}$.

2. Frames in Hilbert spaces.

DEFINITION 2.1 [DS]. A sequence $\{x_n\}$ in a Hilbert space H is a **frame** if there exist numbers $A, B > 0$ such that for all $x \in H$ we have

$$A\|x\|^2 \le \sum_n |\langle x, x_n\rangle|^2 \le B\|x\|^2.$$

The numbers A, B are called the **frame bounds**. The frame is **tight** if $A = B$. The frame is **exact** if it ceases to be a frame whenever any single element is deleted from the sequence.

Frames were first introduced in 1952 by Duffin and Schaeffer in connection with nonharmonic Fourier series [DS].

Since $\sum |\langle x, x_n\rangle|^2$ is a series of positive real numbers, it converges absolutely, hence unconditionally. That is, every rearrangement of the sum also converges, and converges to the same value. Therefore, every rearrangement of a frame is also a frame, and all sums involving frames actually converge unconditionally.

It follows immediately from the definition that frames are complete. That is, the only $x \in H$ orthogonal to every x_n is $x = 0$, or equivalently, the set of finite linear combinations of the x_n is dense in H. The following theorem shows that they also provide a decomposition of the space similar to that of orthonormal bases. Given operators $S, T: H \to H$ we write $S \leq T$ if $\langle Sx, x\rangle \leq \langle Tx, x\rangle$ for all $x \in H$, and we denote by I the identity map on H, i.e., $Ix = x$ for all $x \in H$.

THEOREM 2.2 [DS]. *Given a sequence $\{x_n\}$ in a Hilbert space H, the following two statements are equivalent:*

(1) *$\{x_n\}$ is a frame with bounds A, B.*
(2) *$Sx = \sum\langle x, x_n\rangle x_n$ is a bounded linear operator with $AI \leq S \leq BI$, called the* **frame operator** *for $\{x_n\}$.*

Proof. 2 ⇒ 1. Follows from $\langle Ix, x\rangle = \|x\|^2$ and $\langle Sx, x\rangle = \sum |\langle x, x_n\rangle|^2$.
1 ⇒ 2. S is well-defined and continuous since

$$\begin{aligned}
\|Sx\|^2 &= \sup_{\|y\|=1} |\langle Sx, y\rangle|^2 \\
&= \sup_{\|y\|=1} \left|\sum_n \langle x, x_n\rangle\langle x_n, y\rangle\right|^2 \\
&\leq \sup_{\|y\|=1} \left(\sum_n |\langle x, x_n\rangle|^2\right)\left(\sum_n |\langle x_n, y\rangle|^2\right) \\
&\leq \sup_{\|y\|=1} B\|x\|^2 \cdot B\|y\|^2 \\
&= B^2\|x\|^2.
\end{aligned}$$

The relations $AI \leq S \leq BI$ then follow immediately from the definition of frames. □

We say that a mapping $U: H \to H$ is **invertible**, or a **topological isomorphism**, if U is linear, bijective, continuous, and U^{-1} is continuous.

COROLLARY 2.3 [DS].

(1) *S is invertible and $B^{-1}I \leq S^{-1} \leq A^{-1}I$.*
(2) *$\{S^{-1}x_n\}$ is a frame with bounds $1/B, 1/A$, called the* **dual frame** *of $\{x_n\}$.*
(3) *Every $x \in H$ can be written*

$$x = \sum_n \langle x, S^{-1}x_n\rangle x_n = \sum_n \langle x, x_n\rangle S^{-1}x_n.$$

Proof. 1. Since $AI \leq S \leq BI$ we have $\|I - \frac{1}{B}S\| \leq \|I - \frac{A}{B}I\| = \frac{B-A}{B} < 1$. Elementary Hilbert space results therefore imply that $\frac{1}{B}S$ is invertible, so S itself must be invertible. Since S^{-1} is a positive operator and commutes with both I

and S, we can multiply through by S^{-1} in the equation $AI \leq S \leq BI$ to obtain $B^{-1}I \leq S^{-1} \leq A^{-1}I$.

2. Since S^{-1} is positive it is self-adjoint. Therefore,

$$\begin{aligned}\sum_n \langle x, S^{-1}x_n\rangle S^{-1}x_n &= \sum_n \langle S^{-1}x, x_n\rangle S^{-1}x_n \\ &= S^{-1}\Big(\sum_n \langle S^{-1}x, x_n\rangle x_n\Big) \\ &= S^{-1}S(S^{-1}x) \\ &= S^{-1}x.\end{aligned}$$

The result then follows from the fact that $B^{-1}I \leq S^{-1} \leq A^{-1}I$ and Theorem 2.2 part 2.

3. Simply expand $x = S(S^{-1}x)$ and $x = S^{-1}(Sx)$. □

COROLLARY 2.4. *If $\{x_n\}$ is a tight frame, i.e., if $A = B$, then*

(1) *$S = AI$.*

(2) *$S^{-1} = A^{-1}I$.*

(3) *Every $x \in H$ can be written $x = A^{-1}\sum\langle x, x_n\rangle x_n$.*

PROPOSITION 2.5 [DS]. *Given a frame $\{x_n\}$ and given $x \in H$ define $a_n = \langle x, S^{-1}x_n\rangle$, so $x = \sum a_n x_n$. If it is possible to find other scalars c_n such that $x = \sum c_n x_n$ then we must have*

$$\sum_n |c_n|^2 = \sum_n |a_n|^2 + \sum_n |a_n - c_n|^2.$$

Proof. Follows from $\sum |a_n|^2 = \langle x, S^{-1}x\rangle = \sum c_n \bar{a}_n$. □

THEOREM 2.6 [DS]. *The removal of a vector from a frame leaves either a frame or an incomplete set. In fact,*

$$\begin{aligned}\langle x_m, S^{-1}x_m\rangle \neq 1 &\Rightarrow \{x_n\}_{n\neq m} \text{ is a frame;} \\ \langle x_m, S^{-1}x_m\rangle = 1 &\Rightarrow \{x_n\}_{n\neq m} \text{ is incomplete.}\end{aligned}$$

Proof. Fix m, and define $a_n = \langle x_m, S^{-1}x_n\rangle = \langle S^{-1}x_m, x_n\rangle$. We know that $x_m = \sum a_n x_n$, but we also have $x_m = \sum c_n x_n$ where $c_m = 1$ and $c_n = 0$ for $n \neq m$. Using Proposition 2.5 we find that

$$\sum_{n\neq m} |a_n|^2 = \frac{1 - |a_m|^2 - |a_m - 1|^2}{2} < \infty.$$

Suppose now that $a_m = 1$. Then $\sum_{n\neq m} |a_n|^2 = 0$, so $a_n = \langle S^{-1}x_m, x_n\rangle = 0$ for $n \neq m$. That is, $S^{-1}x_m$ is orthogonal to x_n for every $n \neq m$. But $S^{-1}x_m \neq 0$ since $\langle S^{-1}x_m, x_m\rangle = a_m = 1 \neq 0$. Therefore $\{x_n\}_{n\neq m}$ is incomplete in this case.

On the other hand, suppose $a_m \neq 1$. Then $x_m = \frac{1}{1-a_m}\sum_{n\neq m} a_n x_n$, so for $x \in H$ we have

$$|\langle x, x_m\rangle|^2 = \left|\frac{1}{1-a_m}\sum_{n\neq m} a_n\langle x, x_n\rangle\right|^2 \leq \frac{1}{|1-a_m|^2}\Big(\sum_{n\neq m}|a_n|^2\Big)\Big(\sum_{n\neq m}|\langle x, x_n\rangle|^2\Big).$$

Therefore

$$\sum_n |\langle x, x_n\rangle|^2 = |\langle x, x_m\rangle|^2 + \sum_{n\neq m}|\langle x, x_n\rangle|^2 \leq C\sum_{n\neq m}|\langle x, x_n\rangle|^2,$$

where $C = 1 + |1-a_m|^{-2}\sum_{n\neq m}|a_n|^2$. It follows immediately that $\{x_n\}_{n\neq m}$ is a frame with bounds $A/C, B$. □

COROLLARY 2.7 [**DS**]. *If $\{x_n\}$ is an exact frame then $\{x_n\}$ and $\{S^{-1}x_n\}$ are* **biorthogonal**, *i.e.,*

$$\langle x_m, S^{-1}x_n\rangle = \delta_{mn} = \begin{cases} 1, & \text{if } m = n; \\ 0, & \text{if } m \neq n. \end{cases}$$

DEFINITION 2.8. A sequence $\{\varphi_n\}$ in a Hilbert space H is a **basis** for H if for every $x \in H$ there exist *unique* scalars c_n such that

$$x = \sum_n c_n\varphi_n.$$

The basis is **bounded** if $0 < \inf\|\varphi_n\| \leq \sup\|\varphi_n\| < \infty$. The basis is **unconditional** if the series $\sum c_n\varphi_n$ converges unconditionally, that is, every permutation of the series converges.

In finite dimensional spaces, a series converges unconditionally if and only if it converges absolutely. In infinite dimensional spaces, absolute convergence still implies unconditional convergence but the reverse need not be true. In Hilbert spaces, all bounded unconditional bases are **equivalent** to orthonormal bases. That is, if $\{\varphi_n\}$ is a bounded unconditional basis then there is an orthonormal basis $\{e_n\}$ and a topological isomorphism $U: H \to H$ such that $\varphi_n = Ue_n$ for all n (see [Y]).

It is easy to see that inexact frames are not bases. However, we do have the following characterization of exact frames.

THEOREM 2.9 [**DS; Y**]. *A sequence $\{x_n\}$ in a Hilbert space H is an exact frame for H if and only if it is a bounded unconditional basis for H.*

Proof [**H**]. ⇒. Assume $\{x_n\}$ is an exact frame with bounds A, B. Then $\{x_n\}$ and $\{S^{-1}x_n\}$ are biorthogonal, so for m fixed we have

$$A\,\|S^{-1}x_m\|^2 \leq \sum_n |\langle S^{-1}x_m, x_n\rangle|^2 = |\langle S^{-1}x_m, x_m\rangle|^2 \leq \|S^{-1}x_m\|^2\|x_m\|^2.$$

Therefore $\|x_m\|^2 \geq A$. Also,

$$\|x_m\|^4 = |\langle x_m, x_m\rangle|^2 \leq \sum_n |\langle x_m, x_n\rangle|^2 \leq B\,\|x_m\|^2,$$

so $\|x_m\|^2 \leq B$. Thus the sequence $\{x_n\}$ is bounded in norm. By Corollary 2.3, $x = \sum \langle x, S^{-1}x_n\rangle x_n$ for all $x \in H$, and the biorthogonality of $\{x_n\}$ and $\{S^{-1}x_n\}$ implies that this representation is unique, so $\{x_n\}$ is a basis for H. Since every permutation of a frame is also a frame, we conclude that the basis is unconditional.

$\Leftarrow$. Assume $\{x_n\}$ is a bounded unconditional basis for H. Then there is an orthonormal basis $\{e_n\}$ and a topological isomorphism $U\colon H \to H$ such that $Ue_n = x_n$ for all n. Given $x \in H$ we therefore have

$$\sum_n |\langle x, x_n\rangle|^2 = \sum_n |\langle x, Ue_n\rangle|^2 = \sum_n |\langle U^*x, e_n\rangle|^2 = \|U^*x\|^2.$$

But

$$\frac{\|x\|}{\|U^{*-1}\|} \leq \|U^*x\| \leq \|U^*\|\,\|x\|,$$

so $\{x_n\}$ forms a frame. It is clearly exact since the removal of any vector from a basis leaves an incomplete set. □

3. Weyl-Heisenberg frames.

DEFINITION 3.1. Given $g \in L^2(\mathbf{R})$ and $a, b > 0$, we say that (g, a, b) **generates a W-H frame** for $L^2(\mathbf{R})$ if $\{T_{na}E_{mb}g\}_{m,n\in\mathbf{Z}}$ is a frame for $L^2(\mathbf{R})$. The function g is referred to as the **mother wavelet**. Together, the numbers a, b are the **frame parameters**. Individually, a is the **shift parameter** and b is the **modulation parameter**.

THEOREM 3.2 [DGM]. *Assume $g \in L^2(\mathbf{R})$ and $a, b > 0$ satisfy:*

(1) *There exist constants A, B such that*

$$0 < A = \operatorname*{ess\,inf}_{x\in\mathbf{R}} \sum_n |g(x - na)|^2 \leq \operatorname*{ess\,sup}_{x\in\mathbf{R}} \sum_n |g(x - na)|^2 = B < \infty;$$

(2) *g has compact support, with $\operatorname{supp}(g) \subset I \subset \mathbf{R}$, where I is some interval of length $1/b$.*

Then (g, a, b) generates a W-H frame for $L^2(\mathbf{R})$ with frame bounds $b^{-1}A, b^{-1}B$.

Proof. Fix n, and observe that the function $f\cdot\overline{T_{na}g}$ is supported in $I_n = I+na = \{x + na : x \in I\}$, which is an interval of length $1/b$. From the Plancherel formula for Fourier series we therefore have

$$\begin{aligned}\sum_n\sum_m |\langle f, T_{na}E_{mb}g\rangle|^2 &= \sum_n\sum_m |\langle f\cdot\overline{T_{na}g}, E_{mb}\rangle|^2 \\ &= \sum_n b^{-1}\int_{I_n} |f(x)|^2\,|g(x - na)|^2\,dx \\ &= b^{-1}\int_{\mathbf{R}} |f(x)|^2 \sum_n |g(x - na)|^2\,dx,\end{aligned}$$

from which the result follows. □

By taking functions f to be analyzed which have compact support, one can prove:

PROPOSITION 3.3 [D1]. *Whether g has compact support or not, it is necessary that condition (1) of Theorem 3.2 hold in order that $\{T_{na}E_{mb}g\}$ make a frame.*

REMARK 3.4. It is easy to see that if g satisfies condition (2) of Theorem 3.2 and if $ab > 1$ then $\sum_n |g(x-na)|^2$ is not bounded below, so that g cannot generate a frame. In fact, the set $\{T_{na}E_{mb}g\}$ is not even complete, since $\cup_n \operatorname{supp}(T_{na}g)$ does not cover $\mathbf{R}$. This is a simple illustration of a more general phenomenon discussed in Section 4, namely that if $ab > 1$ then $\{T_{na}E_{mb}g\}$ is not complete in $L^2(\mathbf{R})$ for *any* $g \in L^2(\mathbf{R})$.

PROPOSITION 3.5. *If (g, a, b) generates a W-H frame for $L^2(\mathbf{R})$, then $(\hat{g}, b, a)$ generates a W-H frame for $L^2(\hat{\mathbf{R}})$.*

Proof. Follows immediately from $(T_{na}E_{mb}g)^\wedge = E_{-na}T_{-mb}\hat{g}$. □

REMARK 3.6. It follows from Propositions 3.3 and 3.5 that if (g, a, b) generates a frame for $L^2(\mathbf{R})$ then both g and $\hat{g}$ are bounded functions.

Theorem 3.2 used Fourier series arguments to derive conditions under which a compactly supported function g could be a mother wavelet. If g does not have compact support then this argument breaks down. It turns out, however, that one can perform a similar calculation if g is not compactly supported.

THEOREM 3.7 [D1]. *Assume $g \in L^2(\mathbf{R})$ and $a > 0$ satisfy:*

(1) *There exist constants A, B such that*

$$0 < A = \operatorname*{ess\,inf}_{x\in\mathbf{R}} \sum_n |g(x-na)|^2 \le \operatorname*{ess\,sup}_{x\in\mathbf{R}} \sum_n |g(x-na)|^2 = B < \infty;$$

(2) $\lim_{b\to 0} \sum_{k\neq 0} \beta(k/b) = 0$, *where*

$$\beta(s) = \operatorname*{ess\,sup}_{x\in\mathbf{R}} \sum_n |g(x-na)|\,|g(x-s-na)|.$$

Then there exists a number $b_0 > 0$ such that (g, a, b) generates a W-H frame for $L^2(\mathbf{R})$ for each $0 < b < b_0$.

REMARK 3.8. Condition (2) of Theorem 3.7 is purely a decay condition on g and is satisfied, for example, if

$$|g(x)| \le C(1+|x|^2)^{-1}$$

for some $C < \infty$ and for all $x \in \mathbf{R}$.

EXAMPLE 3.9. Given any $a, b > 0$ with $ab < 1$ it is easy to construct a smooth mother wavelet which generates a W-H frame for those parameters [D1; DGM]. For example, assume $1/2b \le a < 1/b$ and set $\lambda = 1/b - a$. Since $\lambda > 0$ we can define

$$g(x) = a^{-1/2} \begin{cases} 0, & x \le 0; \\ v(x/\lambda), & 0 \le x \le \lambda; \\ 1, & \lambda \le x \le a; \\ \left(1 - v(\frac{x-a}{\lambda})^2\right)^{1/2}, & a \le x \le 1/b; \\ 0, & 1/b \le x; \end{cases}$$

where $v \in C^\infty(\mathbf{R})$ is any function such that $v(x) = 0$ if $x \le 0$, $v(x) = 1$ if $x \ge 1$, and $0 < v(x) < 1$ if $0 < x < 1$. It is easy to see then that $g \in C^\infty(\mathbf{R})$, $\|g\|_2 = 1$, $\mathrm{supp}(g) \subset [0, 1/b]$, and $\sum |g(x - na)|^2 \equiv 1/a$. It therefore follows from Theorem 3.2 that g generates a tight W-H frame with frame bound $A = B = \frac{1}{ab}$.

4. The Zak transform. The Zak transform, also known as the Weil-Brezin map, has been used explicitly and implicitly in many mathematical and signal processing articles. Its history may even extend as far back as Gauss. Zak studied this operator beginning in the 1960's, in connection with solid state physics [Z].

DEFINITION 4.1. The **Zak transform** of a function f is (formally)

$$Zf(t, \omega) = |a|^{1/2} \sum_{k \in \mathbf{Z}} f(ta - ka)\, e^{-2\pi i k \omega} \qquad \text{for } t \in \mathbf{R},\ \omega \in \hat{\mathbf{R}},$$

where $a \in \mathbf{R} \backslash \{0\}$ is fixed.

Zf is defined pointwise at least for continuous functions with compact support. Formally, we have the **quasiperiodicity relations**

$$\begin{aligned} Zf(t+1, \omega) &= e^{-2\pi i \omega} Zf(t, \omega) \\ Zf(t, \omega + 1) &= Zf(t, \omega) \end{aligned}$$

Therefore, the values of $Zf(t, \omega)$ for $(t, \omega) \in \mathbf{R} \times \hat{\mathbf{R}}$ are completely determined by its values in the unit square, i.e., for $(t, \omega) \in Q = [0, 1) \times [0, 1)$.

DEFINITION 4.2. We define $L^2(Q)$ to be the Hilbert space

$$L^2(Q) = \left\{F : \|F\|_2 = \left(\int_0^1 \int_0^1 |F(t, \omega)|^2\, d\omega\, dt\right)^{1/2} < \infty\right\}.$$

We denote by $E_{(m,n)}$ the two-dimensional exponential $E_{(m,n)}(t, \omega) = e^{2\pi i m t}\, e^{2\pi i n \omega}$. Recall that the set of two-dimensional exponentials $\{E_{(m,n)}\}_{m,n \in \mathbf{Z}}$ forms an orthonormal basis for $L^2(Q)$.

THEOREM 4.3 [J]. *The Zak transform is a unitary map of $L^2(\mathbf{R})$ onto $L^2(Q)$, i.e., Z is a linear bijective isometry.*

In general, if Zf is continuous on Q it need not be true that Zf is continuous on $\mathbf{R} \times \hat{\mathbf{R}}$. For example, consider the function f such that Zf is identically 1 on Q. Moreover, one can prove the following.

THEOREM 4.4 [J]. *If $f \in L^2(\mathbf{R})$ and Zf is continuous on $\mathbf{R} \times \hat{\mathbf{R}}$ then Zf has a zero in Q.*

The unitary nature of the Zak transform allows us to translate conditions on frames for $L^2(\mathbf{R})$ into conditions for frames for $L^2(Q)$, where things are frequently easier to deal with.

PROPOSITION 4.5. *Given functions $g_n \in L^2(\mathbf{R})$. Then $\{g_n\}$ is complete/a frame/ an exact frame/an orthonormal basis for $L^2(\mathbf{R})$ if and only if $\{Zg_n\}$ is complete/a frame/an exact frame/an orthonormal basis for $L^2(Q)$.*

The Zak transform is particularly good for analyzing W-H frames when $ab = 1$.

THEOREM 4.6 [J]. *Assume $g \in L^2(\mathbf{R})$ and $a, b > 0$ with $ab = 1$. Then*

$$Z(T_{na}E_{mb}g) = E_{(m,n)}Zg.$$

COROLLARY 4.7 [D1; B]. *Given $g \in L^2(\mathbf{R})$ and $a, b > 0$ with $ab = 1$. Then $\{T_{na}E_{mb}g\}_{m,n\in\mathbf{Z}}$ is a complete set in $L^2(\mathbf{R})$ if and only if $Zg \neq 0$ a.e. in Q.*

COROLLARY 4.8 [D1; H]. *Given $g \in L^2(\mathbf{R})$ and $a, b > 0$ with $ab = 1$. Then the following statements are equivalent, i.e., each implies the other.*

(1) $0 < A \leq |Zg(t, \omega)|^2 \leq B < \infty$ a.e. in Q.

(2) *$\{T_{na}E_{mb}g\}_{m,n\in\mathbf{Z}}$ is a frame for $L^2(\mathbf{R})$ with frame bounds A, B.*

(3) *$\{T_{na}E_{mb}g\}_{m,n\in\mathbf{Z}}$ is an exact frame for $L^2(\mathbf{R})$ with frame bounds A, B.*

Proof. 1 ⇒ 2. Assume 1 holds. By Proposition 4.5 and Theorem 4.6, it suffices to show that $\{E_{(m,n)}Zg\}$ is a frame for $L^2(Q)$ with frame bounds A, B. So, choose any $F \in L^2(Q)$. As Zg is a bounded function, we have that $F \cdot \overline{Zg} \in L^2(Q)$. But $\{E_{(m,n)}\}$ is an orthonormal basis for $L^2(Q)$, so we have

$$\sum_{m,n} |\langle F, E_{(m,n)}Zg\rangle|^2 = \sum_{m,n} |\langle F \cdot \overline{Zg}, E_{(m,n)}\rangle|^2 = \|F \cdot \overline{Zg}\|_2^2,$$

from which the result follows.

2 ⇒ 1. Assume 2 holds. By Proposition 4.5 and Theorem 4.6, $\{E_{(m,n)}Zg\}$ is therefore a frame for $L^2(Q)$ with frame bounds A, B. But $\{E_{(m,n)}\}$ is an orthonormal basis for $L^2(Q)$, so $\sum |\langle F, E_{(m,n)}Zg\rangle|^2 = \|F \cdot \overline{Zg}\|_2^2$, as above. Thus $A\|F\|^2 \leq \|F \cdot \overline{Zg}\|_2^2 \leq B\|F\|_2^2$ for all $F \in L^2(Q)$, which implies easily that $A \leq |Zg|^2 \leq B$ a.e.

2 ⇒ 3. Assume 2 holds, so $\{E_{(m,n)}Zg\}$ is a frame for $L^2(Q)$. By 1 ⇔ 2 we know that Zg is bounded both above and below. Therefore, the mapping U defined by $UF = F \cdot Zg$ is a topological isomorphism of $L^2(Q)$ onto itself. Recall from Section 2 that exact frames are bounded unconditional bases, and that bounded unconditional bases are equivalent to orthonormal bases. Since $\{E_{(m,n)}Zg\}$ is obtained from the orthonormal basis $\{E_{(m,n)}\}$ by the topological isomorphism U, we see that $\{E_{(m,n)}Zg\}$ is a bounded unconditional basis, hence an exact frame. □

COROLLARY 4.9 [**D1**; **B**]. *Given $g \in L^2(\mathbf{R})$ and $a, b > 0$ with $ab = 1$. Then $\{T_{na}E_{mb}g\}_{m,n\in\mathbf{Z}}$ is an orthonormal basis for $L^2(\mathbf{R})$ if and only if $|Zg(t,\omega)| = 1$ a.e. in Q.*

The preceding results give us hope that we can easily find good orthonormal bases for $L^2(\mathbf{R})$, since all we need do is find some nice function whose Zak transform has absolute value 1. Let us look at some examples.

EXAMPLE 4.10 [**DGM**; **BGZ**]. The Zak transform of the Gaussian function $g(x) = e^{-\pi x^2}$ is continuous and has a single zero in Q. Therefore the Weyl-Heisenberg states $\{T_{na}E_{mb}g\}_{m,n\in\mathbf{Z}}$ for $ab = 1$ are complete in $L^2(\mathbf{R})$ but do not form a frame. However, one can show that g does generate a frame for other values of ab, in particular, for $ab = 1/2$.

EXAMPLE 4.11. Let $a = b = 1$ and $g = \chi_{[0,1)}$. The Zak transform of g is $Zg(t,\omega) \equiv 1$ for $(t,\omega) \in Q$. Therefore g generates a W-H frame with $A = B = 1$. In fact, this W-H frame is an orthonormal basis for $L^2(\mathbf{R})$ by Corollary 4.9.

This last example is a bit unpleasant since g is not smooth. This will introduce discontinuities even when analyzing functions which are smooth. The following theorem, due to Balian, Coifman, and Semmes, shows that something like this always happens when $ab = 1$, namely, either g is not smooth or it does not decay very fast.

THEOREM 4.12 [**D1**]. *Given $g \in L^2(\mathbf{R})$ and $a, b > 0$ with $ab = 1$. If (g, a, b) generates a W-H frame, then either $xg(x) \notin L^2(\mathbf{R})$ or $g' \notin L^2(\mathbf{R})$.*

Thus W-H frames with $ab = 1$ are bases for $L^2(\mathbf{R})$ but are not very nice. It can be shown that all W-H frames with $ab < 1$ are inexact, and we indicate below that it is impossible to construct a W-H frame when $ab > 1$. Thus $ab = 1$ is a sort of "critical value" for W-H frames. Daubechies explains this as a "Nyquist density", as occurs in information processing [D1].

For the case $ab = N > 1$, an easy calculation gives $Z(T_{na}E_{mb}g) = E_{(mN,n)}Zg$. But $\{E_{(mN,n)}\}$ is only a part of the orthonormal basis $\{E_{(m,n)}\}$, so it is easy to prove that $E_{(mN,n)}Zg$ is incomplete in $L^2(Q)$, no matter what $g \in L^2(\mathbf{R})$ is chosen. The proof can be adapted to cover the case $ab > 1$ and rational. For $ab > 1$ and irrational, the result follows by computing the coupling constant of a certain Von Neumann algebra [D1; R].

5. Affine frames.

DEFINITION 5.1. We define

$$H_+^2(\mathbf{R}) = \{f \in L^2(\mathbf{R}) : \mathrm{supp}(\hat{f}) \subset [0,\infty)\},$$
$$H_-^2(\mathbf{R}) = \{f \in L^2(\mathbf{R}) : \mathrm{supp}(\hat{f}) \subset (-\infty,0]\}.$$

These are Hilbert spaces, with the same inner products as the $L^2(\mathbf{R})$ inner product, and with norms

$$\|f\|_{H_+^2} = \left(\int_0^\infty |\hat{f}(\gamma)|^2\,d\gamma\right)^{1/2} \quad \text{and} \quad \|f\|_{H_-^2} = \left(\int_{-\infty}^0 |\hat{f}(\gamma)|^2\,d\gamma\right)^{1/2}.$$

Moreover, $H^2_+(\mathbf{R})$ and $H^2_-(\mathbf{R})$ are closed subspaces of the Hilbert space $L^2(\mathbf{R})$ and each is the orthogonal complement of the other.

DEFINITION 5.2. Given $g \in H^2_+(\mathbf{R})$, $a > 1$, and $b > 0$, we say that (g, a, b) **generates an affine frame** for $H^2_+(\mathbf{R})$ if $\{D_{a^n} T_{mb} g\}_{m,n \in \mathbf{Z}}$ is a frame for $H^2_+(\mathbf{R})$. The function g is referred to as the **mother wavelet**. The numbers a, b together are the **frame parameters**, a being the **dilation parameter**, and b the **shift parameter**.

We make similar definitions for affine frames for $H^2_-(\mathbf{R})$ and $L^2(\mathbf{R})$ and remark that it is sometimes necessary to take two mother wavelets in order to form a frame for $L^2(\mathbf{R})$ (cf. Theorem 5.4).

THEOREM 5.3 [DGM]. *Let $g \in L^2(\mathbf{R})$ be such that* $\operatorname{supp}(\hat{g}) \subset [l, L]$, *where $0 \le l < L < \infty$, and let $a > 1$ and $b > 0$ be such that:*

(1) There exist constants A, B such that

$$0 \;<\; A \;=\; \operatorname*{ess\,inf}_{\gamma \ge 0} \sum_n |\hat{g}(a^n\gamma)|^2 \;\le\; \operatorname*{ess\,sup}_{\gamma \ge 0} \sum_n |\hat{g}(a^n\gamma)|^2 \;=\; B \;<\; \infty;$$

(2) $(L - l) \le 1/b$.

Then $\{D_{a^n} T_{mb} g\}$ is a frame for $H^2_+(\mathbf{R})$ with bounds $b^{-1}A, b^{-1}B$.

Proof. Fix $n \in \mathbf{Z}$. Then by condition (2), the function $D_{a^n}\hat{f} \cdot \bar{\hat{g}}$ is supported in $I = [l, l + 1/b]$, which is an interval of length $1/b$. The Plancherel formula for Fourier series therefore implies that

$$\begin{aligned}
\sum_n \sum_m |\langle f, D_{a^n} T_{mb} g\rangle|^2 &= \sum_n \sum_m |\langle D_{a^n}\hat{f} \cdot \bar{\hat{g}}, E_{-mb}\rangle|^2 \\
&= \sum_n b^{-1} \int_I |D_{a^n}\hat{f}(\gamma) \cdot \overline{\hat{g}(\gamma)}|^2 \, d\gamma \\
&= \sum_n b^{-1} \int_0^\infty |\hat{f}(\gamma)|^2 \, |\hat{g}(a^n\gamma)|^2 \, d\gamma \\
&= b^{-1} \int_0^\infty |\hat{f}(\gamma)|^2 \cdot \sum_n |\hat{g}(a^n\gamma)|^2 \, d\gamma,
\end{aligned}$$

from which the result follows. □

A similar theorem can be formulated for $H^2_-(\mathbf{R})$. It is easy to see that if we combine a frame for $H^2_+(\mathbf{R})$ with one for $H^2_-(\mathbf{R})$ then we obtain a frame for $L^2(\mathbf{R})$.

THEOREM 5.4. *Let $g_1, g_2 \in L^2(\mathbf{R})$ satisfy* $\operatorname{supp}(\hat{g}_1) \subset [-L, -l]$ *and* $\operatorname{supp}(\hat{g}_2) \subset [l, L]$, *where $0 \le l < L < \infty$, and let $a > 1$, $b > 0$ be such that:*

(1) There exist constants A, B such that

$$\begin{aligned}
0 \;<\; A \;&=\; \min\left\{ \operatorname*{ess\,inf}_{\gamma \le 0} \sum_n |\hat{g}_1(a^n\gamma)|^2, \; \operatorname*{ess\,inf}_{\gamma \ge 0} \sum_n |\hat{g}_2(a^n\gamma)|^2 \right\} \\
&\le\; \max\left\{ \operatorname*{ess\,sup}_{\gamma \le 0} \sum_n |\hat{g}_1(a^n\gamma)|^2, \; \operatorname*{ess\,sup}_{\gamma \ge 0} \sum_n |\hat{g}_2(a^n\gamma)|^2 \right\} \;=\; B \;<\; \infty;
\end{aligned}$$

(2) $(L - l) \le 1/b$.

Then the collection of functions $\{D_{a^n}T_{mb}g_1, D_{a^n}T_{mb}g_2\}$ *is a frame for* $L^2(\mathbf{R})$ *with bounds* $b^{-1}A, b^{-1}B$.

In analogy with Theorem 3.7, the following theorem gives a condition on g so that (g,a,b) generates an affine frame for $L^2(\mathbf{R})$ for some frame parameters. In particular, g need not have a compactly supported Fourier transform. Note that unlike Theorem 5.4, this theorem requires only one mother wavelet.

THEOREM 5.5 **[D1]**. *Assume* $g \in L^2(\mathbf{R})$ *and* $a > 1$ *satisfy:*

(1) *There exist numbers* A, B *such that*

$$0 < A = \operatorname*{ess\,inf}_{\gamma\in\hat{\mathbf{R}}} \sum_n |\hat{g}(\gamma)|^2 \leq \operatorname*{ess\,sup}_{\gamma\in\hat{\mathbf{R}}} \sum_n |\hat{g}(\gamma)|^2 = B < \infty;$$

(2) $\lim_{b\to 0} \sum_{k\neq 0} \beta(k/b)^{1/2}\beta(-k/b)^{1/2} = 0$, *where*

$$\beta(s) = \operatorname*{ess\,sup}_{|\gamma|\in[1,a]} \sum_n |\hat{g}(a^n\gamma)|\,|\hat{g}(a^n\gamma - s)|.$$

Then there exists a number $b_0 > 0$ *such that* $\{D_{a^n}T_{mb}g\}$ *is a frame for* $L^2(\mathbf{R})$ *for each* $0 < b < b_0$.

When $a = 2$ an improved version of Theorem 5.5 holds in which the function β is replaced by a new function β_1 which takes into account possible cancellations which may arise from the phase portion of $\hat{g}$ and which are lost in the function β. This improved theorem is especially useful in analyzing the Meyer wavelet (described below). The theorem is due to Tchamitchian.

THEOREM 5.6 **[D1]**. *Let* $g \in L^2(\mathbf{R})$ *and* $a = 2$ *satisfy the hypotheses of Theorem 5.5, and assume* $b > 0$. *If* $\{D_{2^n}T_{mb}g\}$ *is a frame for* $L^2(\mathbf{R})$ *with frame bounds* A', B' *then*

$$A' \geq b^{-1}\left(A - 2\sum_{l=0}^{\infty} \beta_1\left(\tfrac{2l+1}{b}\right)^{1/2} \beta_1\left(-\tfrac{2l+1}{b}\right)^{1/2}\right)$$

and

$$B' \leq b^{-1}\left(B + 2\sum_{l=0}^{\infty} \beta_1\left(\tfrac{2l+1}{b}\right)^{1/2} \beta_1\left(-\tfrac{2l+1}{b}\right)^{1/2}\right),$$

where

$$\beta_1(s) = \operatorname*{ess\,sup}_{\gamma\in\hat{\mathbf{R}}} \sum_m \left|\sum_{j\geq 0} \hat{g}(2^{m+j}\gamma)\, \bar{\hat{g}}(2^j(2^m\gamma + s))\right|$$

and A, B *are as in Theorem 5.5.*

EXAMPLE 5.7. Let $\hat{g}_1 = \chi_{(-2,-1]}$ and $\hat{g}_2 = \chi_{[1,2)}$, and take $a = 2$ and $b = 1$. Then $\{D_{2^n}T_m g_1, D_{2^n}T_m g_2\}$ is a tight affine frame for $L^2(\mathbf{R})$, and in fact is an orthonormal basis for $L^2(\mathbf{R})$. Note, however, that the elements of this orthonormal basis are not smooth. In the next example we discuss the Meyer wavelet, which is a C^∞ function that generates an affine orthonormal basis for $L^2(\mathbf{R})$.

As we saw in Section 4, a W-H frame forms a basis for $L^2(\mathbf{R})$ if and only if $ab = 1$. Moreover, W-H frames for this critical value are composed of functions which are either not smooth or do not decay quickly. Y. Meyer showed that a very different situation holds for the affine case when he exhibited a C^∞ function with compactly supported Fourier transform which generates an affine orthonormal basis for $L^2(\mathbf{R})$.

DEFINITION 5.8 [**D2**]. The **Meyer wavelet** is the function $\psi \in L^2(\mathbf{R})$ defined by:

$$\hat{\psi}(\gamma) = e^{i\gamma/2}\omega(|\gamma|),$$

where

$$\omega(\gamma) = \begin{cases} 0, & \gamma \leq 1/3; \\ \sin \frac{\pi}{2} v(3\gamma - 1), & 1/3 \leq \gamma \leq 2/3; \\ \cos \frac{\pi}{2} v(\frac{3\gamma}{2} - 1), & 2/3 \leq \gamma \leq 4/3; \\ 0, & \gamma \geq 4/3; \end{cases}$$

and $v \in C^\infty(\mathbf{R})$ is such that $v(\gamma) = 0$ for $\gamma \leq 0$ and $\gamma \geq 1$, $0 \leq v(\gamma) \leq 1$ for $\gamma \in [0,1]$, and $v(\gamma) + v(1-\gamma) = 1$ for $\gamma \in [0,1]$.

The last condition on v is crucial in showing that ψ generates an orthonormal basis for $L^2(\mathbf{R})$. Note that $\hat{\psi}$ is compactly supported and lies in both the negative and positive portions of the real line. Therefore, we should only need dilations and translations of ψ in order to form a frame for $L^2(\mathbf{R})$, not two functions ψ_1, ψ_2 as in Theorem 5.4 or Example 5.7.

LEMMA 5.9. *Let ψ be as above, and let β_1 be as in Theorem 5.6. Then*

(1) $\|\psi\|_2 = 1$.
(2) $\sum_n |\hat{\psi}(2^n\gamma)|^2 \equiv 1$.
(3) $\beta_1(k) = 0$ *for every odd* $k \in \mathbf{Z}$.

From Theorem 5.6 we therefore have that ψ generates a tight affine frame for $L^2(\mathbf{R})$ for the parameters $a = 2$, $b = 1$. As pointed out before, ψ actually generates an orthonormal basis for $L^2(\mathbf{R})$. Meyer, Lemarie, Daubechies, et al., have developed the concept of "multiscale analysis" to understand more fully the phenomena exhibited by the Meyer wavelet [D2; LM].

REFERENCES

[AT] L. AUSLANDER AND R. TOLIMIERI, *Radar ambiguity functions and group theory*, SIAM J. Math. Anal., 16 (1985), pp. 577–599.

[BGZ] H. BACRY, A. GROSSMANN, AND J. ZAK, *Proof of the completeness of lattice states in the kq representation*, Physical Review B, 12 (1975), pp. 1118–1120.

[B] J. BENEDETTO, *Gabor representations and wavelets*, AMS Contemporary Mathematics Series (to appear).

[D1] I. Daubechies, *Frames of coherent states, phase space localisation, and signal analysis*, (to appear).

[D2] ———, *Orthonormal bases of compactly supported wavelets*, (to appear).

[DGM] I. Daubechies, A. Grossmann, and Y. Meyer, *Painless nonorthogonal expansions*, J. Math. Phys., 27 (1986), pp. 1271–1283.

[DS] R. J. Duffin and A. C. Schaeffer, *A class of nonharmonic Fourier series*, Trans. Am. Math. Soc., 72 (1952), pp. 341–366.

[H] C. Heil, *Ph.D. Thesis*, University of Maryland (1989).

[HW] C. Heil and D. Walnut, *Continuous and discrete wavelet transforms*, (to appear).

[J] A. J. E. M. Janssen, *Bargmann transform, Zak transform, and coherent states*, J. Math. Phys., 23 (1982), pp. 720–731.

[LM] P. Lemarié and Y. Meyer, *Ondelettes et bases hilbertiennes*, Rev. Mat. Iberoamericana, 2 (1986), pp. 1–18.

[R] M. Rieffel, *Von Neumann algebras associated with pairs of lattices in Lie groups*, Math. Ann., 257 (1981), pp. 403–418.

[Y] R. M. Young, *An Introduction to Nonharmonic Fourier Series*, Academic Press, 1980.

[Z] J. Zak, *Finite translations in solid state physics*, Phys. Rev. Lett., 19 (1967), pp. 1385–1397.

LINEAR AND POLYNOMIAL METHODS IN MOTION ESTIMATION

THOMAS S. HUANG* AND ARUN NETRAVALI†

Abstract. almost all motion estimation problems can be formulated in terms of the solution of a set of simultaneous polynomial equations; a few of them can be formulated in terms of linear equations. With respect to these linear and polynomial formulations, many open questions remain. These include: (i) How to find minimal formulations - in the linear case, the smallest number of equations, in the polynomial case, the smallest total degree? (ii)How to find all the real solutions of a polynomial system? (iii) How to find robust algorithms in the presence of data noise? The main goal of this paper is to stimulate interest of the mathematics community. It is hoped that mathematicians will collaborate with computer vision researchers to answer some of these open questions.

I. Introduction and Problem Statement. A straightforward formulation of motion estimation problems leads to the solution of simultaneous transcendental equations. The solution is typically carried out by iterative methods, which may diverge or may converge to a wrong local minimum unless one has a very good initial guess solution. Fortunately, in almost all case, it is possible to formulate the problem in such a way that polynomial equations result. In some case, it is even possible to obtain linear equations. However, many open mathematical questions remain with regard to these linear and polynomial methods. The goal of the present paper is to raise some of the open questions in the context of a specific motion estimation problem, that of two-view motion determination from point correspondences. The hope is that the paper will stimulate interest of the mathematics community so that they may contribute to answering some of these open questions. For a more general review paper on motion analysis, see Ref 1.

We first state the two-view motion determination problem. The basic imaging geometry is shown in Fig. 1. A pinhole camera model is used. Two images are taken of a moving rigid object at time instants t_1 and t_2, respectively. By processing these two images, one aims to determine the motion (in a sense specified below) of the object from t_1 to t_2. The solution approach consist of two steps: (i) Extract and match feature points (e.g., corners of man-made objects) over the two images. (ii) From the image coordinates of the corresponding points, solve equations to determine the motion parameters. In this paper, we are concerned with step (ii).

*Coordinated Science Laboratory, University of Illinois, 1101 W. Springfield Avenue, Urbana, IL 61801

†AT & T Bell Laboratories, 600 Mountain Avenue, Murray Hill, NJ 07974

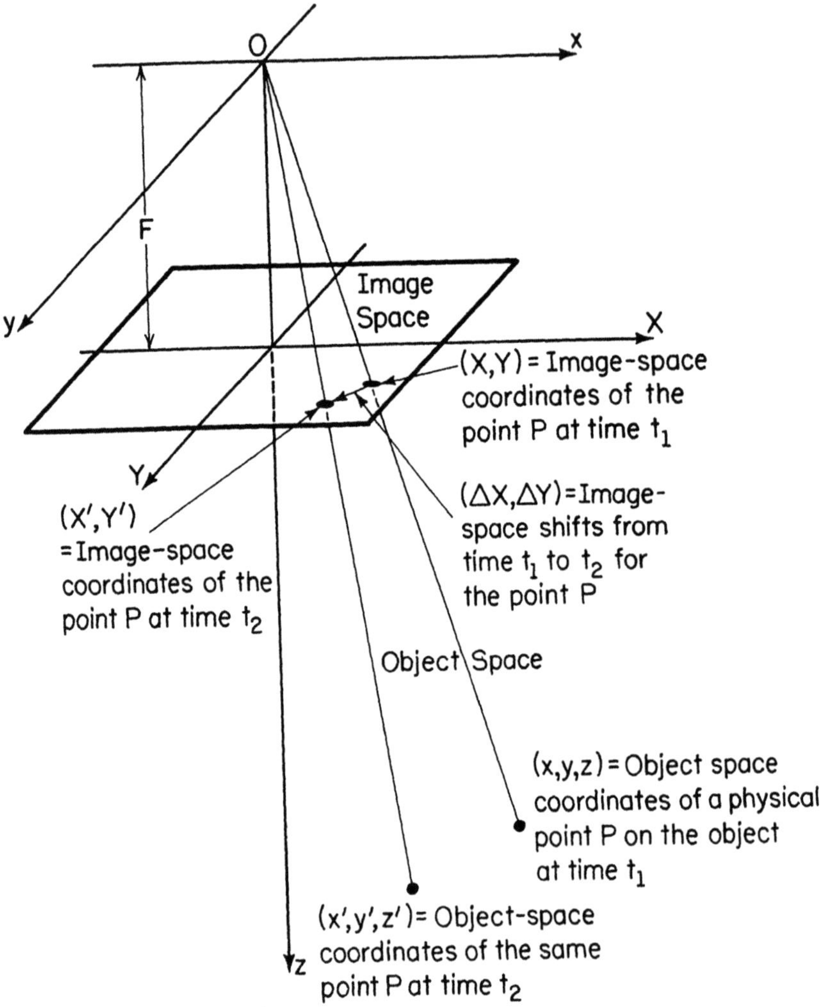

Figure 1

Referring to Figure 1, we shall use the following notation:

$p = (x, y, z)$ = coordinates of a 3–D point at time instant t_1

$p' = (x', y', z')$ = coordinates of the same 3–D point at t_2

$P = (X, Y, 1)$ = coordinates of the image point at t_1

$P' = (X', Y', 1)$ = coordinates of the image point at t_2

without loss of generality, we have set the focal length F to 1. And we shall use a subscript i to denote the ith point.

From kinematics,

$$p_i' = R\, p_i + t \tag{1}$$

where R is a 3×3 rotation matrix (orthonormal and det $(R) = +1$)

$$R = \begin{bmatrix} r_{11} & r_{12} & r_{13} \\ r_{21} & r_{22} & r_{23} \\ r_{31} & r_{32} & r_{33} \end{bmatrix} \tag{2}$$

t is a translation vector (or 3×1 column matrix)

$$t = \begin{bmatrix} t_x \\ t_y \\ t_z \end{bmatrix} \tag{3}$$

and p_i' and p_i are considered as 3×1 column matrices.

From simple geometry, the object point and image point coordinates are related by:

$$\begin{aligned} X_i &= x_i/z_i, \; Y_i = y_i/z_i \\ X_i' &= x_i'/z_i', \; Y_i' = y_i'/z_i' \end{aligned} \tag{4}$$

The mathematical problem can now be stated precisely:

Given $P_i \longleftrightarrow P_i', i = 1, 2, \ldots, n$

Find $R, \dfrac{t}{\|t\|}; \dfrac{z_i}{\|t\|}$.

Note that with a single camera t and z_i can be determined only to within a scale factor. The reason is that: If the object is moved d times farther away from the camera and enlarged d times (i.e., multiply all z_i by d), also the translation t is multiplied by d, then one will get exactly the same two images independent of d.

II. Linear Method.

A. Transcendental Equations [3]

From Eqs (1) and (4) (omitting subscripts),

$$\begin{aligned} X' &= \frac{r_{11}\, X + r_{12}\, Y + r_{13} + t_x/z}{r_{31}\, X + r_{32}\, Y + r_{33} + t_z/z} \\ Y' &= \frac{r_{21}\, X + r_{22}\, Y + r_{23} + t_y/z}{r_{31}\, X + r_{32}\, Y + r_{33} + t_z/z} \end{aligned} \tag{5}$$

Eliminating z,

$$\begin{aligned} &(t_x - t_z\, X')\, \{(r_{31}\, X + r_{32}\, Y + r_{33}Y' - (r_{21}\, X + r_{22}Y + r_{23})\} \\ &= (t_y - t_z\, Y')\, \{(r_{31}\, X + r_{32}\, Y + r_{33})X' - (r_{11}\, X + r_{12}\, Y + r_{13})\} \end{aligned} \tag{6}$$

Expressing r_{ij} in terms of the 3 Euler angles [2], Eq. 6 becomes transcendental in these angles and linear and homogeneous in t_x, t_y, and t_z. With five point correspondences, we have five equations in five unknowns (setting, e.g., $t_z = 1$). Iterative methods can be used to solve these equations. However, nothing can be said about the number of solutions, and convergence to the correct solutions are by no means ensured.

B. Linear Methods [4,5]

It turns out that with eight or more point correspondences, a linear algorithm is possible. Geometrically, it is obvious from Figure 1 that the three vectors Rp, p', and t are coplanar. Thus

$$p' \cdot t \times Rp = 0 \tag{7}$$

which can be written, in matrix form, as

$$p'^T Ep = 0 (T \text{ denotes transposition}) \tag{8}$$

where

$$E \triangleq \begin{bmatrix} e_1 & e_2 & e_3 \\ e_4 & e_5 & e_6 \\ e_7 & e_8 & e_9 \end{bmatrix} = GR \tag{9}$$

$$G \triangleq \begin{bmatrix} 0 & -t_z & t_y \\ t_z & 0 & -t_x \\ -t_y & t_x & 0 \end{bmatrix} \tag{10}$$

Dividing (8) by zz', we get

$$P'^T EP = 0 \tag{11}$$

or

$$[X'Y'1]\ E \begin{bmatrix} X \\ Y \\ 1 \end{bmatrix}$$

Eq. (11) is linear and homogeneous in the nine unknowns e_i. Given eight or more point correspondences, we can solve the set of linear and homogeneous equation for e_i to within a scale factor. Once E is obtained (to within a scale factor), several algorithms have been developed (none involving the solution of nonlinear equations) to find $R, \frac{t}{\|t\|}$, and $\frac{z_i}{\|t\|}$. See Ref. 1.

C. Open Questions

In this paper we consider only one specific motion estimation problem, that of two–view motion determination using point correspondences. there are many other problems. E.g.: (i) Three-view analysis. (ii) Using other features such as straight edges, curves, and corners (a corner consists of a corner point and the edges emanating from it). (iii) Optic flow based methods (see Ref. 1). Of these numerous motion estimation problems, linear algorithms have been developed only for a few. A major open question is whether we can come up with a *systematic* procedure of transforming a nonlinear problem into a linear one by defining intermediate variables. We want to do it in such a way that the number of intermediate variables is as small as possible.

Another important question is noise sensitivity. Because of image sampling and other factors, the image coordinates of the feature points cannot be measured exactly. This introduces errors in the estimated motion parameters and range values. Empirically, it has been found that linear algorithms are more sensitive to noise than nonlinear algorithms. Theoretical analysis is needed to make a quantitative comparison.

III. Polynomial Methods. In the following sections (with the exception of A), results are described briefly and without derivations. The readers are referred to the references for details.

A. Using Rigidity Constraints [6]

From rigidity constraints, z_i and $z_i'(i = 1, 2, \ldots, n)$ are determined. Then R, t are obtained by solving Eq. (1).

The points p_i are on a rigid object, if and only if the distance between any pair of them is the same at t_1 and t_2:

$$\|p_i - p_j\|^2 = \|p_i' - p_j'\|^2, \text{ for all } i, j \tag{12}$$

For n point correspondences ($n \geq 2$) there are 6+3 (N-4)=(3N-6) independent such equations. Eq. (12) can be written as

$$(x_i - x_j)^2 + (y_i - y_j)^2 + (z_i - z_j)^2 = (x_i' - x_j')^2 + (y_i' - y_j')^2 + (z_i' - z_j')^2$$

or

$$\begin{aligned} &(X_i z_i - X_j z_j)^2 + (Y_i z_i - Y_j z_j)^2 + (z_i - z_j)^2 \\ &= (X_i' z_i' - X_j' z_j')^2 + (Y_i' z_i' - Y_j' z_j')^2 + (z_i' - z_j')^2 \end{aligned} \tag{13}$$

The unknows are z_i, z_j, z_i', z_j'. The equation is 2nd-order and homogeneous in the unknowns.

With five point correspondences, we have ten unknowns and nine homogeneous equations and can attempt to solve the equations to find the unknows to within a scale factor. Note that the total degree of the set of nine 2nd-order polynomial equations is $2^9 = 512$.

B. Using Quaternions [7]

Rotation can be represented by a unit quaternion. Equivalently, one can scale the quaternion to make the scalar component 1:

$$q = [1, q_x, q_y, q_z] \tag{14}$$

If we can find q, then we can renormalize it to a unit quaternion.

Given five point correspondences, it is possible, by eliminating t and z_i, to obtain three 4th-order polynomial equations in the three unknowns q_x, q_y, and q_z. The total degree is $4^3 = 64$.

C. Using the E-Matrix [8]

With five point correspondences, we can write five linear homogeneous equations (11) in the nine unknowns e_i. The fact that the matrix E is decomposable into a skew symmetrical matrix post-multiplied by a right-handed orthonormal matrix can be expressed in terms of three homogeneous polynomial equations in e_i. These equations are of degrees 3, 4, and 4, respectively. Taken together, we have eight homogeneous equations in the nine unknowns e_i. The total degree of the system is $3 \times 4 \times 4 = 48$.

D. Using Projective Geometry [9]

By using results and techniques of algebraic projective geometry, it is possible to formulate a single 10th-order polynomial equation in one unknown. However, the derivation of the equation was extremely tedious, and a symbolic mathematics package, MAPLE, had to be used.

E. Open Questions

From Sections A–D, we see that the number of equations equals to the number of unknowns when we have five point correspondences. Since the equations are polynomial, the number of solutions is finite. From Bezout's theorem [10], the number of solutions (generally complex) is equal to the total degree of the polynomial system, if one includes solutions at infinity. We are obviously interested in real solutions. The total degree of a polynomial system from any formulation is an upperbound to the number of real solutions. For our problem (two-view motion/range determination from point correspondences), the smallest upperbound to date is 10 (from Section D). A major question is: For a specific motion estimation problem, how does one derive a set of polynomial equations with the *minimum* total degree? A polynomial system of a smaller total degree not only gives a tighter upperbound to the number of real solutions, but also is obviously easier to solve.

With respect to finding solutions of a set of simultaneous polynomial equations, many questions remain unresolved. First, given a set of M polynomial equations with K unknows (where $K \leq M$), how can one find *all* the real solutions? For $K = 1 = M$, Sturm's method can be used. However, its extension to the general case appears very difficult. One approach is the U-resultants [10], which unfortunately are computationally infeasible except for very simple polynomial systems. Second, a related and simpler question: How can one determine the number of *real* solutions of a polynomial system (without solving it)?

Numerically, the best hope to date of solving a set of polynomial equations (with $K = M$) to find all solutions (generally complex) appears to be the use of homotopy methods [11]. We are currently using this approach to solve the equation set of Section C with encouraging results. However, this approach is still computationally intensive, and may become infeasible if the total degree of the polynomial system is larger than 50 or so. Furthermore, with numerical methods, because of round off and other quantization errors, one can never be completely sure that a specific solution is purely real.

For practical applications, an important third question arises: Can we find computationally efficient methods of estimating all the real solutions in the presence of data noise (which introduces errors into the coefficients of the polynomial equations)? In this connection, a careful noise sensitivity analysis needs to be made for polynomial systems.

Although in the above we have asked the three questions mainly from a numerical point of view, we are indeed interested in some of the deeper questions: (a) How does the number of solutions depend on 3-D point configurations and motion parameters? (b) How does the noise sensitivity depend on 3-D point configurations and motion parameters?

IV. Discussions. Almost all motion estimation problems can be formulated in terms of the solution of a set of simultaneous polynomial equations, while only a few can be solved by linear methods. Even in the case where linear algorithms are available, polynomial methods may still be preferred because they require fewer feature correspondences (which are hard to come by in practice) and are potentially less noise sensitive.

The main purpose of this paper is to arouse the interest of the mathematics community. The relevant areas of mathematics appear to be algebraic and projective geometry. Most researchers in computer vision (one subfield of which is dynamic scene analysis and motion estimation) are not familiar with the tools in these areas. It is hoped that mathematicians will be stimulated by the open questions raised here into collaborating with computer vision researchers to answer some of them.

Acknowledgment. This work was supported in part by the Joint Services Electronics program under Grant N00014–84–C-0149.

REFERENCES

[1] T.S. HUANG, *Motion Analysis*, in *Encyclopedia of Artificial Intelligence*, ed. by S.C. Shapiro, Wiley (1987).

[2] H. GOLDSTEIN, *Classical Mechanics*, Addison-Wesley (1981).

[3] T.S. HUANG AND R.Y. TSAI, *Image Sequence Analysis: Motion Estimation*, in *Image Sequence Analysis*, ed. by T.S. Huang, Springer-Verlag (1981).

[4] H.C. LONGUET-HIGGINS, *A Computer Program for Reconstructing a Scene from Two Projections*, Nature, 293 (September 1981), pp. 133–135.

[5] R.Y. TSAI AND T.S. HUANG, *Uniqueness and Estimation of 3-D Motion Parameters of Rigid Bodies with Curved Surfaces*, IEEE Trans. on PAMI, 6, no. 1 (January 1984), 13–17.

[6] A. MITICHI AND J.K. AGGARWAL, *A Computational Analysis of Time-Varying Images*, in *Handbook of Pattern Recognition and Image Processing*, ed. by T.Y. Young and K.S. Fu, Academic Press (1986).

[7] C. JERIAN AND R. JAIN, *Polynomial Methods for Structure From Motion*, Proc. 2nd International Conference Computer Vision (December 1988), Tarpon Springs, FL.

[8] T.S. Huang and Y.S. Shim, *Linear Algorithm for Motion Estimation: How to Handle Degenerate Cases*, Proc. British Pattern Recognition Association Conference (April 1987), Cambridge, England.

[9] D.D. Faugeras and S.J. Maybank, *Multiplicity of Solutions for Motion Problems*, Technical Report, INRIA, France (1988).

[10] W.V.D. Hodge and D. Pedoe, *Methods of Algebraic Geometry*, 1, Cambridge University Press (1953).

[11] A. Morgan, *Solving Polynomial Systems Using Continuation for Engineering and Science Problems*, Prentice-Hall (1987).

POSITIVE DEFINITE COMPLETIONS: A GUIDE TO SELECTED LITERATURE*

CHARLES R. JOHNSON†

I. Introduction. Our purpose here is to present a personal and current discussion of work on the general problem of completing a "partial Hermitian matrix" to a positive definite matrix. (Such problems arise in a variety of ways and enjoy attractive mathematical structure. This author's interest stems from a long term desire to understand determinantal inequalities.) To this end we describe a general setting for completion problems that permits discussion of very general ones, including contraction completions, ranks of completions, etc. and then go on to focus upon positive definite completions (as in the talk at IMA). A future, broader survey will cover other completion problems and links among them, but we include a rather broad set of references here.

a) THE SETTING. Let S denote a set of items, perhaps a field of scalars, such as the real or complex numbers, or a more arbitrary set. By a *partial matrix* we mean a rectangular array, the entries in certain locations of which are particular, known elements of S, while the remaining entries are unspecified and free to be chosen from S. We refer to the known entries as "specified" in contrast to the remaining unspecified entries, which are conventionally and suggestively denoted by ?'s. For example, if S is the field of rational numbers,

$$\begin{bmatrix} -1 & ? & \frac{1}{2} \\ ? & -2/3 & 3 \end{bmatrix}$$

is an example of a partial matrix. A conventional matrix (all entries specified) is, of course, a special case. Often it will be convenient to refer to a part (submatrix, minor, etc.) of a partial matrix, each of whose entries is specified (or unspecified). In this event the adjective will be extended to the part. For example,

$$[-2/3 \quad 3]$$

is a specified submatrix of the above example of a partial matrix. Also, modifiers applied to the term "partial matrix" will have the meaning naturally suggested by their use relative to an ordinary matrix. For example, an "Hermitian partial matrix", or "partial Hermitian matrix" is a square partial matrix that is Hermitian to the extent of its specified entries; i.e., the set S is the complex field and if the i, j entry is specified, then the j, i entry is also and the two are complex conjugates.

*This work was supported in part by National Science Foundation grant DMS 88 02836. The text closely follows an invited talk given at the Institute for Mathematics and its Applications during a most stimulating workshop on signal processing in July 1988. Part of the preparation was carried out while the author was a guest scientist at NASA/ICASE

†Department of Mathematics, College of William and Mary, Williamsburg, VA 23185

By a *completion* of a partial matrix $A = (a_{ij})$, we mean a conventional matrix $B = (b_{ij})$, of the same dimensions as A and with all entries from the set S, such that if a_{ij} is specified, then $b_{ij} = a_{ij}$. Thus, a completion is just a specification (from S) of all the unspecified entries, resulting in a conventional matrix (over S). A *matrix completion problem*, then, is just the question as to whether a partial matrix has a completion in a certain class or (equivalently) with a certain property of interest. For example, we might ask if a partial Hermitian matrix has a positive definite completion or if a given rectangular partial matrix over the complex numbers has a completion of rank 7. Many such problems carry with them an important inheritance structure. Often, having a completion with certain property requires "obvious" necessary conditions of like kind upon the data (specified entries). For example, since every principal submatrix of a positive definite matrix is positive definite, if a partial Hermitian matrix is to have a positive definite completion, every specified principal submatrix must be positive definite, and, thus, every specified principal minor must be positive. Similarly, for a rectangular, complex partial matrix to have a completion of rank 7, no specified submatrix may have rank greater than 7.

The occurrence of such structure suggests matrix completion problems at another level. On the one hand, a matrix completion problem, as we have described it, in its most general form could be viewed as a request for a decision procedure, which, for each partial matrix, tells whether there is a completion of the desired type. Of course, for the answer to be "yes", any "obvious" necessary conditions must meet. What if they all are? Will the answer then be "yes"? (We distinguish here between obvious necessary conditions, of the sort mentioned above, and possibly more subtle conditions that would characterize completability and yield a complete decision procedure.) Generally, no, but is there a little more, and perhaps natural, structure that might be required by the partial matrices that would yield such a nice situation? A question of this type is the following. Which *patterns* for the specified entries insure that if certain "obvious" necessary conditions are met, then there will be a completion of desired type? This introduces a combinatorial (as opposed to magnitudinal or analytic) aspect into matrix completion problems, which turns out to be rather natural, but we will discuss it further in the context of the specific matrix completion problems that we survey here. Call this the *combinatorial matrix completion problem*; such study has become one major component in emerging work in "combinatorial matrix theory".

b) REMARKS. The setting just described permits an immense variety of specific problems, a nontrivial fraction (but not all) of which are mathematically interesting. Simply pick your favorite set S and a property of interest. Such problems have a variety of applied motivations including seismic reconstruction problems, entropy methods in statistical/physical problems, electrical engineering/systems theory, etc. They also provide a natural vehicle for mathematical understanding of structure; a good example is determinantal inequalities for positive definite matrices.

The general setting we have described is a good deal broader than that which we shall discuss. It includes at least two long standing (and important) lines of research that we mention here but discuss in no further detail. First, completion problems involving familiar combinatorial structures have been studied by combinatorists for some time as a means of better understanding these structures. For example, given a matrix of ±1's, what is the smallest size Hadamard matrix into which it may be embedded; or, given a partial matrix whose specified entries are ±1's may it be completed to an Hadamard matrix? Similarly for Latin squares – which n-by-n partial matrices over the set $\{1, 2, \cdots, n\}$ may be completed to Latin squares. Second, algebraically oriented matrix theorists have been studying for some time the possible invariants that may occur among completions of partial matrices whose specified entries consists of exactly one or two blocks with known invariants. For example, given the Jordan structure of the k-by-k complex matrix A, what are the possible Jordan structures of the n-by-n $(n > k)$ matrix

$$\begin{bmatrix} A & B \\ C & D \end{bmatrix} \quad ?$$

We focus here upon modern work on the most prominent example from an emerging class of completion problems of a different sort. Positive definite completion problems, contraction completion, rank completion problems and others all enjoy some form of the inheritance structure mentioned earlier, as well as other aspects of commonality. The positive definite case is discussed here in some detail, and it is hoped that a future survey will include the others.

II. Positive Definite Completions. This is the most heavily studied of matrix completion problems, and we shall not be able even to mention all the interesting work upon it. The references provide many details and items not given here. The basic problem is to decide whether a given partial Hermitian matrix has a positive definite completion. There is a corresponding, and essentially equivalent, positive semidefinite completion problem. One may then go on and attempt to describe all positive definite (semidefinite) completions (they form a convex set), or to focus upon positive definite completions with additional important features. The obvious necessary condition on the data (positive definite completion case) is that every specified principal minor of the partial Hermitian matrix must be positive, equivalently every specified principal submatrix must be positive definite. We refer to a partial Hermitian matrix meeting this minimal necessary condition as *partial positive definite*. The resulting combinatorial positive definite completion problem asks which patterns for the specified entries ensure that a partial positive definite matrix has a positive definite completion.

Call a square partial matrix *banded* if all entries within some band-width (symmetric w.r.t. the main diagonal) are specified and all entries outside this band-width are unspecified. An early positive definite completion result is the Caratheodory-Fejer theorem from function theory. In our language, it states that a (banded) Toeplitz (equal entries along bands) partial positive definite matrix always has a Toeplitz positive definite completion. This result was notably generalized in the stimulating paper [DG1]. Also translated to our language, [DG1] included the following three major conclusions.

THEOREM *DG*. *Let H be a complex banded partial positive definite matrix. Then, (1) H has a (complex) positive definite completion. (2) Among the positive definite completions of H, there is a unique one with maximum determinant. (3) Among positive definite completions, the determinant maximizer is the unique one whose inverse is banded (at most same band-width) in the usual sense (all entries equal to zero outside some bands).*

This result, then, says that at least the banded patterns are among those sought in the combinatorial version of the positive definition completion problem. It should be noted that theorem DG remains valid when S is the real, or even rational, field as well as the complex field, and that the authors also prove the more general result in which A is "block banded" (obvious interpretation). This could, though, just be viewed as the banded case with the set S consisting of matrices of various sizes).

The special case of the positive definite completion problem in which there is just one (two symmetrically placed) unspecified entry is worth noting. Because the general problem is unchanged under permutation similarity, this is the special case of the case studied in [$DG1$] in which the missing entry lies in the upper right (and lower left) corner. This case exhibits, in a simple way, many features of the problem, and is a substantial piece of the solution of more general problems. Consider the n-by-n partitioned partial Hermitian matrix

$$H = \begin{bmatrix} a & b^* & x \\ b & A & c \\ \overline{x} & c^* & d \end{bmatrix}$$

in which the maximal specified principal submatrices

$$B = \begin{bmatrix} a & b^* \\ b & A \end{bmatrix} \quad \text{and} \quad C = \begin{bmatrix} A & c \\ c^* & d \end{bmatrix}$$

are positive definite, and $x \in \mathbb{C}$ is the unspecified entry. Using Sylvester's determinantal identity [HJ], for example, we then have

$$\det H = \frac{\det B \det C - \left| \det \begin{bmatrix} b^* & x \\ A & c \end{bmatrix} \right|^2}{\det A}$$

$$= \frac{\det B \det C - \left| \det \begin{bmatrix} b^* & 0 \\ A & c \end{bmatrix} + (-1)^n x \det A \right|}{\det A}$$

Since an Hermitian matrix is positive definite if and only if its leading principal minors are positive [HJ] and since B is positive definite, the matrix H is positive definite if and only if $\det H > 0$. Algebraic manipulation of this expression for $\det H$ reveals that $\det H > 0$ if and only if

$$\left| x - \frac{(-1)^{n-1} \det \begin{bmatrix} b^* & 0 \\ A & c \end{bmatrix}}{\det A} \right|^2 < \frac{\det B \det C}{(\det A)^2}$$

Using Schur complements [HJ], we find that

$$\det \begin{bmatrix} b^* & 0 \\ A & c \end{bmatrix} = (-1)^{n-1} b^* A^{-1} c \ \det A.$$

Substitution into the previous inequality then gives that $\det H > 0$ if and only if

$$\left| x - b^* A^{-1} c \right|^2 < \frac{\det B \det C}{(\det A)^2}$$

Furthermore, $\det H$ is maximized exactly when

$$x = b^* A^{-1} c \ .$$

Since $\det B, \det C$ and $\det A > 0$ (the latter because it is a principal submatrix of the positive definite matrix B [HJ]), we may draw the following conclusions.

1) Positive definite completions of H always exist under our assumption that B and C are positive definite.
2) The set of all x that constitute a positive definite completion of H is the open disc with center:

$$\frac{(-1)^{n-1} \det \begin{bmatrix} b^* & 0 \\ A & C \end{bmatrix}}{\det A} = b^* A^{-1} c$$

and radius:

$$\frac{(\det B)^{1/2} (\det C)^{1/2}}{\det A}$$

3) The center of this disc uniquely maximizes $\det H$ among the positive completions of H, the maximum determinant being

$$\frac{\det B \ \det C}{\det A} \quad .$$

Since, by co-factors,

$$\frac{\det \begin{bmatrix} b^* & x \\ A & c \end{bmatrix}}{\det H}$$

is the $1, n$ entry of H^{-1} (if it exists), we also note that

4) the corner entries ($1, n$ and $n, 1$) of H^{-1} are 0 exactly for the x that provides the unique determinant maximizing completion of H among the positive definite completions of H. These corner entries are in just the positions of our unspecified entry x in H.

It should also be noted if H is a partial positive definite *symmetric* matrix over the real field $\mathbf{R}$ (respectively, the rational field $\mathbf{Q}$), then there are positive definite completions over $\mathbf{R}$ (respectively $\mathbf{Q}$) and the determinant maximizing completion is from $\mathbf{R}$ (respectively $\mathbf{Q}$). In the case of the real field $\mathbf{R}$, the set of all positive definite completions is just a line segment with the same center and half-width as in the case of $\mathbf{C}$.

Of course, conclusion (1) of theorem DG may be proven by sequential application (down the bands one-at-a-time, going out) of the one-variable case, and conclusions (3) and (4) in the one variable case are indicative special cases of theorem DG. Sequential application of the one variable case will continue to be important in the general problem.

In further considering the positive definite completion problem, we henceforth assume that all diagonal entries of any partial positive definite matrix under study are specified. For existence of positive definite completions this is no loss of generality, for, if not, any unspecified diagonal entires may be chosen arbitrarily large, leaving the question of a positive definite completion to rest upon that maximal partial principal submatrix all of whose diagonal entries are specified. For determinant maximization this is an obviously necessary regularity condition because the determinant of a positive definite matrix is monotone increasing in its diagonal entries.

Conclusions (2) and (3) of theorem DG may be generalized to *any* partial positive definite matrix (independent of pattern) and proven in a simple way using optimization. This was first noted in [GJSW].

THEOREM 1. *Let H be a partial positive definite matrix over $\mathbf{C}$ all of whose diagonal entries are specified, and suppose that H has a positive definite completion. Then, among all positive definite completions of H there is a unique one with maximum determinant. Among positive definite completions the determinant maximizer is characterized by having a zero entry in the **inverse** in every position in which H has an unspecified entry.*

Proof Outline. Consider the set of all positive *semi*definite completions of H. This set is nonempty (because of the assumption that H has a positive definite completion), closed and bounded (no off-diagonal entry can be too large because the diagonal entries are specified and every 2-by-2 principal submatrix must be positive semidefinite). In fact it is also convex. Since det is a continuous function on this compact set, the maximum of det is attained on it by Weierstrass' theorem. Any maximizing matrix must be in the (relative) interior of this set, because the boundary consists of positive semidefinite matrices that are not positive definite and therefore have determinant 0 (and there are matrices with positive determinant, in the interior, per assumption of the existence of positive definite completions).

The key to both uniqueness and the inverse 0-pattern is the not sufficiently well known fact that the determinant function is strictly log concave over the positive definite matrices. This may be proven via simultaneous diagonalization by congruence of two positive definite matrices, as in [HJ], which reduces the question to the diagonal case. Since strictly log concave functions behave as strictly concave functions with respect to optimization, we conclude that there must be a *unique* determinant maximizing positive definite completion, which must occur at a (*unique*) critical point (because it lies in the interior of the constraint set). To see what it means to be a critical point we differentiate the determinant with respect to an unspecified entry h_{ij} of H:

$$\frac{\partial}{\partial h_{ij}} \det H = (-1)^{i+j} 2 \det H(i/j),$$

in which $H(i/j)$ denotes the submatrix of H resulting from deleting row i and column j. Thus,

$$\det H(i/j) = 0$$

for the determinant maximizing completion if the i,j entry is unspecified. But in view of the cofactor version for the inverse and the symmetry of H, we must have

$$H_{ij}^{-1} = H_{ji}^{-1} = 0 \ ,$$

as $\det H(i,j) = 0$ is the numerator of H_{ji}^{-1} (and H^{-1} exists for any positive definite completion). This explains the unique *inverse* zero pattern of the unique determinant maximizing completion. Strictly speaking, this argument is valid only in the case of real partial positive definite matrices (in which case it produces a real solution), but can be adjusted, as in [GJSW], to cover the complex case. □

Before discussing existence of positive definite completions, we mention an example that shows that not all patterns for the specified entries of a partial positive definite matrix insure a positive definite completion. Thus, the answer to the combinatorial positive definite completion question lies somewhere between the banded (or block banded) patterns of [DG1] and all patterns.

Consider the partial positive *semi*definite matrix

$$H = \begin{bmatrix} 1 & 1 & ? & -1 \\ 1 & 1 & 1 & x \\ ? & 1 & 1 & 1 \\ -1 & \overline{x} & 1 & 1 \end{bmatrix}$$

with an unspecified entry in the 1,3 and 3,1 positions and another (denoted by x) in the 2,4 and 4,2 positions. It is a simple exercise that the patterns that insure completability to a positive semidefinite matrix of a partial positive semidefinite matrix are the same as those that insure completability of a partial positive definite matrix to a positive definite. In any even we consider this partial positive semidefinite example for simplicity, but it could be converted to a partial positive definite one with the same point by adding small positive ϵ's to the diagonal entries.

In order to complete H to a positive semidefinite matrix, x must be chosen so that every principal submatrix in which it lies is positive semidefinite. Notice that

x completes two different maximal 3-by-3 principal submatrices of H, the ones lying in rows and columns 2,3,4 and 1,2,4:

$$\underset{\{2,3,4\}}{\begin{bmatrix} 1 & 1 & x \\ 1 & 1 & 1 \\ \bar{x} & 1 & 1 \end{bmatrix}} \quad \text{and} \quad \underset{\{1,2,4\}}{\begin{bmatrix} 1 & 1 & -1 \\ 1 & 1 & x \\ -1 & \bar{x} & 1 \end{bmatrix}} .$$

A simple calculation, using the one variable case, shows that for the former to be positive semidefinite we must have $x = 1$ and for the latter we must have $x = -1$. Not both these objectives may be achieved simultaneously. The problem, of course, is that x completes more than one different maximal principal submatrix and this happens to impose conflicting requirements upon x. It would not have helped to focus upon the unspecified entry.

In order to discuss the existence of positive definite completions and describe a complete solution to the combinatorial positive definite completion problem we need a brief digression into relevant graph theory. First of all, it is natural to describe the *pattern* of the specified (and, therefore, unspecified) entries of an n-by-n partial Hermitian matrix $H = (h_{ij})$ with an undirected (because h_{ij} is specified when h_{ji} is) graph on n vertices. In this graph we place an edge joining vertex i and vertex j if and only if h_{ij} is specified. Because the diagonal entries are all assumed specified, we ignore loops at the vertices. For example, the graph associated with the partial Hermitian matrix

$$\begin{bmatrix} 5 & 2 & ? & 1 & -1 \\ 2 & 6 & ? & 2 & ? \\ ? & ? & 3 & i & ? \\ 1 & 2 & -i & 4 & ? \\ -1 & ? & ? & ? & 6 \end{bmatrix}$$

is

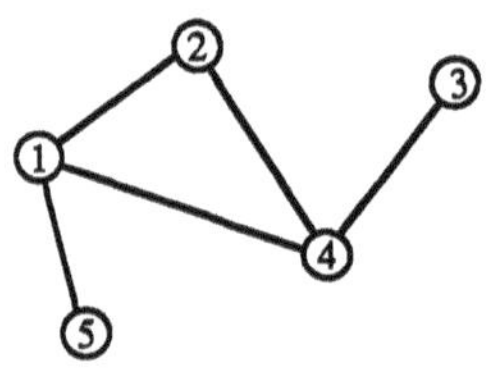

In any undirected graph, we say that 2 vertices are *adjacent* if they are joined by an edge. A *path* just a sequence of edges concatenated at vertices (i.e. a sequence of adjacent vertices);

a *circuit* is a path with an "initial" and "terminal" vertex that coincide;

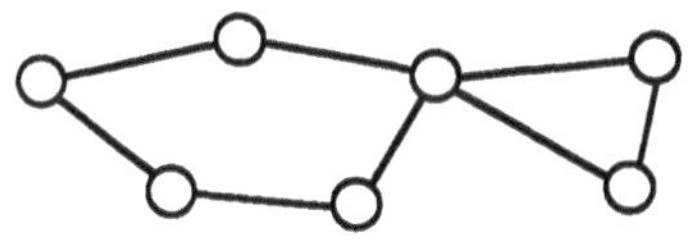

and a *simple circuit* is one that has no other self intersections

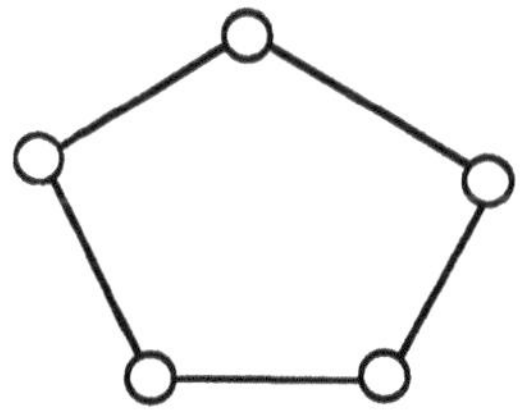

A *chord* of a circuit is just an edge joining two vertices (of the circuit) that are not adjacent in the circuit. A simple circuit is *minimal* if no proper subset of its vertices are themselves the vertices of a circuit. (Equivalently, the graph contains no chords for the circuit.) Note that an edge by itself does not constitute a circuit. The key notion is that of a chordal graph.

DEFINITION. An undirected graph is called *chordal* if it has no minimal simple circuits of length (measured in number of edges) 4 or more. Equivalently, any simple circuit of length at least 4 must have a chord.

Chordal graphs turn out to be a very nicely structured, heavily studied class of graphs that arise in many ways. For example, they have previously arisen in numerical linear algebra in the analysis of Gaussian elimination; the usual undirected graph of a (combinatorially symmetric) "sparse" matrix is chordal if and only if the calculations in Gaussian elimination may be ordered so as to avoid any intermediate fill-in (independent of the numerical values of the nonzero entries). A standard reference is [G]. Chordal graphs are also nice from the point of view of computation. Most tasks may be carried out algorithmically in low polynomial order for chordal graphs, including many that are known to be difficult or *np*-complete for general graphs. Chordal graphs also form quite a large class that includes the "trees" and the graphs from banded (and block banded) pattern but also much more. The simplest example of a chordal graph that is not associated with a (block) banded pattern for any numbering of the vertices is the tree

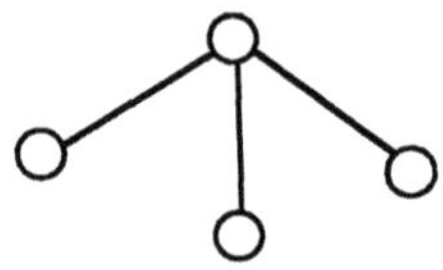

See [JR3] for a detailed discussion of where the block banded patterns sit within the chordal graphs.

We return to a few remaining graph theoretic concepts. An (undirected) graph is called *complete* if every pair of vertices is adjacent. An (induced) *subgraph* of a given graph is simply that graph resulting from retaining all edges both of whose vertices are contained in some selected subset of the vertices. A *clique* of a graph is an induced subgraph that is complete, and a clique is called *maximal* if its vertices do not constitute a proper subset of a clique. A (vertex) *separator* of two vertices i, j of a given graph is an induced subgraph whose removal leaves i an j but no path joining them; an i, j-separator is called *minimal* if it does not properly contain an i, j-separator.

It is now informative to return to the example partial positive semidefinite matrix that did not have a positive semidefinite completion:

$$H = \begin{bmatrix} 1 & 1 & ? & -1 \\ 1 & 1 & 1 & ? \\ ? & 1 & 1 & 1 \\ -1 & ? & 1 & 1 \end{bmatrix}$$

The graph of the specified entries of H is

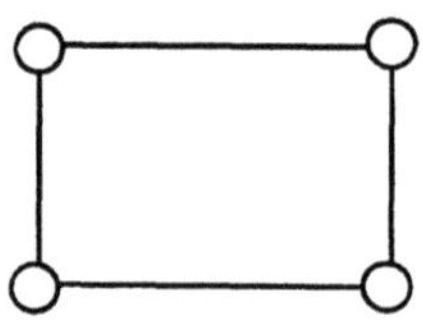

which is the simplest example of a nonchordal graph. When we tried to add the 2,4 entry (edge) and remain partial positive semidefinite

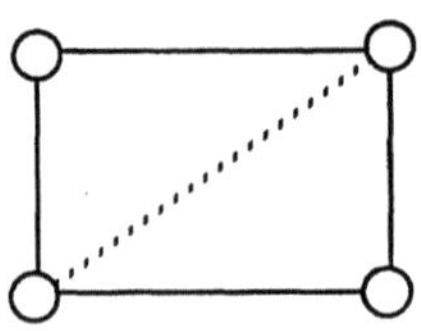

the two conflicting 3-by-3 principal submatrices corresponded to the two triangles

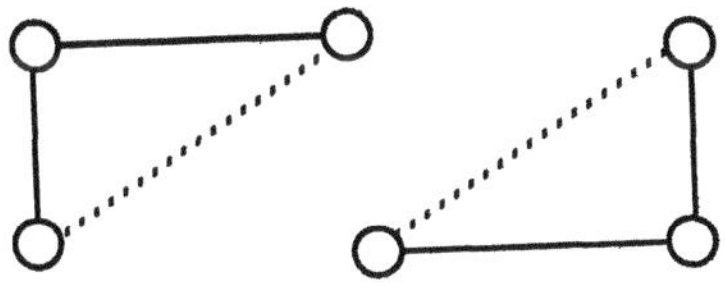

with the common (dotted) edge being a chord of the simple circuit of length 4. This is correctly suggestive of the general result proven in [GJSW].

THEOREM 2. *Every partial positive definite matrix with undirected graph G for its specified entries has a positive definite completion if and only if G is chordal.*

Proof Outline. Since any principal submatrix of a positive definite matrix is positive definite, any partial principal submatrix of a partial positive definite matrix (that has a positive definite completion) must itself have a positive definite completion. Since *non*chordal graphs are just those that have induced subgraphs consisting of just one "long" simple circuit, it suffices for necessity go give an example for each simple circuit of length $K \geq 4$ of a partial positive definite matrix that has not positive definite completion. This may be done in the same way as the 4-by-4 example we have discussed.

The key to sufficiency is to reduce the problem of finding a completion to a sequence of one-variable problems (already discussed). This may be done using a "chordal ordering" of the edges *missing* from G. It may be shown that for any chordal graph, the missing edges may be ordered so that their sequential addition results in a sequence of chordal graphs. Such an ordering is not unique (and there may be quite a few), but not every ordering will necessarily work. The addition of each such edge completes exactly one new maximal clique (principal submatrix) and gives a single one-variable problem to be solved. Its solution gives a new partial positive definite matrix with one more entry. Since the new pattern is chordal, this process may be repeated until its termination in a positive definite completion of the original partial positive definite matrix. It should be noted that this (nonunique) one-step-at-a-time procedure, which is so closely linked to chordality, is both conceptually and computationally simple. It is also a natural paradigm for other completion problems and has further uses in the positive definite case. □

We conclude this section on the positive definite problem by mentioning some related results and making several remarks,

In [JR1] the possible inertias (distribution of positive, negative and zero eigenvalues) of Hermitian completions of partial Hermitian matrices are studied. Again the inertias of the maximal specified principal submatrices impose a constraint, this time through the interlacing inequalities. Under a necessary technical assumption about nonvanishing of certain key principal minors (which was automatic in the partial positive definite case), all inertias allowed by interlacing can be achieved in the chordal case. It follows from the positive definite results that in the chordal case there is a completion whose minimum eigenvalue is equal to the smallest of the minimum eigenvalues of all specified principal submatrices (this, of course, the most it could be). The inertia results may be viewed as a generalization, but how large the minimum eigenvalue can be is an open question in the general (not necessarily chordal) case. Results somehow relating it to the graph structure and the smallest eigenvalue of any specified principal submatrix would be most welcome.

In [JR3] the case of partial matrices in which the set S is a ring of functions on a set x is considered. In the chordal case, if the partial matrix is partial positive definite for each element of X there is, of course, a positive definite completion point-wise. Conditions on the ring that allow a point-wise positive definite completion with functions from the ring are studied. It is likely that stronger results of this type could be obtained.

Several authors have sought "nice" descriptions of the set of all positive (semi-) definite completions of a partial positive definite matrix with given pattern. A natural strategy is to try to generalize the disc that occurs in the one variable case. Some examples include [GKW, W2] among others listed in the references. This continues to be an area of active research that generally involves the pattern of unspecified entries.

Finally, it is natural to ask what can be said about existence of positive definite completions beyond the chordal case. It is clear that if only partial positive definite is assumed the question becomes one of numerical values if the graph is not chordal. Two natural approaches may be suggested but neither has yet produced a fully satisfactory solution; so this is an area in which further work should be done.

First, since the chordal structure is so nice, one might write down the conditions associated with retaining partial positive definiteness as certain entries are added until a chordal pattern is achieved. Using the one-variable case and noting the *set* of maximal principal submatrices completed at each step, this may be done. The result is a list of "quadratic" (in each variable) inequalities that must be simultaneously satisfied. This gives a theorem that has two unfortunate drawbacks. First, minimal chordal supergraphs are not unique and, for a general graph, are not computationally easy to find. Second, in practice, feasibility of an arbitrary list of quadratic inequalities may be difficult to determine. More special structure of the problem needs to be exploited.

A second approach is to add to the list of "obvious" necessary conditions, hopefully until it becomes sufficient. The positive definiteness of specified principal submatrices is an especially obvious condition, but there are others. According to the Schur Product theorem [HJ], the Hadamard, or entry-wise, product of any two positive definite matrices A and B, denoted $A \circ B$, must be positive definite. In particular, if H is a partial Hermitian matrix and P is any positive definite matrix with 0's in all positions in which H has unspecified entries, then $H \circ P$, an ordinary matrix, must be positive definite if H is to have a positive definite completion. This adds many necessary conditions, some of which are surely redundant. For example, for the minimal nonchordal pattern

$$\begin{bmatrix} * & * & ? & * \\ * & * & * & ? \\ ? & * & * & * \\ * & ? & * & * \end{bmatrix}$$

it may be shown that only one new condition need be added to check for a positive semidefinite completion, namely positive semidefiniteness of Hadamard multiplication by

$$P = \begin{bmatrix} 1 & 1 & 0 & 1 \\ 1 & 2 & 1 & 0 \\ 0 & 1 & 1 & -1 \\ 1 & 0 & -1 & 2 \end{bmatrix}$$

(other choices for P also work). In particular, the example

$$H = \begin{bmatrix} 1 & 1 & ? & -1 \\ 1 & 1 & 1 & ? \\ ? & 1 & 1 & 1 \\ -1 & ? & 1 & 1 \end{bmatrix}$$

has no positive semidefinite completion because $\det(H \circ P) < 0$. It is not yet clear whether this approach can be pursued to produce a finite set of necessary and sufficient conditions for any given pattern. This is a focus of research with W. Barrett, and if so, it would provide a nicer solution than the one mentioned above.

III. Maximum Determinant/Maximum Entropy Completions. If a partial positive definite matrix has positive definite completions, the one with maximum determinant has attracted the most attention. (We know it is unique and has a striking inverse zero pattern by theorem 1.) This is due to an important role in applications and the attractive mathematical structure associated with it. The determinant maximizing completion is often called the *maximum entropy* completion because of its role in statistical entropy methods. Specific applications have come from seismic reconstruction and other missing data problems, image enhancement and the so-called model reduction problem among others. Discussion of some of these may be found among the references.

This author's interest in the determinant maximizing completion stems primarily from a long term interest in determinantal inequalities for positive definite matrices. Here we elaborate further on the determinant maximizer, formulae for its determinant and its role in determinantal inequalities. Many results still focus upon the chordal case in which the entries and determinant of the maximizing completion are rational in the specified entires. As before, we continue to assume that all diagonal entries of a partial positive definite matrix are specified.

We begin with a trivial observation and then show how it can be put to use. If the maximum determinant over all positive definite completions of a given partial positive definite matrix A may be expressed in terms of the specified entries, then a (sharp) upper bound for all determinants of positive definite matrices B that agree with A in its specified entries results. This idea may be used to show that either completion result (theorem DG or theorems 1 and 2) implies the classical determinantal inequality of Hadamard. Consider the simplest partial positive definite matrix

$$A = \begin{bmatrix} a_{11} & ? & \dots & ? \\ ? & a_{22} & \ddots & \vdots \\ \vdots & \ddots & \ddots & ? \\ ? & \dots & ? & a_{nn} \end{bmatrix}$$

in which only the diagonal entries $a_{11}, a_{22}, \dots a_{nn} > 0$ are specified. According to either completion result, there is a unique determinant maximizing completion of $\widehat{A}$ of A, and its *inverse* must be diagonal (all off-diagonal entries 0, because the off-diagonal entries of A are unspecified). But, if $\widehat{A}^{-1}$ is diagonal, then $\widehat{A}$ must be also, and it follows that

$$\widehat{A} = \begin{bmatrix} a_{11} & 0 & \dots & 0 \\ 0 & a_{22} & & \vdots \\ \vdots & & \ddots & 0 \\ 0 & \dots & 0 & a_{nn} \end{bmatrix}$$

is the determinant maximizing completion of A. Since $\det \widehat{A} = a_{11}a_{22}\cdots a_{nn}$, we have

$$\det B \leq a_{11}a_{22}\cdots a_{nn}$$

if B is any positive definite matrix with diagonal entries $a_{11}, \cdots, a_{nn}$. This is just Hadamard's inequality, and equality occurs exactly when B is diagonal.

More complicated examples require more sophisticated methodology to express the maximum determinant in terms of the data. As chance would have it, such methodology was being developed in [BJ1, BJ2, JB] independent of the work on completions. This work focussed upon simple formulae for the determinant of a nonsingular matrix A based upon some information about the 0-pattern of A^{-1}, and any 0-pattern could be at least partly exploited. To give an example we need some standard notation.

For index sets $\alpha \subseteq \{1, \cdots, m\}$ and $\beta \subseteq \{1, \cdots, n\}$, we denote the submatrix of an m-by-n (partial or ordinary) matrix lying in the rows indicated by α and the columns indicated by β as

$$A[\alpha, \beta] \ ,$$

Alternatively, the complementary submatrix in which the rows α and columns β are deleted is denoted by

$$A(\alpha, \beta) \ .$$

When A is square ($m = n$), we abbreviate the principal submatrices $A[\alpha, \alpha]$ to

$$A[\alpha]$$

and $A(\alpha, \alpha)$ to

$$A(\alpha).$$

A simple, but good, example of the formulae for $\det A$ given the 0-pattern of A^{-1} is the following. Suppose that $A^{-1} = B = (b_{ij})$ is tridiagonal, i.e. $b_{ij} = 0$ whenever $|i - j| > 1$. Then, if $A = (a_{ij})$,

$$\det A = \frac{\det A\big[\{1,2\}\big] \det A\big[\{2,3\}\big] \cdots \det A\big[\{n-1,n\}\big]}{a_{22}a_{33} \cdots a_{n-1,n-1}}$$

The numerator is the product of the 2-by-2 principal minors of A occurring consecutively down the main diagonal (there are $n-1$ of these) and the denominator is the product of the principal minors in which the numerator terms overall (the $n-2$ diagonal entries 2 through $n-1$, in this case), and all of this hinges upon the inverse B being tridiagonal. This is indicative of a fairly general, though more complicated theory developed in [BJ1, BJ2, JB]. It follows by the same logic used in our proof of Hadamard's inequality that for any n-by-n positive definite matrix $A = (a_{ij})$, we have the universal inequality

$$\det A \leq \frac{\det A\big[\{1,2\}\big] \det A\big[\{2,3\}\big] \cdots \det A\big[\{n-1,n\}\big]}{a_{22}a_{33} \cdots a_{n-1,n-1}}$$

Moreover, equality occurs in this inequality exactly when A^{-1} is tridiagonal, and given the diagonal and sub-and super-diagonal, there is a unique matrix that achieves equality. The reasoning is this. Consider an n-by-n positive definite matrix A. Forget the entries outside the sub-and superdiagonals. The resulting partial positive definite matrix has positive definite completions. The (unique) determinant maximizing one has tridiagonal inverse (theorem 1), but the determinant of this one (which bounds all others) is as indicated $\big($[BJ1]$\big)$. This idea was used to exhibit very broad generalizations of the so-called Hadamard-Fischer inequalities in [JB],

Instead of summarizing the ideas of [BJ1, BJ2], we describe three different ways of describing the maximum determinant over all positive definite completions of a given partial positive definite matrix in the most general case in which we know such completions exist, namely when the graph of the specified entries is chordal. Though we do not pursue it here, bounds for the maximum determinant may be given in general, using chordal supergraphs. Simple, exact descriptions of the maximum determinant for a general partial positive definite matrix that has positive definite completions are yet to be found. It is clear that they are no longer rational in the specified entries, and it may be difficult to give such a nice description.

It should be kept in mind that any formula for the maximum determinant over positive definite completions gives another general determinantal inequality for positive definite matrices.

In order to proceed, we need a few more graph theoretical ideas. A *tree* is a connected (undirected) graph on n vertices that has the fewest possible edges, namely $n - 1$. A *spanning tree* T of a connected graph G is simply a tree on the same vertices as G whose set of edges is a subset of that of G. Given a family of subsets $C_1, C_2, \cdots, C_m \subseteq \{1, 2, \cdots, n\}$, the ***intersection graph*** of this family is just the graph whose vertices are $C_1, \cdots, C_m$ with an (undirected) edge between C_i and C_j if and only if $C_i \cap C_j \neq \emptyset$. We say that a spanning tree of the intersection graph of $C_1, C_2, \cdots, C_m$ satisfies the ***intersection property*** if whenever C_k lies on the (unique) path between C_i and C_j, then $C_i \cap C_j \subseteq C_k$. A natural intersection graph to associate with a given graph G is the one in which $C_1, \cdots, C_m$ are the vertex sets of the maximal cliques of G. We call this graph $G_{\mathcal{C}}$. As with much of the development of completion problems for positive definite matrices, further understanding of the structure of chordal graphs is useful. An important idea is the following fact, which may be found in [BJL].

THEOREM 3. *Let G be a connected, undirected graph and let $G_{\mathcal{C}}$ be the intersection graph of the maximal cliques of G. Then, there is a spanning tree of $G_{\mathcal{C}}$ satisfying the intersection property if and only if G is chordal.*

As with the chordal orderings of the missing edges of a chordal graph, the spanning tree guaranteed by theorem 3 need not be unique. However, the important expressions we derive from it are unique.

We may now mention the first of three (theorems 4, 5, 6) presentations of the maximal determinant for completions of a partial positive definite matrix, the graph of whose specified entries is chordal. For further details and links with determinantal formulae and inequalities, see [BJ1, BJ2, BJL, JB].

THEOREM 4. *Let G be a connected, chordal graph and suppose A is a partial positive definite matrix, the graph of whose specified entries is G. The maximum determinant occurring among all positive definite completions of A is then given by*

$$\frac{\prod_{k=1}^{m} \det A[C_k]}{\prod \det A[C_i \cap C_j]}$$

Here, $C_1, \cdots, C_m$ are the maximal cliques of G, and for any spanning tree T of the intersection graph $G_{\mathcal{C}}$ of the maximal cliques that satisfies the intersection property, the product in the deonominator is taken over all pairs i, j such that $\{C_i, C_j\}$ is an edge of T. Furthermore, there is a unique positive definite completion of A that attains this maximum, and the undirected graph of its inverse is contained in G. This completion is the only one with the latter property.

The example given earlier in this section involving a tridiagonal inverse is the special case of theorem 3 in which G is just a path from 1 to n.

Since virtually all computational tasks associated with chordal graphs may be carried out efficiently (including those necessary for theorem 4), theorem 4 does provide a computationally effective, as well as theoretically useful, description of the maximum determinant. The denominator terms turn out to be the minimal vertex separators, and, together with their multiplicities (as well as the numerator terms, which are maximal cliques) they are unique, even though the necessary spanning tree is not. However, an alternate presentation that requires no "external" (such as the special spanning tree) is a natural goal. It turns out that there is one that is attractive by its elementary nature. It has the flavor of inclusion/exclusion. If we begin, as in theorem 4 with numerator terms corresponding to maximal cliques and then divide by terms corresponding to *all* pairwise intersections, we have put too much in the denominator. If we pay back by putting all 3-way intersections in the numerator, we have gone too far on top. Continuing with odd order intersections in the numerator and even order in the denominator turns out to terminate with a correct ratio. It can be shown that in the case of a chordal pattern, cancellation occurs to leave just the expression given in theorem 4.

THEOREM 5. *Let G be a connected chordal graph and suppose A is a partial positive definite matrix, the graph of whose specified entries is G. Let $C_1, \cdots, C_m$ be the index sets of the maximal cliques of G. The maximum determinant occurring among all positive definite completions of A is then given by*

$$\frac{\prod\limits_{k \text{ odd}} \prod\limits_{|S|=k} \det A\left[\bigcap\limits_{i \in S} C_i\right]}{\prod\limits_{k \text{ even}} \prod\limits_{|S|=k} \det A\left[\bigcap\limits_{i \in S} C_i\right]}.$$

It is useful to illustrate theorem 5 with an example that relates it to theorem 4 by exhibiting the cancellation pattern that reduces the expression of the former to the latter.

Example. Consider the chordal graph

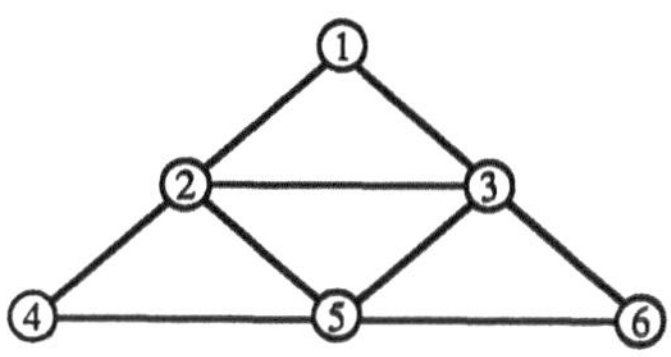

with maximal cliques $C_1 = \{1,2,3\}, C_2 = \{2,3,5\}, C_3 = \{2,4,5\}$ and $C_4 = \{3,5,6\}$. Then, in this case the expression of theorem 4 is

$$\frac{\prod\limits_{k \text{ odd}} \prod\limits_{|S|=k} \det A\left[\bigcap\limits_{i \in S} C_i\right]}{\prod\limits_{k \text{ even}} \prod\limits_{|S|=k} \det A\left[\bigcap\limits_{i \in S} C_i\right]}.$$

$$= \frac{\det A[C_1] \cdot \det A[C_2] \cdot \det A[C_3] \cdot \det A[C_4]}{\det A|C_1 \cap C_2] \cdot \det A[C_1 \cap C_3] \cdot \det A[C_1 \cap C_4] \cdot \det A[C_2 \cap C_3] \cdot \det A[C_2 \cap C_4] \cdot \det .}$$

$$\cdot \frac{\det A[C_1 \cap C_2 \cap C_3] \cdot \det A[C_1 \cap C_2 \cap C_4] \cdot \det A[C_1 \cap C_3 \cap C_4]. \det A[C_2 \cap C_3 \cap C}{\det A|C_1 \cap C_2 \cap C_3 \cap C_4]}$$

$$= \frac{\det A[C_1] \cdot \det A[C_2] \cdot \det A[C_3] \cdot \det A[C_4] \cdot \det A[\{2\}] \cdot \det A[\{3\}] \cdot \det A[\phi] \cdot \det A[\{5}{\det A[\{2,3\}] \cdot \det A[\{2\}] \cdot \det A[\{3\}] \cdot \det A[\{2,5\}] \cdot \det A[\{3,5\}] \cdot \det A[\{5\}] \cdot \det A[\phi}$$

$$= \frac{\det A[C_1] \cdot \det A[C_2] \cdot \det A[C_3] \cdot \det A[C_4]}{\det A[C_1 \cap C_2] \cdot \det A[C_2 \cap C_3] \cdot \det A[C_2 \cap C_4]}$$

The latter is the expression of theorem 5 corresponding to the unique (in this case) spanning tree of $G_{\mathcal{C}}$ satisfying the intersection property

$$\begin{array}{c} C_1 \\ | \\ C_2 \\ / \quad \backslash \\ C_3 \quad C_4 \end{array}$$

While there is great redundancy among numerator and denominator terms in theorem 5, this expression has the virtue of being independent of the tree T; only a knowledge of the set of maximal cliques is needed.

A feature that the presentations of theorems 4 and 5 lack is that they do not explicitly describe the entries of the determinant maximizing completion. A third, and perhaps even more elementary description of the maximum determinant, is a surprisingly simple algorithm that actually produces the maximizing matrix in the chordal case. The very nice surprise is that the sequence of one variable determinant maximizing problems associated with a chordal ordering of the missing edges actually produces the global maximum. This aspect of chordal structure likely has further implications in continuous optimization as well as those that are recognized in discrete optimization. That this local/global property holds for *some* chordal orderings was first noticed in [JR3] and that it actually holds for *any* chordal ordering was shown in [BJL].

THEOREM 6. *Let G be a chordal graph and let A be a partial positive definite matrix the graph of whose specified entries is G. Let F be a chordal graph obtained by adding one edge u, v to G and let C be the unique maximal clique of F containing the new edge. If $L = (l_{ij})$ is the unique determinant maximizing positive definite completion of $A[C]$ and $M = (m_{ij})$ is the unique determinant maximizing completion of A, then*

$$l_{uv} = m_{uv} \ .$$

It follows that the global maximum M may be found by a sequence of problems of one variable type (described in detail in section II). This has important implications. In the chordal case not only the maximum determinant but also the entries of the maximizing completion are rational in the specified entries. This means that if S is the real or rational field, instead of the complex field, the determinant maximizing completion will remain in the field. This does not remain valid beyond the chordal case.

Example [JB]. Consider the partial positive definite matrix

$$\begin{bmatrix} 2 & -1 & x & -1 \\ -1 & 2 & -1 & y \\ \overline{x} & -1 & 2 & -1 \\ -1 & \overline{y} & -1 & 2 \end{bmatrix}$$

Of course, here the graph of the specified entries is not chordal. Calculation based upon theorem 2 yields that the maximum determinant is $24\sqrt{3}\ - 36$, which is attained at the completion $x = y = \sqrt{3} - 1$, neither of which is not rational in the data.

It is, however, a consequence of uniqueness (theorem 2) that real data results in a real-entried determinant maximizing matrix, regardless of pattern.

See [JB] for more details about the generality of determinantal inequalities for positive definite matrices resulting from the chordal case.

Acknowledgements. The author gratefully acknowledges helpful conversations with Wayne Barrett, Israel Koltracht, Hanoch LevAri, Michael Lundquist, Leiba Rodman and Douglas Shier during the preparation of this manuscript.

REFERENCES

[AHMR] J. AGLER, W. HELTON, S. MCCULLOUGH AND L. RODMAN, *Positive Semidefinite Matrices with a Given Sparsity Pattern*, Linear Algebra Appl. 107 (1988), 101-149.

[B] J.P. BURG, *Maximum Entropy Spectral Analysis*, Ph.D. Thesis (Dept. of Geophysics), Stanford University, Stanford, CA (1975).

[BEGL] A. BEN-ARTZI, R.L. ELLIS, I. GOHBERG AND D.C. LAY, *The Maximal Distance Problem and Band Sequences*, Linear Algebra Appl. 87 (1987), 93–112.

[BF] W. BARRETT AND P. FEINSILVER, *Inverse of Banded Matrices*, Linear Algebra Appl. 41 (1981) 111–130.

[BJ1] W. BARRETT AND C.R. JOHNSON, *Determinantal Formulae for Matrices with Sparse Inverses*, Linear Algebra Appl. 56 (1984), 73–88.

[BJ2] W. BARRET AND C.R. JOHNSON, *Determinantal Formulae for Matrices with Sparse Inverses, II: Asymmetric Zero Patterns*, Linear Algebra Appl. 81 (1986), 237–261.

[BJL] W. BARRETT, C.R. JOHNSON AND M. LUNDQUIST, *Determinantal Formulae for Matrix Completions Associated with Chordal Graphs*, Linear Algebra Appl., (to appear).

[BJOV] W. BARRET, C.R. JOHNSON, D. OLESKY AND P. VAN DEN DRIESSCHE, *Inherited Matrix Entries: Principal Submatrices of the Inverse*, SIAM J. Alg. Disc. Meth. 8 (1987), 313–322.

[CJRW] N. COHEN, C.R. JOHNSON, L. RODMAN AND H. WOERDEMAN, *Ranks of Completions of Partial Matrices*, submitted.

[D] C. DAVIS, *Completing a Matrix so as to Minimize its Rank*, Operator Theory: Advances and Appl. 29 (1988), 87–95.

[DG1] H. DYM AND I. GOHBERG, *Extensions of Band Matrices with Band Inverses*, Linear Algebra Appl. 36 (1981), 1–24.

[DG2] H. Dym, I. Gohberg, *A new class of contractive interpolents and maximum entropy principles*, preprint.

[DKW] C. Davis, W.M. Kahan and H.F. Weinberger, *Norm-preserving Dilations and their Applications to Optimal Error Bounds*, SIAM J. Numerical Anal. 19 (1982), 445-469.

[EGL1] R.L. Ellis, I. Gohberg and D.C. Lay, *Band Extensions, Maximum Entropy and the Permanence Principle*, in *Maximum Entropy and Bayesian Methods in Applied Statistics*. ed. J. Justice. Cambridge University Press, Cambridge (1986)

[EGL2] R. Ellis, I. Gohberg and D. Lay, *Invertible Self-adjoint Extensions of Band Matrices and their Entropy*, SIAM J. Alg. Disc. Meth. 8 (1987), 483–500.

[EGL3] R. Ellis, I. Gohberg and D.C. Lay, *On Negative Eigenvalues of Self-adjoint Extensions of Band Matrices*, submitted.

[G] M. Golumbic, *Algorithmic Graph Theory and Perfect Graphs*, Academic Press, New York (1980).

[GJSW] R. Grone, C.R. Johnson, E. Sá and H. Wolkowicz, *Positive Definite Completions of Partial Hermitian Matrices*, Linear Algebra Appl. 58 (1984).

[GKW] I. Gohberg, M.A. Kaashoek and H. Woerdeman, *The Band Method for Positive and Contractive Extension Problems*, preprint from the Summer Institute in Operator Theory, Operator Algebras and Applications, Durham, N.H., July (1988).

[GR] I. Gohberg, S. Rubinstein, *Proper Contractions and their Unitary Minimal Completions, Operator Theory: Advances and Applications vol. 33.*

[HJ] R. Horn and C.R. Johnson, *Matrix Analysis*, Cambridge University Press, N.Y. (1985).

[HL] D. Hartfiel and R. Loewy, *A Determinant Version of the Frobenius-König Theorem*, Lin. and Multilin. Alg. 16 (1984), 155–165.

[J] C.R. Johnson, *Optimization, Matrix Inequalities and Matrix Completions*, in *Operator Theory, Analytic Functions, Matrices and Electrical Engineering*. ed. J. William Helton. CBMS Regional Conference Series in Mathematics 68, American Mathematical Society, Providence, RI (1987)

[JB] C.R. Johnson and W. Barrett, *Spanning Tree Extensions of the Hadamard-Fischer Inequalities*, Linear Algebra Appl. 66 (1985), 177–193.

[JR1] C.R. Johnson and L. Rodman, *Inertia Possibilities for Completions of Partial Hermitian Matrices*, Linear and Multilinear Algebra 16 (1984), 179–195.

[JR2] C.R. Johnson and L. Rodman, *Completion of Partial Matrices to Contractions*, J. Func. Anal. 69 (1986), 260–267.

[JR3] C.R. Johnson and L. Rodman, *Chordal Inheritance Principles and Positive Definite Completions of Partial Matrices over Function Rings*, Birkhäuser-Verlag, Basel (1988), in *Contributions to operator Theory and its Applications.* Proc. of the Mesu Conference on Operator Theory and Functional Analysis

[JR4] C.R. Johnson and L. Rodman, *Completion of Toeplitz Partial Contractions*, SIAM J. Matrix Anal. Appl. 9 (1988), 195–167.

[KW] M.A. Kaashoek and H.J. Woerdeman, *Unique Minimal Rank Extensions of Triangular Operators*, J. Math Anal. Appl. 131 (1988) 501–516.

[NDD] H. Nelis, E. Deprettere and P. Dewilde, *Approximate Inversion of Positive Definite Matrices, Specified on a Multiple Band*, preprint (1988).

[P] S. Parrot, *On a Quotient Norm and St.-Nagy-Foias Lifting Theorem*, J. Funct. Anal. 30 (1978), 311–328.

[W1] H. Woerdeman, *The Lower Order of Lower Triangular Operators and Minimal Rank Extensions*, Integral Equations and Operator Theory 10 (1987), 859–879.

[W2] H. Woerdeman, *Strictly Contractive and Positive Definite Completions for Block Matrices*, Rapport WS-337, Urije Universiteit, Amsterdam, November (1987).

[W3] H. Woerdeman, *Minimal Rank Completions for Block Matrices*, Linear Algebra Appl., (to appear).

AN EXISTENCE THEOREM AND LATTICE APPROXIMATIONS FOR A VARIATIONAL PROBLEM ARISING IN COMPUTER VISION*

SANJEEV R. KULKARNI†, SANJOY MITTER‡, THOMAS J. RICHARDSON‡

Abstract. A variational method for the reconstruction and segmentation of images was recently proposed by Mumford and Shah [15]. In this paper we treat two aspects of the problem. The first concerns existence of solutions, and the second concerns discrete approximations. Discrete versions of this problem have been proposed and studied in [5,12,14,15]. However, it seems that these discrete versions do not properly approximate the continuous problem in the sense that their solutions may not converge to a solution of the continuous problem as the lattice spacing tends to zero. Thus, these discrete formulations in the limit fail to capture properties of the continuous formulation (such as rotation invariance).

Here we consider the use of an alternate lattice approximation for the boundaries of the image and Minkowski content as a cost term for the boundaries. Several properties of Minkowski content are derived. These are used to show that partially discrete versions of the variational problem possess some desirable convergence properties. Specifically, under suitable conditions, solutions to the discrete problem converge in the continuum limit to a solution of the continuous problem, thereby retaining (in the limit) the advantages of the continuous problem. We also present an existence result that is applicable to both discrete and continuous versions of the problem.

1. Introduction. A variational approach to the problem of reconstructing and segmenting an image degraded by noise was recently proposed by Mumford and Shah in [15] (see also Blake and Zisserman [4,5]). The method involves minimizing a cost functional over a space of boundaries with suitably smooth functions within the boundaries. Specifically, if g represents the observed image defined on $\Omega \subset \mathbf{R}^2$, then a reconstructed image f and its associated edges Γ are found by minimizing

$$E(f,\Gamma) = c_1 \iint_{\Omega} (f-g)^2 \, dx\, dy + c_2 \iint_{\Omega\setminus\Gamma} \| \nabla f \|^2 \; dx\, dy + c_3 L(\Gamma) \tag{1}$$

where c_1, c_2, c_3 are constants, $\|\cdot\|$ denotes the Euclidean norm and $L(\Gamma)$ denotes the length of Γ. An interesting special case of this problem is obtained if f is restricted to be constant within connected components of $\Omega\setminus\Gamma$. In this case, the optimal value of f on a connected component of $\Omega\setminus\Gamma$ is simply the mean of g over the connected component. Hence, the solution depends only on Γ and is obtained by minimizing

$$E(\Gamma) = c_1 \sum_{i=1}^{k} \iint_{\Omega_i} (g - \overline{g}_i)^2 \, dx\, dy + c_3 L(\Gamma) \tag{2}$$

where $\Omega_1, \ldots, \Omega_k$ are the connected components of $\Omega\setminus\Gamma$, and $\overline{g}_i$ is the mean of g over Ω_i.

---

*This research was supported in part by the U.S. Army Research Office, contract DAAL03-86-K-0171 (Center for Intelligent Control Systems) and by the Department of the Navy for SDIO.

†Center for Intelligent Control Systems, M.I.T., 35-423, Cambridge, MA, 02139 and M.I.T./Lincoln Laboratory, 244 Wood St., Lexington, MA 02173.

‡Center for Intelligent Control Systems, 35-308, M.I.T., Cambridge, MA, 02139.

Discrete versions of these problems have also been proposed [5,15]. In these discrete problems, the original image g is defined on a subset of the lattice $\frac{1}{n}\mathbf{Z}^2$ with lattice spacing $\frac{1}{n}$. The reconstructed image f is defined on the same lattice, while the boundary Γ consists of a subset of line segments joining neighboring points of the dual lattice. For the discrete problem, f and Γ are found by minimizing

$$E(f,\Gamma) = c_1 \sum_{i \in \Omega} \frac{1}{n^2}(f_i - g_i)^2 + c_2 \sum_{\substack{i,i' \in \Omega \\ \text{adjacent} \\ \overline{ii'} \cap \Gamma = \emptyset}} (f_i - f_{i'})^2 + c_3 L(\Gamma) \tag{3}$$

Similar discrete problems arise in the context of using Markov random fields for problems in vision as proposed by Geman and Geman [12] and studied by Marroquin [14] and others.

The continuous formulation has some distinct advantages over the discrete formulation. For example, the continuous problem is invariant under arbitrary rotations and translations. Also, results from the calculus of variations can be applied in the continuous case. In fact, such methods have yielded interesting results concerning the properties of the minimizing f and Γ [16,22,23]. However, since analytic solutions are not available, the problem must eventually be digitized to obtain numerical solutions. The discrete problem has the advantages of being more directly amenable to computer implementations, particularly with parallel algorithms or hardware. A desirable property of any discrete version of a continuous problem would be for solutions of the discrete problem to converge to solutions of the continuous problem in the continuum limit. In the examples above, one would like convergence of the discrete solutions as the lattice spacing tends to zero. It seems that this is not the case for the problems as defined above. Thus, these discrete formulations in the limit fail to capture properties of the continuous formulation. In particular, the discrete problem in the limit is generally not rotationally invariant, and the analytical results concerning solutions to the continuous problem are not applicable.

In this paper we consider modifications to both the cost functional and the discretization procedure which ensure convergence in the continuum limit. For the cost functional, we propose the use of Minkowski content as the penalty term for the boundaries instead of Hausdorff measure which has been previously used [1,2,17]. For the discretization procedure, we consider only digitizing the boundary. The observed and reconstructed images are still defined on continuous domains. Also, the discrete boundary consists of a union of closed lattice squares rather than a union of line segments. In Section 2 we introduce some preliminary definitions and results from geometric measure theory, and in Section 3 some additional properties of Minkowski content are derived. Section 4 gives an existence result applicable to the problems of interest and Section 5 contains results on the application of these ideas to the variational problem.

2. Metrics and Measures on the Space of Boundaries. In this section we introduce a variety of notions useful in dealing with the 'boundaries' or 'edges' of an image. The 'image' is usually a real valued function defined on a bounded open set $\Omega \subset \mathbf{R}^2$, although some of the results consider the more general case of $\Omega \subset \mathbf{R}^n$. A *boundary* generally refers to a closed subset of $\overline{\Omega}$. However, sometimes the boundary may be restricted to have certain additional properties such as having a finite number of connected components. A topology on the space of boundaries is required for the notion of convergence, and a measure of the 'cost' of a boundary is required for the variational problem.

For $A \subset \mathbf{R}^n$, the *δ-neighborhood* of A will be denoted by $A^{(\delta)}$ and is defined as

$$A^{(\delta)} = \{x \in \mathbf{R}^n : \inf_{y \in A} \|x - y\| < \delta\}$$

The notion of distance between boundaries which we will use is the Hausdorff metric $d_H(\cdot, \cdot)$ defined as

$$d_H(A_1, A_2) = \inf\{\rho : A_1 \subset A_2^{(\rho)} \text{ and } A_2 \subset A_1^{(\rho)}\}$$

It is elementary to show that $d_H(\cdot, \cdot)$ is in fact a metric on the space of all non-empty compact subsets of $\mathbf{R}^n$. An important property of this metric is that it induces a topology which makes the space of boundaries compact.

THEOREM 1. *Let $\mathcal{C}$ be an infinite collection of non-empty closed subsets of a bounded closed set $\overline{\Omega}$. Then there exists a sequence $\{\Gamma_n\}$ of distinct sets of $\mathcal{C}$ and a non-empty closed set $\Gamma \subset \overline{\Omega}$ such that $\Gamma_n \to \Gamma$ in the Hausdorff metric.*

Proof: See [10], Theorem 3.16. □

For the 'cost' of a boundary, the usual notion of length cannot be applied to highly irregular boundaries. Hence a measure on the space of boundaries which generalizes the usual notion of length is desired. A variety of such measures for subsets of $\mathbf{R}^n$ have been investigated. (e.g., see [10]). Perhaps the most widely used and studied are Hausdorff measures [10,11,19].

For a non-empty subset A of $\mathbf{R}^n$, the *diameter* of A is defined by $|A| = \sup\{\|x - y\| : x, y \in A\}$. Let

$$\omega_s = \frac{\Gamma(\frac{1}{2})^s}{\Gamma(\frac{s}{2} + 1)}$$

where $\Gamma(\cdot)$ is the usual Gamma function. For integer values of s, ω_s is the volume of the unit ball in $\mathbf{R}^s$. For $s > 0$ and $\delta > 0$ define

$$\mathcal{H}^s_\delta(A) = 2^{-s}\omega_s \inf\{\sum_{i=1}^{\infty} |U_i|^s : A \subset \bigcup_{i=1}^{\infty} U_i, \; |U_i| \leq \delta\}$$

The *Hausdorff s-dimensional measure* of A is then given by

$$\mathcal{H}^s(A) = \lim_{\delta \to 0} \mathcal{H}^s_\delta(A) = \sup_{\delta > 0} \mathcal{H}^s_\delta(A)$$

Note that the factor $2^{-s}\omega_s$ in the definition of $H^s_\delta(\cdot)$ is included for proper normalization. With this definition, for integer values of s Hausdorff measure gives the desired value on sets where the usual notions of length, area, and volume apply.

Many properties of Hausdorff measure can be found in [10,11,19]. The following definitions are required to state several useful properties. A *curve* $\Gamma \subset \mathbf{R}^n$ is the image of a continuous injection $g : [0,1] \to \mathbf{R}^n$. The *length* of a curve Γ is defined as

$$L(\Gamma) = \sup\{\sum_{i=1}^{m} \|g(t_i) - g(t_{i-1})\| \ : \ 0 = t_0 < t_1 < \cdots < t_m = 1\}$$

and Γ is said to be *rectifiable* if $L(\Gamma) < \infty$. Finally, a compact connected set is called a *continuum*.

THEOREM 2. *If $\Gamma \subset \mathbf{R}^n$ is a curve, then $\mathcal{H}^1(\Gamma) = L(\Gamma)$.*

Proof: See [10] Lemma 3.2. □

THEOREM 3. *If Γ is a continuum with $\mathcal{H}^1(\Gamma) < \infty$, then Γ consists of a countable union of rectifiable curves together with a set of $\mathcal{H}^1$-measure zero.*

Proof: See [10], Theorem 3.14. □

THEOREM 4. *If $\{\Gamma_n\}$ is a sequence of continuua in $\mathbf{R}^n$ that converges (in Hausdorff metric) to a compact set Γ, then Γ is a continuum and $\mathcal{H}^1(\Gamma) \leq \liminf_{n\to\infty} \mathcal{H}^1(\Gamma_n)$.*

Proof: See [10], Theorem 3.18. □

Theorem 4 asserts that $\mathcal{H}^1$-measure is lower-semicontinuous on the set of connected boundaries with respect to the Hausdorff metric. In what follows we extend this result to a cost term for boundaries which depends on the number of connected components. Specifically, we define $\nu(\Gamma) = \mathcal{H}^1(\Gamma) + F(\#(\Gamma))$ where $\#(\Gamma)$ denotes the number of connected components of Γ, and F is any non-decreasing function such that $\lim_{n\to\infty} F(n) = \infty$.

THEOREM 5. *$\#(\cdot)$ and $\nu(\cdot)$ are lower-semicontinuous on the space of boundaries with respect to the Hausdorff metric.*

Proof: We will use the following notation,

$$\mathrm{r}(A_1, A_2) = \sup_{x\in A_1} \inf_{y\in A_2} \|x - y\|$$

Suppose $\Gamma_n \to \Gamma$. First we show $\#(\cdot)$ is a lower-semicontinuous function on the space of boundaries. Assume $\#(\Gamma) = c < \infty$. There exists an open cover of Γ consisting of c disjoint open sets $G_1, G_2, \ldots, G_c$ such that $\Gamma \cap G_i \neq \emptyset$, $\forall i$. Γ is closed so $\exists \delta > 0$ such that $\forall i$, $\mathrm{r}(\Gamma \cap G_i, \mathbf{R}^2 \backslash G_i) > \delta$. Since $\mathrm{r}(\Gamma, \Gamma_n) \to 0$, for n sufficiently large $\Gamma_n \subset \cup_i G_i$ and $\Gamma_n \cap G_i \neq \emptyset$. Thus $\liminf_{n\to\infty} \#(\Gamma_n) \geq c$. If $\#(\Gamma) = \infty$ then we can repeat this argument for any c and the result follows.

Now we proceed to show $\nu(\Gamma) \leq \liminf_{n\to\infty} \nu(\Gamma_n)$. Assume (without loss of generality) that $\{\nu(\Gamma_n)\} \leq K$, for some $K < \infty$. It follows that $\#(\Gamma_n)$ is uniformly bounded, by $M < \infty$ say, and by the result above, $\#(\Gamma) \leq M$. Since the connected components of Γ are thus separated pairwise by some finite distance the result follows once we show it for connected Γ.

Assume Γ is connected. Let $\delta_n = d_H(\Gamma_n, \Gamma)$. Suppose Γ_n has more than one connected component and let C be one connected component of Γ_n. If for some $\epsilon > 0$, $d(C, \Gamma_n \backslash C) = 2(\delta_n + \epsilon)$, then $\{x : d(x, C) < \delta_n + \epsilon\}$ and $\{x : d(x, \Gamma_n \backslash C) < \delta_n + \epsilon\}$ are two disjoint open sets both containing points of Γ and whose union covers Γ. This contradicts the connectedness of Γ. Thus we can find $x \in C$ and $y \in \Gamma_n \backslash C$ such that $\|x - y\| \leq 2\delta_n$. Consider the straight line segment from x to y. It connects C to some other connected component of Γ_n. Since C was an arbitrary connected component of Γ_n we can find a similar straight line segment from each connected component of Γ_n joining it to some other component. Now if we add all the line segments to Γ_n the number of connected components is reduced to $M/2$ or fewer, the Hausdorff measure will increase by at most $2M\delta_n$ and we will have $d_H(\Gamma_n, \Gamma) \leq 2\delta_n$. Let p be the smallest integer such that $2^p \geq M$, then by repeating the above argument p times we get a modified, connected Γ_n such that its Hausdorff measure is at most $(2pM)\delta_n$ larger than before and $d_H(\Gamma_n, \Gamma) \leq 2^p\delta_n$. Thus the modified Γ_n still converge to Γ and since they are connected we can apply Theorem 4 to get, in terms of the original sequence

$$\mathcal{H}^1(\Gamma) \leq \liminf_{n\to\infty} \mathcal{H}^1(\Gamma_n) + 2pM\delta_n = \liminf_{n\to\infty} \mathcal{H}^1(\Gamma_n).$$

The first result implies lower semicontinuity of F and together with the above result we get lower semicontinuity of ν. □

In view of the fact that the $\mathcal{H}^1$ measure can be discontinuous on the space of boundaries with the topology induced by the Hausdorff metric, a discretization procedure for the continuous problem which will be convergent is not immediately attainable. Here we consider the use of an alternate notion for the cost of boundaries and a modified discretization process (discussed in Section 5).

To measure the cost of the boundaries, we suggest the use of Minkowski content [11]. Let $\mu(\cdot)$ denote Lebesgue measure in $\mathbf{R}^n$. For any $A \subset \mathbf{R}^n$, $0 \leq s \leq n$, and $\delta > 0$, define

$$\mathcal{M}^s_\delta(A) = \frac{\mu(A^{(\delta)})}{\delta^{n-s}\omega_{n-s}}$$

As in the definition of Hausdorff measure, the term ω_{n-s} is included for proper normalization. Recall that $A^{(\delta)}$ is the δ-neighborhood of A — i.e. those points within distance δ of A. Equivalently, $A^{(\delta)}$ is the Minkowski set sum of A and the open ball of radius δ; or in the terminology of mathematical morphology [21] it is the dilation of A with the open ball of radius δ. In general, $\lim_{\delta\to 0} \mathcal{M}^s_\delta(A)$ may not exist (for an example see [11], Section 3.2.40). However, *lower* and *upper Minkowski contents* can be defined by

$$\mathcal{M}^s_*(A) = \liminf_{\delta\to 0^+} \mathcal{M}^s_\delta(A)$$

and

$$\mathcal{M}^{*s}(A) = \limsup_{\delta \to 0+} \mathcal{M}^s_\delta(A)$$

respectively. If these two values agree (i.e if $\lim_{\delta\to 0} \mathcal{M}^s_\delta(A)$ exists) then the common value is simply called the *s-dimensional Minkowski content* and is denoted by $\mathcal{M}^s(A)$.

3. Properties of Minkowski Content. In this section we develop several properties of Minkowski content some of which will be used in Section 5. The results can roughly be categorized as properties of δ-neighborhoods, continuity and regularity properties of Minkowski content, and relationships between Minkowski content and Hausdorff measure.

First, we state two elementary properties. Two sets A_1, A_2 are said to be *positively separated* if

$$d(A_1, A_2) \equiv \inf\{\|a_1 - a_2\| : a_1 \in A_1, a_2 \in A_2\} > 0$$

The sets $A_1, A_2, \dots, A_m$ are called positively separated if $\min_{i\neq j} d(A_i, A_j) > 0$. The first property is that $\mathcal{M}^s$ is additive on positively separated sets, i.e. if $A_1, A_2, \dots, A_m$ are positively separated then $\mathcal{M}^s(\cup_{i=1}^m A_i) = \sum_{i=1}^m \mathcal{M}^s(A_i)$. This follows from the fact that for sufficiently small δ, the δ-neighborhoods of the A_i are disjoint. The second property is that for any set A, $A^{(\delta)} = \overline{A}^{(\delta)}$ and so $\mathcal{M}^s_\delta(A) = \mathcal{M}^s_\delta(\overline{A})$ for every $\delta > 0$, where $\overline{A}$ denotes the closure of A. Clearly $A^{(\delta)} \subset \overline{A}^{(\delta)}$. On the other hand, if $x \in \overline{A}^{(\delta)}$ then $\|x - y\| = \eta < \delta$ for some $y \in \overline{A}$. But $\|y - a\| < \delta - \eta$ for some $a \in A$, so that $\|x - a\| \le \|x - y\| + \|y - a\| < \delta$. Hence, $x \in A^{(\delta)}$ and so the result follows.

The following two lemmas give properties of δ-neighborhoods which will be useful in showing continuity properties of Minkowski content. $B_r(x)$ and $\overline{B}_r(x)$ denote the open and closed balls, respectively, of radius r centered at x.

LEMMA 1. *$\mu(\partial\Gamma^{(\delta)}) = 0$ for every $\Gamma \subset \mathbf{R}^2$.*

Proof: Let $\Gamma \subset \mathbf{R}^2$ and let $E = \partial\Gamma^{(\delta)}$. The Lebesgue density of E at x, $D_\mu(E, x)$, is defined as

$$D_\mu(E, x) = \lim_{r\to 0} \frac{\mu(E \cap B_r(x))}{\mu(B_r(x))}$$

when the limit exists. We will show that the Lebesgue density of E is less than 1 for all $x \in E$. Hence, $\mu(E) = 0$ will follow from the Lebesgue Density Theorem.

Let $x \in E = \partial\Gamma^{(\delta)}$. Then for each $r > 0$, there exists $c(r) \in \Gamma$ with $\|x - c(r)\| < \delta + r^2$. If $w \in B_\delta(c(r))$ then $w \notin E$, so that

$$\mu(E \cap B_r(x)) \le \mu(B_r(x)) - \mu(B_r(x) \cap B_\delta(c(r)))$$

The circle of radius δ centered at $c(r)$ intersects the circle of radius r centered at x in two points which determine a chord C. Let S denote the segment of $B_r(x)$

determined by C, θ the central angle at x subtended by C, and a the distance from x to C. Then

$$\mu(B_r(x) \cap B_\delta(c(r))) \geq \mu(S) = \frac{1}{2}r^2(\theta - \sin\theta)$$

and

$$\lim_{r\to 0} \theta = \lim_{r\to 0} 2\cos^{-1}(\frac{d}{r}) = \lim_{r\to 0} 2\cos^{-1}(\frac{r^2 + 2\delta r^2 + r^4}{2r(\delta + r^2)}) = \pi$$

Therefore,

$$D_\mu(E,x) = \lim_{r\to 0} \frac{\mu(E \cap B_r(x))}{\mu(B_r(x))} \leq \lim_{r\to 0} \frac{\mu(B_r(x)) - \mu(S)}{\mu(B_r(x))} = \lim_{r\to 0}(1 - \frac{1}{2\pi}(\theta - \sin\theta)) = \frac{1}{2} \quad \square$$

LEMMA 2. *If $\Gamma_n \to \Gamma$ in Hausdorff metric, then $\Gamma_n^{(\delta)} \to \Gamma^{(\delta)}$.*

Proof: Let $\epsilon > 0$. Since $\Gamma_n \to \Gamma$, $\exists\, N < \infty$ such that $d_H(\Gamma_n, \Gamma) < \epsilon\ \forall n > N$. If $x \in \Gamma^{(\delta)}$ then $x = a + \rho$ with $a \in \Gamma$ and $\| \rho \| < \delta$. For all $n > N$, there exists $a_n \in \Gamma_n$ with $\| a - a_n \| < \epsilon$. Then $x_n \equiv a_n + \rho \in \Gamma_n^{(\delta)}$ and $\| x - x_n \| = \| a - a_n \| < \epsilon$. Hence, $\Gamma^{(\delta)} \subset (\Gamma_n^{(\delta)})^{(\epsilon)}$. Similarly, $\Gamma_n^{(\delta)} \subset (\Gamma^{(\delta)})^{(\epsilon)}$. Thus, $d_H(\Gamma_n^{(\delta)}, \Gamma^{(\delta)}) < \epsilon\ \forall n > N$. $\square$

Two continuity properties of $\mathcal{M}_\delta^s$ may now be deduced. These follow directly from the corresponding continuity properties of Lebesgue measure on δ-neighborhoods.

THEOREM 6. *If $\Gamma_n \to \Gamma$ in Hausdorff metric then $\mu(\Gamma_n^{(\delta)}) \to \mu(\Gamma^{(\delta)})$ and so $\mathcal{M}_\delta^s(\Gamma_n) \to \mathcal{M}_\delta^s(\Gamma)$. I.e., $\mathcal{M}_\delta^s(\Gamma)$ is continuous in Γ with respect to Hausdorff metric.*

Proof: Since $\Gamma_n \to \Gamma$, by Lemma 2 we have $\Gamma_n^{(\delta)} \to \Gamma^{(\delta)}$. Let $\epsilon > 0$. Then there exists $N < \infty$ such that $\Gamma_n^{(\delta)} \subset (\Gamma^{(\delta)})^{(\epsilon)}\ \forall n \geq N$. Therefore, $\sup_{n \geq N} \mu(\Gamma_n^{(\delta)}) \leq \mu(\Gamma^{(\delta+\epsilon)})$. As $\epsilon \downarrow 0$, $\Gamma^{(\delta+\epsilon)} \downarrow \overline{\Gamma^{(\delta)}}$ so that $\limsup_{n\to\infty} \mu(\Gamma_n^{(\delta)}) \leq \mu(\overline{\Gamma^{(\delta)}})$. Then by Lemma 1 it follows that $\limsup_{n\to\infty} \mu(\Gamma_n^{(\delta)}) \leq \mu(\Gamma^{(\delta)})$.

Let K be a compact subset of $\Gamma^{(\delta)}$. Since $\{B_\delta(x) : x \in \Gamma\}$ is an open cover of K, there exists a finite subcover $B_\delta(x_1), \ldots, B_\delta(x_m)$. Let $\epsilon > 0$. Since $\Gamma_n \to \Gamma$, there exists $N < \infty$ such that $\forall n \geq N$ we can find $y_{n,1}, \ldots, y_{n,m} \in \Gamma_n$ with $\|y_{n,i} - x_i\| < \epsilon$ for $i = 1, \ldots, m$. Then $\mu(B_\delta(x_i) \setminus B_\delta(y_{n,i})) < f(\epsilon) = \mu(B_1 \setminus B_2) \leq 2\delta\epsilon$ where B_1 and B_2 are balls of radius δ whose centers are ϵ apart. Therefore,

$$\mu(\Gamma_n^{(\delta)}) \geq \mu(\bigcup_{i=1}^m B_\delta(y_{n,i})) > \mu(K) - mf(\epsilon) \quad \forall n \geq N$$

and so $\inf_{n \geq N} \mu(\Gamma_n^{(\delta)}) > \mu(K) - mf(\epsilon)$. Since $\epsilon > 0$ is arbitrary and $f(\epsilon) \to 0$ as $\epsilon \to 0$ we have $\liminf_{n\to\infty} \mu(\Gamma_n^{(\delta)}) \geq \mu(K)$. Finally, since this is true for every compact $K \subset \Gamma^{(\delta)}$, we have $\liminf_{n\to\infty} \mu(\Gamma_n^{(\delta)}) \geq \sup_{K \subset \Gamma^{(\delta)}} \mu(K) = \mu(\Gamma^{(\delta)})$.

Thus,

$$\liminf_{n\to\infty} \mu(\Gamma_n^{(\delta)}) = \limsup_{n\to\infty} \mu(\Gamma_n^{(\delta)}) = \lim_{n\to\infty} \mu(\Gamma_n^{(\delta)}) = \mu(\Gamma^{(\delta)}) \quad \square$$

PROPOSITION 1. *$\mathcal{M}^s_\delta(\Gamma)$ is continuous in δ for all $\delta > 0$.*

Proof: As $\eta \uparrow \delta$, we have $\Gamma^{(\eta)} \uparrow \Gamma^{(\delta)}$ so that $\mu(\Gamma^{(\eta)}) \uparrow \mu(\overline{\Gamma^{(\delta)}})$. As $\eta \downarrow \delta$, we have $\Gamma^{(\eta)} \downarrow \overline{\Gamma^{(\delta)}}$. Then by Lemma 1, $\mu(\Gamma^{(\eta)}) \downarrow \mu(\overline{\Gamma^{(\delta)}}) = \mu(\Gamma^{(\delta)})$. Thus, $\lim_{\eta\to\delta} \mu(\Gamma^{(\eta)}) = \mu(\Gamma^{(\delta)})$. □

All the results given so far in this section were proved for $\Gamma \subset \mathbf{R}^2$. However, these results and proofs can easily be extended to $\mathbf{R}^n$.

We now state a result given in Federer [11] relating Minkowski content to Hausdorff measure. A subset Γ of $\mathbf{R}^n$ is called *m-rectifiable* if there exists a Lipschitzian function mapping a bounded subset of $\mathbf{R}^m$ onto Γ.

THEOREM 7. *If Γ is a closed m-rectifiable subset of $\mathbf{R}^n$ then $\mathcal{M}^m(\Gamma) = \mathcal{H}^m(\Gamma)$.*

Proof: See [11] Theorem 3.2.39. □

We will present a proof of Theorem 7 in the restricted case of 1-dimensional measure in $\mathbf{R}^2$ (i.e., $m = 1, n = 2$), which is stated as Theorem 8. The basic idea of our proof is contained in the proof of Proposition 4. This idea will be used again in the proof of Theorem 9 on the Γ-convergence of Minkowski content, which is true only for 1-dimensional measures, and this motivates including the proof.

The following two preliminary results give upper and lower bounds on $M^1_\delta(\Gamma)$ for rectifiable and connected sets respectively. These two results could be appropriately extended to s-dimensional measure in $\mathbf{R}^n$.

PROPOSITION 2. *If $\Gamma \subset \mathbf{R}^2$ is rectifiable then $\mu(\Gamma^{(\delta)}) \leq 2\delta\mathcal{H}^1(\Gamma) + \pi\delta^2$ and so $\mathcal{M}^1_\delta(\Gamma) \leq \mathcal{H}^1(\Gamma) + \frac{1}{2}\pi\delta$.*

Proof: Since Γ is rectifiable, $\Gamma = \{\gamma(t) : 0 \leq t \leq 1\}$ where $\gamma : [0,1] \to \mathbf{R}^2$ is rectifiable and $\mathcal{H}^1(\Gamma) = \sup\{\sum_{i=1}^m \|\gamma(t_i) - \gamma(t_{i-1})\| : 0 = t_0 < t_1 < \cdots < t_m = 1\}$. For $E = 1, 2, \ldots$ let $\{t_{ij}\}$ be a sequence of dissections such that $\max_i\{\|t_{ij} - t_{i-1,j}\|\} \to 0$ and $\mathcal{H}^1(\Gamma) = \lim_{j\to\infty} \sum_{i=1}^{m(j)} \|\gamma(t_{ij}) - \gamma(t_{i-1,j})\|$ Let $C_j = \cup_{i=1}^{m(j)} S_i$ where S_i is the straight line joining $\gamma(t_{i-1,j})$ and $\gamma(t_{ij})$. Then $\mu(S_{ij}^{(\delta)}) = 2\delta\|\gamma(t_{ij}) - \gamma(t_{i-1,j})\| + \pi\delta^2$, and

$$
\begin{aligned}
\mu(\cup_{i=1}^k S_{ij}^{(\delta)}) &= \mu(\cup_{i=1}^{k-1} S_{ij}^{(\delta)}) + \mu(S_{kj}^{(\delta)}) - \mu(S_{kj}^{\delta} \cap \bigcup_{i=1}^{k-1} S_{ij}^{(\delta)}) \\
&\leq \mu(\cup_{i=1}^{k-1} S_{ij}^{(\delta)}) + \mu(S_{kj}^{(\delta)}) - \pi\delta^2 \\
&= \mu(\cup_{i=1}^{k-1} S_{ij}^{(\delta)}) + 2\delta\|\gamma(t_{kj}) - \gamma(t_{k-1,j})\|
\end{aligned}
$$

By induction on i, we get

$$\mu(C_j^\delta) \leq \sum_{i=1}^{m(j)} 2\delta\|\gamma(t_{ij}) - \gamma(t_{i-1,j})\| + \pi\delta^2$$

Since $C_j \to \Gamma$ in Hausdorff metric, by Theorem 6

$$\mu(\Gamma^{(\delta)}) = \lim_{j\to\infty} \mu(C_j^{(\delta)}) \leq 2\delta\mathcal{H}^1(\Gamma) + \pi\delta^2 \quad \square$$

PROPOSITION 3. *If $\Gamma \subset \mathbf{R}^2$ is connected, then $\mathcal{M}^1_\delta(\Gamma) \geq |\Gamma|$.*

Proof: Let $x, y \in \Gamma$, and let $\epsilon > 0$. Since Γ is connected, we can find $x = x_0, x_1, \ldots, x_k = y$ in Γ with $\|x_i - x_{i-1}\| < \epsilon$ for $1 \leq i \leq k$. Let $P(w)$ denote the point obtained by the orthogonal projection of w onto the straight line T through x and y, and let $p(w)$ be the coordinate of $P(w)$ considering T as the real line with origin at x and positive direction towards y. I.e.,

$$p(w) = \frac{\langle w - x, y - x\rangle}{\|y - x\|}$$

where $\langle \cdot, \cdot \rangle$ denotes the usual inner product. Note that $|p(x_i) - p(x_{i-1})| = \|P(x_i) - P(x_{i-1})\| \leq \|x_i - x_{i-1}\| < \epsilon$. By deleting intermediate points and reordering the indices as necessary, we can assume that $0 = p(x_0) < p(x_1) < \cdots < p(x_k) = \|x - y\|$ and $p(x_i) - p(x_{i-1}) < \epsilon$.

For $u, v \in \mathbf{R}^2$ with $p(u) < p(v)$, let $R(u, v) = \{w \in \mathbf{R}^2 : p(u) < p(w) < p(v)\}$. Then

$$\Gamma^{(\delta)} \supset \cup_{i=0}^k B_\delta(x_i) \supset \cup_{i=1}^k B_\delta(x_i) \cap R(x_i, x_{i-1})$$

Since the $R(x_i, x_{i-1})$ for $i = 1, 2, \ldots, k$ are disjoint,

$$\begin{aligned} \mathcal{M}^1_\delta(\Gamma) &\geq \frac{\mu(\cup_{i=1}^k B_\delta(x_i) \cap R(x_i, x_{i-1}))}{2\delta} \\ &= \frac{1}{2\delta} \sum_{i=1}^k \mu(B_\delta(x_i) \cap R(x_i, x_{i-1})) \\ &\geq \frac{1}{2\delta} \sum_{i=1}^k 2\sqrt{\delta^2 - \epsilon^2}\,(p(x_i) - p(x_{i-1})) \\ &= \|x - y\| \sqrt{1 - \frac{\epsilon^2}{\delta^2}} \end{aligned}$$

Since $\epsilon > 0$ is arbitrary we have $\mathcal{M}^1_\delta(\Gamma) \geq \|x - y\|$. Finally, the result follows since $x, y \in \Gamma$ are arbitrary. □

Using the bounds of Propositions 2 and 3, the following proposition can be shown.

PROPOSITION 4. *If $\Gamma \subset \mathbf{R}^2$ is connected and consists of a countable union of rectifiable curves then $\mathcal{M}^1(\Gamma) = \mathcal{H}^1(\Gamma)$.*

Proof: First, we prove the result when Γ is a rectifiable curve which does not intersect itself. Let $\Gamma = \{\gamma(t) : 0 \leq t \leq 1\}$ where $\gamma : [0, 1] \to \mathbf{R}^2$ is rectifiable and $\gamma(s) \neq \gamma(t)$ if $s \neq t$. Let $0 = t_0 < t_1 < \cdots < t_m = 1$, and for $i = 1, 2, \ldots, m$ let $\Gamma_i = \{\gamma(t) : t_{i-1} < t < t_i\}$. If $K_i \subset \Gamma_i$ $i = 1, 2, \ldots, m$ are continuua then they are positively separated. Therefore, for sufficiently small δ the $K_i^{(\delta)}$ are disjoint. From Proposition 3 we have

$$\mathcal{M}^1_\delta(\Gamma) \geq \sum_{i=1}^m \mathcal{M}^1_\delta(K_i) \geq \sum_{i=1}^m |K_i|$$

for all sufficiently small δ. Hence,

$$\liminf_{\delta\to 0} \mathcal{M}^1_\delta(\Gamma) \geq \sum_{i=1}^{m} \|\gamma(t_i) - \gamma(t_{i-1})\|$$

and since the dissection $\{t_i\}$ is arbitrary

$$\liminf_{\delta\to 0} \mathcal{M}^1_\delta(\Gamma) \geq \mathcal{H}^1(\Gamma)$$

On the other hand, from Proposition 2, $\mathcal{M}^1_\delta(\Gamma) \leq \mathcal{H}^1(\Gamma) + \frac{1}{2}\pi\delta$ so that

$$\limsup_{\delta\to 0} \mathcal{M}^1_\delta(\Gamma) \leq \mathcal{H}^1(\Gamma)$$

Thus,

$$\mathcal{M}^1(\Gamma) = \lim_{\delta\to 0} \mathcal{M}^1_\delta(\Gamma) = \mathcal{H}^1(\Gamma)$$

Now, suppose $\Gamma = \cup_{i=1}^{\infty} C_i$ is connected where the C_i are rectifiable curves. By decomposing the C_i as necessary, we can assume that they are not self-intersecting and that C_i intersects C_j in at most a finite number of points for $i \neq j$. Then $\mathcal{H}^1(\Gamma) = \sum_{i=1}^{\infty} \mathcal{H}^1(C_i)$. Let $E_k = \cup_{i=1}^{k} C_i$. Then by a dissection argument similar to that used above we get

$$\liminf_{\delta\to 0} \mathcal{M}^1_\delta(E_k) \geq \sum_{i=1}^{k} \mathcal{H}^1(C_i)$$

and so

$$\liminf_{\delta\to 0} \mathcal{M}^1_\delta(\Gamma) \geq \sup_k \liminf_{\delta\to 0} \mathcal{M}^1_\delta(E_k) \geq \mathcal{H}^1(\Gamma)$$

Also, from Proposition 2 and the fact that Γ is connected we have

$$\begin{aligned} \mu(E_k^{(\delta)}) &= \mu(E_{k-1}^{(\delta)}) + \mu(C_k^{(\delta)}) - \mu(E_{k-1}^{(\delta)} \cap C_k^{(\delta)}) \\ &\leq \mu(E_{k-1}^{(\delta)}) + \mu(C_k^{(\delta)}) - \pi\delta^2 \leq \mu(E_{k-1}^{(\delta)}) + 2\delta\mathcal{H}^1(C_k) \end{aligned}$$

By induction we get

$$\mu(E_k^{(\delta)}) \leq 2\delta \sum_{i=1}^{k} \mathcal{H}^1(C_k) + \pi\delta^2$$

for every integer k. Since $E_k^{(\delta)}$ is an increasing sequence of sets with $\Gamma^{(\delta)} = \cup_{i=k}^{\infty} E_k$ we have

$$\mu(\Gamma^{(\delta)}) = \lim_{k\to\infty} \mu(E_k^{(\delta)}) \leq 2\delta\mathcal{H}^1(\Gamma) + \pi\delta^2$$

Thus

$$\limsup_{\delta\to 0} \mathcal{M}^1_\delta(\Gamma) \leq \mathcal{H}^1(\Gamma)$$

and so the result follows. □

The next inequality gives bounds for s-dimensional Minkowski content in $\mathbf{R}^2$ which are valid for *every* subset of $\mathbf{R}^2$. This could also be appropriately extended to $\mathbf{R}^n$. Here, we use the notation

$$\mathcal{H}^s_{\delta,2\delta}(\Gamma) = 2^{-s}\omega_s \inf\{\sum_{i=1}^{\infty} |U_i|^s \ : \Gamma \subset \bigcup_{i=1}^{\infty} U_i,\ \delta \leq |U_i| \leq 2\delta\}$$

PROPOSITION 5. *For every* $\Gamma \subset \mathbf{R}^2$ *and* $0 \leq s \leq 2$,

$$\frac{2^{s-1}}{\omega_s\omega_{2-s}}\mathcal{H}^s_\delta(\Gamma) \leq \mathcal{M}^s_\delta(\Gamma) \leq \frac{16}{\omega_s\omega_{2-s}}\mathcal{H}^s_{\delta,2\delta}(\Gamma)$$

and so

$$\frac{2^{s-1}}{\omega_s\omega_{2-s}}\mathcal{H}^s(\Gamma) \leq \mathcal{M}^s_*(\Gamma) \leq \liminf_{\delta\to 0} \frac{16}{\omega_s\omega_{2-s}}\mathcal{H}^s_{\delta,2\delta}(\Gamma)$$

where $\mathcal{H}^s_{\delta,2\delta}(\Gamma) = 2^{-s}\omega_s \inf\{\sum_{i=1}^{\infty} |U_i|^s \ : \Gamma \subset \bigcup_{i=1}^{\infty} U_i,\ \delta \leq |U_i| \leq 2\delta\}$

Proof: Consider the closed lattice squares formed by the points $\frac{1}{\sqrt{2}\delta}\mathbf{Z}^2$. Form a cover $\{U_i\}$ of Γ by taking all lattice squares whose intersection with Γ is non-empty. Then $\{U_i\}$ is a δ-cover of Γ and $\bigcup_i U_i \subset \overline{\Gamma^{(\delta)}}$. Hence,

$$\begin{aligned}\frac{2^s}{\omega_s}\mathcal{H}^s_\delta(\Gamma) &\leq \sum_i |U_i|^s = \frac{2}{\delta^{2-s}}\sum_i \left(\frac{\delta}{\sqrt{2}}\right)^2 = \frac{2}{\delta^{2-s}}\mu\Big(\bigcup_i U_i\Big)\\ &\leq \frac{2}{\delta^{2-s}}\mu(\overline{\Gamma^{(\delta)}}) = \frac{2}{\delta^{2-s}}\mu(\Gamma^{(\delta)}) = 2\omega_{2-s}\mathcal{M}^s_\delta(\Gamma)\end{aligned}$$

To show the second part of the first inequality, let $\{U_i\}$ be any cover of Γ with $\delta \leq |U_i| \leq 2\delta$. Without loss of generality, we assume that $U_i \cap \Gamma$ is non-empty for each i. Select $x_i \in \Gamma \cap U_i$. Then $\cup_i \overline{B}_{|U_i|}(x_i) \supset \cup_i U_i \supset \Gamma$ so that $\cup_i \overline{B}_{2|U_i|}(x_i) \supset \Gamma^{(\delta)}$ since $|U_i| \geq \delta$. Therefore,

$$\mathcal{M}^s_\delta(\Gamma) \leq \frac{\mu(\cup_i \overline{B}_{2|U_i|}(x_i))}{\delta^{2-s}\omega_{2-s}} \leq \frac{\sum_i 4\pi|U_i|^2}{\delta^{2-s}\omega_{2-s}} \leq \frac{4\pi}{\omega_{2-s}}\sum_i \frac{|U_i|^2}{(\frac{|U_i|}{2})^{2-s}} = \frac{2^{4-s}}{\omega_{2-s}}\sum_i |U_i|^s$$

and so

$$\mathcal{M}^s_\delta(\Gamma) \leq \frac{2^{4-s}}{\omega_{2-s}} \inf\{\sum_{i=1}^{\infty} |U_i|^s \ : \Gamma \subset \bigcup_{i=1}^{\infty} U_i,\ \delta \leq |U_i| \leq \delta\} = \frac{16}{\omega_s\omega_{2-s}}\mathcal{H}^s_{\delta,2\delta}(\Gamma) \quad \square$$

Note that the definition of $\mathcal{H}^s_{\delta,2\delta}$ is similar to Hausdorff measure, except that the diameter of the covering sets is bounded below as well as above. Hence, its value may be quite different from Hausdorff measure. As an aside, one consequence of the above proposition is the known result that the *Minkowski dimension* of a set is greater than or equal to its *Hausdorff dimension* [9,13].

We can now prove the following special case of Theorem 7.

THEOREM 8. *If $\Gamma \subset \mathbf{R}^2$ is a compact set with a finite number of connected components then* $\mathcal{M}^1(\Gamma) = \mathcal{H}^1(\Gamma)$.

Proof: Since the connected components of Γ are compact, disjoint, and finite in number, they are positively separated. By additivity of both $\mathcal{M}^1$ and $\mathcal{H}^1$, we need only consider the case in which Γ has one connected component. Hence, we assume that Γ is a continuum. If $\mathcal{H}^1(\Gamma) = \infty$ then $\mathcal{M}^1(\Gamma) = \infty$ from Proposition 5. Therefore, we can assume that $\mathcal{H}^1(\Gamma) < \infty$.

Then from Lemma 3.12 of [10], Γ is arcwise connected. Since Γ is compact, we can define a sequence of curves C_j inductively as follows (as in the proof of Lemma 3.13 of [10]). Let C_1 be a curve in Γ joining two of the most distant points of Γ. Given $C_1, C_2, \ldots, C_j$, let $x \in \Gamma$ be at a maximum distance from $\cup_{i=1}^{j} C_i$ and let d_j denote this maximum distance. If $d_j = 0$ then the procedure terminates and we let $C_i = \emptyset$ for $i \geq j+1$. Otherwise, let C_{j+1} be a curve in Γ joining x and $\cup_{i=1}^{j} C_i$ that is disjoint from $\cup_{i=1}^{j} C_i$ except for an endpoint.

Let $E_k = \cup_{j=1}^{k} C_j$. It is shown in [10] (proof of lemma 3.13) that $\mathcal{H}^1(\Gamma) = \mathcal{H}^1(\cup_{i=1}^{\infty} E_k)$. Also,

$$\sum_{j=1}^{\infty} d_j \leq \sum_{j=1}^{\infty} \mathcal{H}^1(C_j) = \mathcal{H}^1(\Gamma) < \infty$$

so that $d_j \to 0$. This implies that $E_k = \cup_{j=1}^{k} C_j \to E$ in Hausdorff metric as $k \to \infty$ and so $\overline{\cup_{k=1}^{\infty} E_k} = \Gamma$. Hence, from Proposition 10 and using the fact that $\mathcal{M}^1(A) = \mathcal{M}^1(\overline{A})$ for any A, we get

$$\mathcal{H}^1(\Gamma) = \mathcal{H}^1(\cup_{k=1}^{\infty} E_k) = \mathcal{M}^1(\cup_{k=1}^{\infty} E_k) = \mathcal{M}^1(\overline{\cup_{k=1}^{\infty} E_k}) = \mathcal{M}^1(\Gamma) \quad \square$$

Note that $\mathcal{M}^1$ and $\mathcal{H}^1$ do not agree on all compact sets. An example of a compact set on which they disagree is given in [11] (Section 3.2.40).

The final result shown in this section is that Minkowski content possesses a useful type variational convergence property known as Γ-convergence (or epi-convergence). This notion of convergence, introduced by De Giorgi [6,7,8] and independently by Attouch [3], is useful in problems involving the convergence of functionals. The result on Γ-convergence will be used in Section 5 to prove some convergence properties of solutions to certain variational problems. Given a topological space (X, τ), and functions $F_n, F : X \to \mathbf{R} \cup \{-\infty, +\infty\}$, the sequence $\{F_n\}$ is said to be Γ-*convergent* (or *epi-convergent*) to F at $x \in X$ if the following two conditions hold:

(i) for every sequence $\{x_n\}$ converging to x in (X, τ), $F(x) \leq \liminf_{n\to\infty} F_n(x_n)$, and

(ii) there exists a sequence $\{x_n\}$ converging to x in (X, τ) such that $F(x) \geq \limsup_{n\to\infty} F_n(x_n)$.

We will show that for every sequence $\delta_n \to 0$, $\mathcal{M}^1_{\delta_n}$ is Γ-convergent to $\mathcal{M}^1$ on the space of compact subsets of $\mathbf{R}^2$ with a bounded number of connected components and with the topology induced by the Hausdorff metric.

First, we need the following lemma as stated in [10].

LEMMA 3. *Let $\mathcal{C}$ be a collection of balls contained in a bounded subset of $\mathbf{R}^n$. Then there exists a finite or countably infinite disjoint subcollection $\{B_i\}$ such that*

$$\bigcup_{B\in\mathcal{C}} B \subset \bigcup_i B'_i$$

where B'_i is the ball concentric with B_i and of three times the radius.

Proof: See [10], Lemma 1.9.

Now the the Γ-convergence of Minkowski content can be shown.

THEOREM 9. *For every sequence $\delta_n \to 0^+$, $\mathcal{M}^1_{\delta_n}$ is Γ-convergent to $\mathcal{M}^1$ on the space of compact subsets of $\mathbf{R}^2$ with a bounded number of connected components and with the topology induced by the Hausdorff metric. I.e., let $\Gamma \subset \mathbf{R}^2$ be compact with $\#(\Gamma) \le M < \infty$, and let $\delta_n > 0$ satisfy $\lim_{n\to\infty}\delta_n = 0$. Then the following two conditions hold:*

(i) *For every sequence of compact sets $\Gamma_n \subset \mathbf{R}^2$ with $\Gamma_n \to \Gamma$ in Hausdorff metric and $\#(\Gamma_n) \le M$ $\forall n$ we have*

$$\mathcal{M}^1(\Gamma) \le \liminf_{n\to\infty} \mathcal{M}^1_{\delta_n}(\Gamma_n)$$

(ii) *There exists a sequence of compact sets $\Gamma_n \subset \mathbf{R}^2$ with $\Gamma_n \to \Gamma$ in Hausdorff metric and $\#(\Gamma_n) \le M$ $\forall n$ such that*

$$\mathcal{M}^1(\Gamma) \ge \limsup_{n\to\infty} \mathcal{M}^1_{\delta_n}(\Gamma_n)$$

Proof: Since $\#(\Gamma) \le M$ we have $\Gamma = \cup_{i=1}^k F_i$ where $k \le M$ and $F_1, F_2, \dots, F_k$ are the connected components of Γ. Since the F_i are compact and disjoint, they are positively separated, i.e there exists $\eta > 0$ such that $F_i^{(\eta)} \cap F_j^{(\eta)} = \emptyset$ for $i \ne j$. Then $\mathcal{M}^1(\Gamma) = \sum_{i=1}^k \mathcal{M}^1(F_i)$, and for sufficiently large n, $\mathcal{M}^1_{\delta_n}(\Gamma_n) = \sum_{i=1}^k \mathcal{M}^1_{\delta_n}(\Gamma_n \cap F_i^{(\eta)})$. Thus, it is sufficient to prove the result under the assumption that Γ is connected.

Suppose $\mathcal{H}^1(\Gamma) = \infty$. Form a δ-covering of Γ by placing a closed ball of radius δ about each point of Γ. Then by Lemma 3, we can find a disjoint subcollection (necessarily finite) of balls such that concentric balls of radius 3δ cover $\Gamma^{(\delta)}$. Let $N(\delta)$ be the number of balls in this finite disjoint subcollection. Then $6\delta N(\delta) \ge \mathcal{H}^1_{6\delta}(\Gamma) \to \infty$ as $\delta \to 0$. Let $\epsilon > 0$. Since $\Gamma_n \to \Gamma$, for sufficiently large n we have $\Gamma_n \cap B_{\frac{\delta}{2}}(x_i) \ne \emptyset$. Also, since $\#(\Gamma_n) \le M$, there is a connected component of $\Gamma_n \cap B_\delta(x_i)$ with diameter greter than or equal to $\frac{\delta}{2}$ for at least $N(\delta) - M$ values of i. Using Proposition 3 and the fact that the balls are positively separated, we have for sufficiently large n

$$\mathcal{M}^1_{\delta_n}(\Gamma_n) \ge \mathcal{M}^1_{\delta_n}(\Gamma_n \cap \cup_{i=1}^{N(\delta)} B_\delta(x_i)) = \sum_{i=1}^{N(\delta)} \mathcal{M}^1_{\delta_n}(\Gamma_n \cap B_\delta(x_i))$$

$$\ge \sum_{i=1}^{N(\delta)} |\Gamma_n \cap B_\delta(x_i)| \ge \frac{\delta}{2}(N(\delta) - M)$$

Since $\delta N(\delta) \to \infty$ as $\delta \to 0$, $\liminf_{n\to\infty} \mathcal{M}^1_{\delta_n}(\Gamma_n) = \infty$, and so the result follows.

Now, suppose $\mathcal{H}^1(\Gamma) < \infty$. From Theorem 3 we have $\Gamma = S \cup (\bigcup_{i=1}^{\infty} C_i)$ where $\mathcal{H}^1(S) = 0$, and C_i are rectifiable curves. From the construction used in the proof of this result (see [10], also part of the proof is reproduced in the proof of Proposition 4), $\mathcal{H}^1(\Gamma) = \sum_{i=1}^{\infty} \mathcal{H}^1(C_i)$ and if $x \in C_i \cap C_j$ then x is an endpoint of at least one of C_i or C_j.

Consider $\cup_{i=1}^{k} C_i$. By decomposing the C_i, we can assume that they are simple curves which meet each other only at endpoints. The C_i are rectifiable curves, so that $C_i : [0,1] \to \mathbf{R}^2$ and

$$\mathcal{H}^1(C_i) = \mathcal{M}^1(C_i) = \sup\{\sum_{j=1}^{m(i)} \|C_i(t_{i,j-1} - C_i(t_{ij})\| : 0 = t_{i0} < t_{i1} < \ldots < t_{i,m(i)} = 1\}$$

For each $i = 1, 2, \ldots, k$, let $0 = t_{i0} < t_{i1} < \cdots < t_{i,m(i)} = 1$, and consider the points $x_{ij} = C_i(t_{ij})$.

The connected components of $\cup_{i=1}^{k} C_i \setminus \{x_{ij}\}$ are given by $G_{ij} = \{C_i(t) : t_{i,j-1} < t < t_{ij}\}$ for $1 \le i \le k$, $1 \le j \le m(i)$. For each i, j, let K_{ij} be a compact subset of G_{ij}. Then the K_{ij} are positively separated since they are a finite collection of disjoint compact sets. Therefore, for some $\eta > 0$, the $\overline{K_{ij}^{(\eta)}}$ are disjoint. Since $\Gamma_n \to \Gamma$ and $\#(\Gamma_n) \le M$, for n sufficiently large $\Gamma_n \cap \overline{K_{ij}^{(\eta)}}$ has a connected component whose diameter approaches the diameter of K_{ij} except for at most M values of i, j. I.e., except for at most M values of i, j, there is a connected component T_{nij} of $\Gamma_n \cap \overline{K_{ij}^{(\eta)}}$ such that for every $\epsilon > 0$ there exists $N > 0$ with $|T_{nij}| > |K_{ij}| - \epsilon$ and $\delta_n < \eta$ for all $n \ge N$. Hence, by Proposition 3, for all $n \ge N$

$$\begin{aligned}\mathcal{M}^1_{\delta_n}(\Gamma_n) &\ge \mathcal{M}^1_{\delta_n}(\Gamma_n \cap \bigcup_{i,j} \overline{K_{ij}^{(\eta)}}) \\ &= \sum_{i=1}^{k} \sum_{j=1}^{m(i)} \mathcal{M}^1_{\delta_n}(\Gamma_n \cap \overline{K_{ij}^{(\eta)}}) \\ &\ge \sum_{i=1}^{k} \sum_{j=1}^{m(i)} (|K_{ij}| - \epsilon) - M(\max_{i,j}\{|K_{ij}|\})\end{aligned}$$

and so

$$\liminf_{n\to\infty} \mathcal{M}^1_{\delta_n}(\Gamma_n) \ge \sum_{i=1}^{k} \sum_{j=1}^{m(i)} |K_{ij}| - M(\max_{i,j}\{|K_{ij}|\})$$

Taking the sup over the compact sets K_{ij} gives

$$\begin{aligned}\liminf_{n\to\infty} \mathcal{M}^1_{\delta_n}(\Gamma_n) &\ge \sup_{K_{ij}}\{\sum_{i=1}^{k} \sum_{j=1}^{m(i)} |K_{ij}| - M(\max_{i,j}\{|K_{ij}|\})\} \\ &= \sum_{i=1}^{k} \sum_{j=1}^{m(i)} \|C_i(t_{i,j-1}) - C_i(t_{ij})\| - M(\max_{i,j}\{\|C_i(t_{i,j-1}) - C_i(t_{ij})\|\})\end{aligned}$$

Then, taking the sup over the t_{ij} gives

$$\liminf_{n\to\infty} \mathcal{M}^1_{\delta_n}(\Gamma_n) \geq \sum_{i=1}^{k} \mathcal{H}^1(C_i)$$

since $M < \infty$ and $\max_{i,j}\{\|C_i(t_{i,j-1}) - C_i(t_{ij})\|\} \to 0$ as $\max_{i,j}\{\|t_{i,j-1} - t_{ij})\|\} \to 0$. Finally, letting $k \to \infty$ gives

$$\liminf_{n\to\infty} \mathcal{M}^1_{\delta_n}(\Gamma_n) \geq \mathcal{H}^1(\Gamma) = \mathcal{M}^1(\Gamma)$$

which proves (i).

To show (ii), take $\Gamma_n = \Gamma$. From Theorem 8, $\mathcal{M}^1(\Gamma) = \mathcal{H}^1(\Gamma)$ so that in particular $\lim_{\delta\to 0} \mathcal{M}^1_\delta(\Gamma) = \mathcal{M}^1(\Gamma)$ exists. Hence, for every sequence $\delta_n \to 0$, condition (ii) is satisfied by taking $\Gamma_n = \Gamma$. □

Note that Theorem 9 is not true in general if the bound on the number of connected components is dropped. For example, let $r_1, r_2, \ldots$ denote an enumeration of the rationals between 0 and 1. Take $\Gamma_n = \{(r_i, 0) : 1 \leq i \leq n\}$ and $\delta_n = 1/n^2$. Then $\Gamma_n \to \Gamma = \{(x,0) : 0 \leq x \leq 1\}$, but $\mathcal{M}^1_{\delta_n}(\Gamma_n) \leq \frac{1}{2}\pi n \delta_n \to 0$ while $\mathcal{M}^1(\Gamma) = 1$. However, we conjecture that the restriction on the number of connected components can be dropped if we impose the additional assumption that $d_H(\Gamma_n, \Gamma)/\delta_n \to 0$ as $n \to \infty$.

4. An Existence Theorem. In this section we will treat the question of the existence of a minimizing pair (f, Γ) for E. We have already developed some results for the cost associated strictly with the boundary so in this section we will be focusing on the function f. Since it may be desirable to introduce other costs associated with the boundary, we will state assumptions required on the boundaries in order to treat the remainder of the problem rather than quote results from the last section. We mention here however that these assumptions are satisfied by the definitions given in Section 2. Also, we will generalize the functional E. We will use the following set of assumptions on the space of boundaries.

- A1 The space of boundaries is contained in the set of nonempty closed sets in $\mathbf{R}^2$.
- A2 With respect to the topology induced by the Hausdorff metric on the space of boundaries $\nu(\cdot)$ is a nonnegative lower semicontinuous, coercive functional. (I.e. ν bounded sets are compact.)

We now generalize the functional E somewhat in anticipation of other applications. Henceforth E is defined by,

$$E(f, \Gamma) = \int_{\Omega\setminus\Gamma} \Phi(g, f, D^{\alpha_1} f, D^{\alpha_2} f, \ldots, D^{\alpha_s} f) + \nu(\Gamma)$$

and, for convenience we introduce the notation,

$$J(f, \Gamma) = \int_{\Omega\setminus\Gamma} \Phi(g, f, D^{\alpha_1} f, D^{\alpha_2} f, \ldots, D^{\alpha_s} f)$$

$g \in L^\infty(\Omega)$. s is a positive integer. Each α_i is a fixed multi-index, using the notation of [20]. f belongs to the subspace of functions in $L^{p_0}(\Omega\backslash\Gamma)$ whose distributional derivative $D^{\alpha_i} f$ exists as an $L^{p_i}(\Omega\backslash\Gamma)$ function, where each p_i satisfies $1 \le p_i < \infty$ for all $1 \le i \le s$. We will denote this space of functions by $\mathcal{D}(\Omega\backslash\Gamma)$. The following describes the assumptions on Φ.

A3 Φ is a nonnegative real function on $\mathbf{R}^{2+s}$ such that for any fixed domain $\Omega' \subset \Omega$ and fixed $g \in L^\infty(\Omega)$ the functional $\int_{\Omega'} \Phi(g, f, v_1, v_2, \ldots, v_s)$ is a lower semicontinuous, coercive functional on $L^{p_0}(\Omega') \times L^{p_1}(\Omega') \times \ldots \times L^{p_s}(\Omega')$ with respect to the weak (product) topology. Furthermore $\int_\Omega \Phi(g, 0, 0, \ldots, 0) < \infty$.

We note that $(g-f)^2 + v_1^2 + v_2^2$ is such a function with $p_0 = p_1 = p_2 = 2$. The formulation presented in the introduction satisfies these conditions with $\mathcal{D}(\Omega\backslash\Gamma) = W^{1,2}(\Omega\backslash\Gamma)$.

We now introduce a notion of convergence on sequences of pairs $\{(f_n, \Gamma_n)\}$. $(f_n, \Gamma_n) \to (f, \Gamma)$ will imply $\Gamma_n \to \Gamma$ in the topology induced by the Hausdorff metric. Now, for each n if $f_n \in L^p(\Omega\backslash\Gamma_n)$ let $\widehat{f_n} \in L^p(\Omega\backslash\Gamma)$ be defined by extending f_n to Ω, setting it to zero on Γ_n and then restricting it to $\Omega\backslash\Gamma$. By $(f_n, \Gamma_n) \to (f, \Gamma)$ we mean $\Gamma_n \to \Gamma$ in the topology induced by the Hausdorff metric and $\widehat{f_n} \to f$ weakly in $L^{p_0}(\Omega\backslash\Gamma)$ and $\widehat{D^{\alpha_i} f_n} \to D^{\alpha_i} f$ weakly in $L^{p_i}(\Omega\backslash\Gamma)$ for each $1 \le i \le s$.

LEMMA 4. *Under assumptions A1, A2 and A3 we can for any E bounded sequence $\{(f_n, \Gamma_n)\}$ extract a subsequence (also denoted $\{(f_n, \Gamma_n)\}$) such that for some boundary Γ and some $f \in \mathcal{D}(\Omega\backslash\Gamma)$, $(f_n, \Gamma_n) \to (f, \Gamma)$.*

Proof: Assume the conditions of the Lemma and suppose we are given an E bounded sequence. We can assume there is some Γ such that $\Gamma_n \to \Gamma$ since otherwise by assumption A2 we can first extract a subsequence and find a boundary with this property. Since the sequence is E bounded we can conclude from A3 that the sequence $\{\int_{\Omega\backslash\Gamma} \Phi(g, \widehat{f_n}, \widehat{D^{\alpha_1} f_n}, \ldots, \widehat{D^{\alpha_s} f_n})\}$ is bounded. Hence, by A3, we can find functions $f \in L^{p_0}(\Omega\backslash\Gamma), v_1 \in L^{p_1}(\Omega\backslash\Gamma), \ldots, v_s \in L^{p_s}(\Omega\backslash\Gamma)$ and a subsequence (which we still denote the same way) such that, $\widehat{f_n} \to f$ weakly in $L^{p_0}(\Omega\backslash\Gamma)$ and $\widehat{D^{\alpha_i} f_n} \to v_i$ weakly in $L^{p_i}(\Omega\backslash\Gamma)$ for each $1 \le i \le s$. We claim that $f \in \mathcal{D}$ and $D^{\alpha_i} f = v_i$.

Let g be any test function in $\Omega\backslash\Gamma$, i.e. $g \in C_o^\infty(\Omega\backslash\Gamma)$. Consider the subsequence extracted above. Since $d(\text{supp}(g), \Gamma) > 0$ (using A1) it follows that for n sufficiently large $g \in C_o^\infty(\Omega\backslash\Gamma_n)$ and $\widehat{f_n} = f_n$ on supp(g) for any f_n defined on $\Omega\backslash\Gamma_n$. Thus along the subsequence we have,

$$\begin{aligned}\int_{\Omega\backslash\Gamma} v_i g = \lim_{n\to\infty} \int_{\Omega\backslash\Gamma} \widehat{D^{\alpha_i} f_n} g \quad &= \lim_{n\to\infty} \int_{\Omega\backslash\Gamma_n} D^{\alpha_i} f_n g \\ = -\lim_{n\to\infty} \int_{\Omega\backslash\Gamma_n} f_n D^{\alpha_i} g &= -\lim_{n\to\infty} \int_{\Omega\backslash\Gamma} \widehat{f_n} D^{\alpha_i} g \\ = -\int_{\Omega\backslash\Gamma} f D^{\alpha_i} g &\end{aligned}$$

We conclude from this that $D^{\alpha_i} f = v_i$ and hence $f \in \mathcal{D}(\Omega\backslash\Gamma)$. □

COROLLARY. *If the space of boundaries is the space of closed sets in $\overline{\Omega}$ then for any J bounded sequence $\{(f_n, \Gamma_n)\}$ we can extract a subsequence (also denoted $\{(f_n, \Gamma_n)\}$) such that for some boundary Γ and some $f \in \mathcal{D}(\Omega\backslash\Gamma)$, $(f_n, \Gamma_n) \to (f, \Gamma)$.*

Proof: For this case Theorem 1 substitutes for A2, yeilding a Γ and a subsequence such that $\Gamma_n \to \Gamma$. The rest of the proof is the same. □

LEMMA 5. *Let $\{(f_n, \Gamma_n)\}$ be any E bounded sequence such that $(f_n, \Gamma_n) \to (f, \Gamma)$, then under assumptions A1, A2 and A3*

$$E(f, \Gamma) \leq \liminf_{n \to \infty} E(f_n, \Gamma_n)$$

Proof: Let Γ^ϵ be a closed ϵ neighbourhood of Γ, i.e. a closed neighbourhood of Γ such that $r(\Gamma^\epsilon, \Gamma) \leq \epsilon$ and define,

$$E_\epsilon(f, \Gamma') = \int_{\Omega\backslash\Gamma^\epsilon\cup\Gamma'} \Phi(g, f, D^{\alpha_1}f, \ldots, D^{\alpha_s}f) + \nu(\Gamma')$$

For n sufficiently large ($\geq N$ say), $\Gamma_n \subset \Gamma^\epsilon$ and since $\Gamma \subset \Gamma^\epsilon$ we get $\widehat{D^{\alpha_i} f_n}|_{\Omega\backslash\Gamma^\epsilon\cup\Gamma_n} = D^{\alpha_i} f_n|_{\Omega\backslash\Gamma^\epsilon}$. Hence the sequence $\{D^{\alpha_i} f_n|_{\Omega\backslash\Gamma^\epsilon\cup\Gamma_n}\}_{n\geq N}$ converges weakly to $D^{\alpha_i} f|_{\Omega\backslash\Gamma^\epsilon}$ in $L^{p_i}(\Omega\backslash\Gamma^\epsilon)$ and similarly $\{f_n|_{\Omega\backslash\Gamma^\epsilon\cup\Gamma_n}\}_{n\geq N}$ converges weakly to $f|_{\Omega\backslash\Gamma^\epsilon}$ in $L^{p_0}(\Omega\backslash\Gamma^\epsilon)$. We can now write

$$\begin{aligned}
\liminf_{n\to\infty} E_\epsilon(f_n, \Gamma_n) &\geq \liminf_{n\to\infty} \int_{\Omega\backslash\Gamma^\epsilon\cup\Gamma_n} \Phi(g, f_n, D^{\alpha_1}f_n, \ldots, D^{\alpha_s}f_n) + \liminf_{n\to\infty} \nu(\Gamma_n) \\
&\geq \int_{\Omega\backslash\Gamma^\epsilon} \Phi(g, f, D^{\alpha_1}f, \ldots, D^{\alpha_s}f) + \nu(\Gamma) \\
&= E_\epsilon(f, \Gamma)
\end{aligned}$$

where the second inequality follows from A_2 and lower semicontinuity of $\int \Phi$ in the weak topology on $\mathcal{D}(\Omega\backslash\Gamma^\epsilon)$. From the nonnegativity of Φ and the fact that Γ is closed we conclude $\sup_{\epsilon>0} J_\epsilon(\cdot) = J(\cdot)$ and hence

$$\liminf_{n\to\infty} E(f_n, \Gamma_n) \geq E(f, \Gamma) \qquad \square$$

THEOREM 10. *Under assumptions A1, A2 and A3 (and in particular letting ν be defined as in Section 2), there exists a minimizing pair (f, Γ) for the functional E.*

Proof: Apply Lemma 4 to a minimizing sequence, then apply Lemma 5. □

5. Application to Variational Problems in Vision. In this section we apply some results of the previous sections to the variational problem discussed in the introduction. As before, g represents an observed image defined on a bounded open set $\Omega \subset \mathbf{R}^2$, f is the reconstructed image, and Γ are the boundaries of the image. In the variational approach, f and Γ are obtained by minimizing the cost functional (1) or (2). Normally, g is assumed to be in $L^\infty(\Omega)$, Γ is a closed subset of

$\overline{\Omega}$, and f is in the Sobolev space $W^{1,2}(\Omega\backslash\Gamma)$. Under certain regularity assumptions, a number of interesting results concerning the nature of the minimizing f and Γ have been obtained [5,16,18,23]. Also, the existence of a minimizing pair (f, Γ) for various versions of the problem has been shown [1,2,17]. We have included the essence of [17] in Section 4.

Here we are concerned with the behavior of solutions to discrete versions of the problem as the lattice spacing tends to zero. Specifically, we are interested in whether or not solutions to the discrete problem converge to a solution of the continuous problem. It seems that this may not necessarily be the case for the discrete problem of (3). For example, consider the segmentation problem (2) where f is required to be piecewise constant. Take $\Omega = (0,1)\times(0,1)$, $g(x,y) = 0$ for $x < y$ and $g(x,y) = 1$ otherwise, and $4\sqrt{2}c_3 < c_1 < 8c_3$. Then the optimal solution to the discrete problem with sufficiently small lattice spacing seems to be $\Gamma = \emptyset$, while the optimal solution to the continuous problem seems to be $\Gamma = \{(x,x) : 0 \leq x \leq 1\}$.

This problem appears to be a result of the possible strict lower semicontinuity of the length of curves with respect to the Hausdorff metric. E.g., in this case, if $\Gamma = \{(x,x) : 0 \leq x \leq 1\}$ and Γ_n is the discrete approximation to Γ with lattice spacing $1/n$, then $\Gamma_n \to \Gamma$ but $L(\Gamma) = \sqrt{2}$ while $\lim_{n\to\infty} L(\Gamma_n) = 2$. The notion of length in the discrete case does not coincide in the continuum limit with the usual measure of length.

As previously mentioned, it may be possible to resolve this problem by modifying one or more of the topology on the space of boundaries, the cost functional, and the discretization process. Here we consider the use of Minkowski content for the cost of the boundaries and propose a modified discrete version of the problem. Specifically, given an observed image $g \in L^\infty(\Omega)$ we consider the problem of minimizing

$$E_\delta(f,\Gamma) = c_1 \int\int_\Omega (f-g)^2 \, dx\, dy + c_2 \int\int_{\Omega\backslash\Gamma} \| \nabla f \|^2 \, dx\, dy + c_3 \mathcal{M}_\delta(\Gamma)$$

with Γ a closed subset of $\overline{\Omega}$ and $f \in W^{1,2}(\Omega\backslash\Gamma)$. For the discrete version of the problem with lattice spacing $\frac{1}{n}$, we simply restrict Γ to be composed of a union of closed lattice squares whose corners lie on $\frac{1}{n}\mathbf{Z}^2$. However, we still take g and f to be defined on the continuous domains Ω and $\Omega\backslash\Gamma$ respectively. Hence, we have only incorporated a partial discretization, i.e. we have only discretized Γ. However, the primary difficulty in numerical solutions is to properly deal with the boundary. For a fixed Γ, the minimization reduces to a standard variational problem whose Euler-Lagrange equations can be solved by standard algorithms for partial differential equations.

We now give some results concerning the problem of minimizing E_δ.

THEOREM 11. *For every $\delta > 0$, there exists a pair $(f_\delta, \Gamma_\delta)$ which minimizes E_δ.*

Proof: Since we have shown that $\mathcal{M}_\delta$ is continuous (Theorem 6), the existence proof given in Section 4 can be applied. □

Note that for any bounded Γ, $\mathcal{M}_\delta(\Gamma) < \infty$. Hence, a minimizing boundary may quite possibly have nonzero Lebesgue measure.

The next theorem establishes the desirable property of discrete to continuous convergence for E_δ with a fixed $\delta > 0$. We will use the same notion of convergence as used in Section 4. For $f \in W^{1,2}(\Omega \backslash \Gamma_n)$, f and its weak first order derivatives $D_{x_i} f$, $i = 1, 2$, can be considered as functions in $L^2(\Omega \backslash \Gamma)$ by defining them to be zero on Γ_n and restricting. By $(f_n, \Gamma_n) \to (f, \Gamma)$ we mean that $\Gamma_n \to \Gamma$ in Hausdorff metric and that for the modified functions $\widehat{f_n} \to f$, and $\widehat{D_{x_i} f_n} \to D_{x_i} f$, $i = 1, 2$ weakly in $L^2(\Omega \backslash \Gamma)$.

THEOREM 12. *Let $(f^*_{\delta,n}, \Gamma^*_{\delta,n})$ denote a minimizing pair for $E_{\delta,n}$, i.e. for the discrete problem E_δ with lattice spacing $\frac{1}{n}$. Then there exists a subsequence (still denoted $(f^*_{\delta,n}, \Gamma^*_{\delta,n})$) and a pair $(f_\delta, \Gamma_\delta)$ such that $(f^*_{\delta,n}, \Gamma^*_{\delta,n}) \to (f_\delta, \Gamma_\delta)$ and $(f_\delta, \Gamma_\delta)$ minimizes E_δ.*

Proof: The existence of a pair $(f_\delta, \Gamma_\delta)$ with $(f^*_{\delta,n}, \Gamma^*_{\delta,n}) \to (f_\delta, \Gamma_\delta)$ follows from the corollary to lemma 4. We only need show that $(f_\delta, \Gamma_\delta)$ minimizes E_δ.

Let $(f^*_\delta, \Gamma^*_\delta)$ minimize E_δ. For each n, let Λ_n be obtained from Γ^*_δ by taking the smallest cover of Γ^*_δ using the closed lattice squares of the lattice with spacing $\frac{1}{n}$. Let h_n be the restriction of f^*_δ to $\Omega \backslash \Lambda_n$. From Theorem 6, $\lim_{n\to\infty} E_\delta(h_n, \Lambda_n) = E_\delta(f^*_\delta, \Gamma^*_\delta)$. Then, by the lower-semicontinuity of E_δ and the optimality of $(f^*_{\delta,n}, \Gamma^*_{\delta,n})$ for the discrete problem with lattice spacing $\frac{1}{n}$, we have

$$\begin{aligned} E_\delta(f_\delta, \Gamma_\delta) &\le \liminf_{n\to\infty} E_\delta(f_{\delta,n}, \Gamma_{\delta,n}) \le \liminf_{n\to\infty} E_\delta(h_n, \Lambda_n) \\ &= \lim_{n\to\infty} E_\delta(h_n, \Lambda_n) = E_\delta(f^*_\delta, \Gamma^*_\delta) \end{aligned}$$

Therefore, $E_\delta(f_\delta, \Gamma_\delta) = E_\delta(f^*_\delta, \Gamma^*_\delta)$ so that $(f_\delta, \Gamma_\delta)$ minimizes E_δ. □

A natural question at this point concerns the behavior of $(f^*_\delta, \Gamma^*_\delta)$ as $\delta \to 0$. One would like $(f^*_\delta, \Gamma^*_\delta)$ to converge to a minimizing solution of the original cost functional E. We can show a convergence result if the number of connected components of the admissible boundaries is uniformly bounded. Following Section 2, we let the cost term for the boundaries be

$$\nu_\delta(\Gamma) = \mathcal{M}^1_\delta(\Gamma) + F(\#(\Gamma))$$

where $F(k) = 0$ for $k \le M < \infty$ and $F(k) = \infty$ for $k > M$. Let E^M_δ denote the cost functional with the above boundary term, and let E^M denote the cost functional whose boundary term is

$$\nu(\Gamma) = \mathcal{M}^1(\Gamma) + F(\#(\Gamma))$$

By Theorem 8, $\mathcal{M}^1(\Gamma)$ in the equation for $\nu(\Gamma)$ could equivalently be replaced by $\mathcal{H}^1(\Gamma)$. For these variational problems, we have the following convergence result, which essentially follows from the result on the Γ-convergence of Minkowski content (Theorem 9).

THEOREM 13. *Let $(f^*_\delta, \Gamma^*_\delta)$ denote a minimizing pair for E^M_δ, and let $\delta_n \to 0^+$. Then there is a subsequence (which we still denote by δ_n) such that $(f^*_{\delta_n}, \Gamma^*_{\delta_n}) \to (f, \Gamma)$ for some (f, Γ) which minimizes E^M. Furthermore, $E^M_{\delta_n}(f^*_{\delta_n}, \Gamma^*_{\delta_n}) \to E^M(f, \Gamma)$.*

Proof: The existence of a pair (f, Γ) with $(f^*_{\delta_n}, \Gamma^*_{\delta_n}) \to (f, \Gamma)$ follows from corollary to lemma 4. We need to show that (f, Γ) minimizes E_δ and that $E^M_{\delta_n}(f^*_{\delta_n}, \Gamma^*_{\delta_n}) \to E^M(f, \Gamma)$.

This follows from Theorem 9 on the epi-convergence of Minkowski content in the case of a bounded number of connected components. Specifically,

$$E^M(f, \Gamma) \le \liminf_{n\to\infty} E^M_{\delta_n}(f_{\delta_n}, \Gamma_{\delta_n}) \le \liminf_{n\to\infty} E^M_{\delta_n}(f^*, \Gamma^*) = E^M(f^*, \Gamma^*)$$

so that (f, Γ) minimizes E^M.

Also, we have

$$E^M(f, G) = \limsup_{n\to\infty} E^M_{\delta_n}(f, \Gamma) \ge \limsup_{n\to\infty} E^M_{\delta_n}(f_{\delta_n}, \Gamma_{\delta_n})$$

Thus,

$$\limsup_{n\to\infty} E^M_{\delta_n}(f, \Gamma) \le E^M(f, \Gamma) \le \liminf_{n\to\infty} E^M_{\delta_n}(f, \Gamma)$$

and so

$$E^M(f, \Gamma) = \lim_{n\to\infty} E^M_{\delta_n}(f, \Gamma) \quad \square$$

Finally, we give a result concerning the convergence of solutions when the lattice spacing and δ are simultaneously allowed to go to zero. The following theorem guarantees convergence of a subsequence to a solution of the continuous problem if $\delta \to 0$ at a rate slower than the lattice spacing.

THEOREM 14. *Let $\delta_n > 0$ with $\delta_n \to 0$ and let $(f^*_{\delta_n,n}, \Gamma^*_{\delta_n,n})$ denote a minimizing pair for $E^M_{\delta_n,n}$, i.e. for the discrete problem $E^M_{\delta_n}$ with lattice spacing $\frac{1}{n}$. If $n\delta_n \to \infty$ as $n \to \infty$ then there exists a subsequence (still denoted $(f^*_{\delta_n,n}, \Gamma^*_{\delta_n,n})$) and a pair (f, Γ) such that $(f^*_{\delta_n,n}, \Gamma^*_{\delta_n,n}) \to (f, \Gamma)$ and (f, Γ) minimizes E^M.*

Proof: As before, the existence of a pair (f, Γ) with $(f^*_{\delta_n,n}, \Gamma^*_{\delta_n,n}) \to (f, \Gamma)$ follows from corollary to lemma 4 and so we need to show that (f, Γ) minimizes E^M.

Let (f^*, Γ^*) minimize E^M, and for each n let (h_n, Λ_n) be obtained from (f^*, Γ^*) as in the proof of Theorem 12. Namely, Λ_n is the smallest cover of Γ^* using lattice squares of the lattice with spacing $\frac{1}{n}$, and h_n is the restriction of f^* to $\Omega \backslash \Lambda_n$. Then using Theorem 9 and the optimality of $(f^*_{\delta_n,n}, \Gamma^*_{\delta_n,n})$ we have

$$E^M(f, \Gamma) \le \liminf_{n\to\infty} E^M_{\delta_n}(f^*_{\delta_n,n}, \Gamma^*_{\delta_n,n}) \le \liminf_{n\to\infty} E^M_{\delta_n}(h_n, \Lambda_n)$$

Since Λ_n is the minimal cover of Γ^* on the lattice with spacing $\frac{1}{n}$, we have $\Lambda_n \subset (\Gamma^*)^{(\frac{\sqrt{2}}{n})}$ so that

$$\begin{aligned}\liminf_{n\to\infty} \mathcal{M}^1_{\delta_n}(\Lambda_n) &\leq \liminf_{n\to\infty} \frac{\mu((\Gamma^*)^{(\delta_n+\frac{\sqrt{2}}{n})})}{2\delta_n} \\ &= \liminf_{n\to\infty} \frac{\mu((\Gamma^*)^{(\delta_n+\frac{\sqrt{2}}{n})})}{2(\delta_n+\frac{\sqrt{2}}{n})} \frac{\delta_n+\frac{\sqrt{2}}{n}}{\delta_n} \\ &= \lim_{n\to\infty} \frac{\mu((\Gamma^*)^{(\delta_n+\frac{\sqrt{2}}{n})})}{2(\delta_n+\frac{\sqrt{2}}{n})}(1+\frac{\sqrt{2}}{n\delta_n}) = \mathcal{M}^1(\Gamma^*)\end{aligned}$$

It follows that

$$\liminf_{n\to\infty} E^M_{\delta_n}(h_n,\Lambda_n) \leq E^M(f^*,\Gamma^*)$$

Therefore, $E^M(f,\Gamma) \leq E^M(f^*,\Gamma^*)$ and (f,Γ) minimizes E^M. □

REFERENCES

[1] AMBROSIO, L., *Existence Theory for a New Class of Variational Problems*, Center for Intelligent Control Systems Report CICS-P-93, MIT (1988).

[2] AMBROSIO, L., *Variational Problems in SBV*, Center for Intelligent Control Systems Report CICS-P-86, MIT (1988).

[3] ATTOUCH, H., *Variational Convergence for Functions and Operators*, Pitman Publishing Inc., 1984.

[4] BLAKE, A. AND A. ZISSERMAN, *Invariant surface reconstruction using weak continuity constraints*, Proc. IEEE Conf. Computer Vision and Pattern Recognition, Miami (1986), pp. 62–67.

[5] BLAKE, A. AND A. ZISSERMAN, *Visual Reconstruction*, MIT Press, 1987.

[6] DE GIORGI, E., *Γ-convergenza e G-convergenza*, Boll. Un. Mat. Ital. (5), 14-A (1977), pp. 213–220.

[7] DE GIORGI, E., *Convergence problems for functionals and operators. Proc. Int. Meeting on Recent Methods*, in *Nonlinear Analysis, Rene 1978, ed. E. De Giorgi*, Magenes, Mosco Pitagora, Bologna, 1979, pp. 131–188.

[8] DE GIORGI, E., *New problems in Γ-convergence and G-convergence*, Proc. Meeting on Free Boundary Problems, Pavia 1979, Istituto Nazionale di Atta Matematica, Roma, Vol. II (1980), pp. 183–194.

[9] DUBUC, B., C. ROQUES-CARMES, C. TRICOT, AND S.W. ZUCKER, *The variation method: a technique to estimate the fractal dimension of surfaces*, Vol. 845 Visual Communications and Image Processing II (1987), pp. 241–248.

[10] FALCONER, K.J., *The Geometry of Fractal Sets*, Cambridge University Press, 1985.

[11] FEDERER, H., *Geometric Measure Theory*, Springer-Verlag, 1969.

[12] GEMAN, S. AND D.GEMAN, *Stochastic Relaxation, Gibbs Distributions, and the Bayesian Restoration of Images*, IEEE Trans. Pattern Analysis and Machine Intelligence, 6 (1984), pp. 721–741.

[13] MANDELBROT, B.B., *The Fractal Geometry of Nature*, W.H. Freeman and Company, 1982.

[14] MARROQUIN, J.L., *Probabilistic Solution of Inverse Problems. Ph.D. Thesis*, Dept. of E.E.C.S., MIT (1985).

[15] MUMFORD, D. AND J. SHAH, *Boundary detection by minimizing functionals*, Proc. IEEE Conf. Computer Vision and Pattern Recognition, San Francisco (1985), pp. 22–26.

[16] MUMFORD, D. AND J. SHAH, *Optimal Approximations by Piecewise Smooth Functions and Associated Variational Problems*, Center for Intelligent Control Systems Report CICS-P-68 (1988); Also submitted to *Communications on Pure and Applied Mathematics*.

[17] RICHARDSON, T.J., *Existence Result for a Problem Arising in Computer Vision*, Center for Intelligent Control Systems Report CICS-P-63, MIT (1988).

[18] RICHARDSON, T.J., *Recovery of Boundaries by a Variational Method*, Center for Intelligent Control Systems report, MIT (to appear).
[19] ROGERS, C.A., *Hausdorff Measure*, Cambridge University Press, 1970.
[20] RUDIN, W., *Functional Analysis*, McGraw Hill, 1973.
[21] SERRA, J., *Image Analysis and Mathematical Morphology*, Academic Press Inc., 1982.
[22] SHAH, J., *Segmentation by Minimizing Functionals: Smoothing Properties*, to be published.
[23] WANG, Y., unpublished notes.

EXTENSION PROBLEMS UNDER THE DISPLACEMENT STRUCTURE REGIME*

H. LEV-ARI†

Abstract. This paper presents a unified approach to certain function-theoretic and matrix extension problems, which is based on the recently developed concept of matrices with a generalized displacement structure. We show that a variety of function extension problems, including Pade approximation and Caratheodory extension, are equivalent to the problem of extending a finite matrix with a given displacement structure into a larger (possibly infinite) matrix with the same structure. Moreover, such matrix extension problems can be efficiently solved by the same layer-peeling procedure that is used to determine the triangular factorization of matrices with a generalized displacement structure.

In general, the matrix extension problem mentioned above has many solutions. The set of all feasible extensions can be conveniently characterized in terms of the cascade model that is constructed by the layer peeling procedure: a particular extension is obtained by terminating the cascade model (which is the same for all extensions) with an arbitrary termination. Most often, the desired solution corresponds to a matrix extension that has finite rank. This is so, for instance when the Pade approximation is required to have minimal degree, or when the Schur-Caratheodory extension is required to have minimal H^∞ norm (but not so for maximum-entropy extension problems). We show that such finite-rank extensions correspond to choosing lossless terminations for the cascade model, and that the resulting extended matrix is intimately connected with the recently developed notion of generalized Bezoutians.

1. Introduction. Two classical extension problems frequently arise in a variety of signal processing and linear system theory applications. These are:

(i) *The Caratheodory-Fejer problem*: Given $c_1, c_2, \ldots, c_m$ construct a power-series

$$c(z) \ := 1 + \sum_{t=1}^{\infty} c_t z^t$$

such that $Re\ c(z) \geq 0$ for all $|z| < 1$.

(ii) *The Padé (partial realization) problem*: Given $h_0, h_1, \ldots, h_m$ construct a power-series

$$h(z) \ := \sum_{t=0}^{\infty} h_t z^t$$

such that $h(z)$ is rational of degree r, i.e., $h(z) = b(z)/a(z)$ with $deg\ b(z) \leq deg\ a(z) = r$.

---

*Research supported in part by the Air Force Office of Scientific Research, Air Force Systems Command under Contract AF-88-0327, by the Department of the Navy, Office of Naval Research, under Contract N00014-85-K-0612, by the U.S. Army Research Office, under Contract DAAL03-86-K-0045.

†Information Systems Laboratory, Stanford University Stanford, CA 94305

Both extension problems can be solved via a set of linear equations involving a structured matrix, i.e.,

$$\mathbf{T} := \begin{pmatrix} 1 & c_1^* & c_2^* & \cdots \\ c_1 & 1 & c_1^* & \\ c_2 & c_1 & 1 & \\ \vdots & & & \ddots \end{pmatrix}, \qquad \mathbf{T} \geq 0$$

for the Caratheodory-Fejer problem [see, e.g., Akhiezer (1965), Krein and Nudelman (1977), Kalman (1981)], and

$$\mathbf{H} := \begin{pmatrix} h_0 & h_1 & h_2 & \cdots \\ h_1 & h_2 & & \\ h_2 & & & \\ \vdots & & & \ddots \end{pmatrix}, \qquad \mathit{rank}\ \mathbf{H} = \mathit{deg}\ a(z) = r$$

for the partial realization problem [see, e.g., Kalman (1979), Rissanen (1971)]. Alternatively both problems can be solved by recursively constructing a cascade model with a suitable termination [see, e.g., Brent, Gustavson and Yun (1980), Citron, Bruckstein and Kailath (1984)]. The known data (i.e., the known coefficients of $c(z)$ or $h(z)$) determine the input/output signals $u_0(z)$, $v_0(z)$ and these serve to determine the internal parameters of the cascade model (Fig. 1).

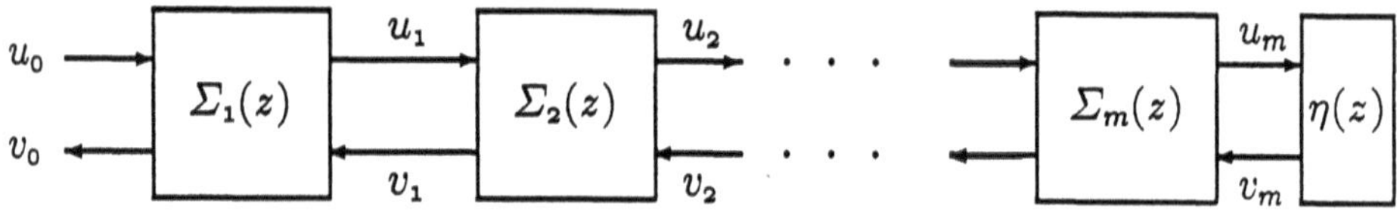

Fig. 1 Terminated Lossless Cascade Network

The procedure employed to construct the cascade model from its input/output signals is known as *layer peeling* [Bruckstein and Kailath (1987)]. The unknown system is identified by layers (also known as sections): the outermost layer is identified first, then peeled off to reveal the input/output signals associated with the next layer. This means that beginning with the given boundary data $u_0(z)$ and $v_0(z)$ one repeats the following steps for $i = 0, 1, \ldots$:

(a) Identify the internal parameters of the i-th layer from $u_i(z)$, $v_i(z)$.

(b) Compute $u_{i+1}(z)$, $v_{i+1}(z)$ via the *chain-scattering matrix* $\Theta_i(z)$, which relates the two peripheries of the i-th layer, i.e.,

$$\begin{pmatrix} u_{i+1}(z) \\ v_{i+1}(z) \end{pmatrix} = \Theta_i(z) \begin{pmatrix} u_i(z) \\ v_i(z) \end{pmatrix}, \quad i \geq 0 \tag{1.1}$$

A simple transformation maps the chain-scattering matrix $\Theta_i(z)$ into the *scattering matrix* $\sum_i(z)$, which relates the outputs of the i-th layer to its inputs, viz.,

$$\begin{pmatrix} u_{i+1}(z) \\ v_i(z) \end{pmatrix} = \Sigma_i(z) \begin{pmatrix} u_i(z) \\ v_{i+1}(z) \end{pmatrix} \quad . \tag{1.2}$$

This layer-peeling procedure can also be interpreted as an efficient recursive procedure for triangular factorization of certain structured matrices. In particular, it is well known that the Schur algorithm produces a triangular factorization of Toeplitz matrices [see, e.g., Bareiss (1969), Rissanen (1973), Morf (1974), Le Roux and Gueguen (1977)], and of certain related matrices [Kailath and Lev-Ari (1984), Kailath, Bruckstein and Morgan (1986)], while the Lanczos algorithm produces the triangular factorization of Hankel matrices [see, e.g., Rissanen (1973), Kung (1977), Citron, Bruckstein and Kailath (1984), Bruckstein and Kailath (1987)].

The foregoing interpretation has motivated us in the past to introduce a notion we called *generalized displacement structure*. Matrices that have this structure can be efficiently factored by a layer-peeling procedure, which generates a sequence of chain-scattering matrices $\Theta_i(z)$, thereby associating a cascade model with each structured matrix [Lev-Ari and Kailath (1986)]. Moreover, this association determines a *nested bi-unique correspondence* between the coefficients $\{K_i \ ; \ 0 \le i \le m\}$ of the cascade model (i.e., the parameters of the chain-scattering matrices $\Theta_0(z), \Theta_1(z), \ldots, \Theta_m(z)$), and the elements of the leading principal submatrix

$$\mathbf{R}_{0:m} := \{r_{t,s} \ ; \ 0 \le t,s \le m\} \tag{1.3a}$$

of the associated structured matrix

$$\mathbf{R} := \{r_{t,s} \ ; \ 0 \le t,s < \infty\} \ . \tag{1.3b}$$

This means that the (algebraic) structure of the matrix $\mathbf{R}$ is captured by the internal structure of the chain-scattering matrices $\Theta_i(z)$: a sequence of such matrices uniquely determines a matrix $\mathbf{R}$ with a specified displacement structure. Notice that $\Theta_i(z)$ may involve several parameters, so K_i is, in general, a *vector*.

We can now recast the classical extension problems as particular cases of the following matrix extension problem:

> Given $\mathbf{R}_{0:m} := \{r_{t,s} \ ; \ 0 \le t,s \le m\}$ and a specified displacement structure, construct a *structured Hermitian matrix* $\mathbf{R} := \{r_{t,s}; 0 \le t,s < \infty\}$ that matches the given $\mathbf{R}_{0:m}$.

This structured matrix extension problem can be solved by constructing the cascade model that corresponds to the specified displacement structure and determining its first $m+1$ coefficients. Because of the nested bi-unique correspondence,

$$\mathbf{R}_{0:m} \longleftrightarrow K_{0:m} := \{K_i \ ; \ 0 \le i \le m\} \tag{1.4}$$

our extension problem reduces to the selection of the missing parameters $K_{m+1:\infty}$ or, equivalently, to the selection of a termination $\eta(z)$ that represents the effect of the cascade model corresponding to the parameters $K_{m+1:\infty}$ (Fig. 1).

The advantage of solving the extension problem in terms of the cascade coefficients K_i rather than directly in terms of the elements of $\mathbf{R}$ is that the latter are (non-trivially) constrained by the displacement structure characterization, whereas the former are subject to simple norm constraints. For instance, the cascade model associated with Toeplitz matrices involves scalar coefficients k_i, usually known as *reflection coefficients*, whose magnitudes are bounded by unity, viz., $|k_i| \leq 1$.

Since the termination $\eta(z)$ represents the coefficients $K_{m+1:\infty}$, the foregoing discussion implies that every choice of $\eta(z)$ results in a feasible extension of $\mathbf{R}_{0:m}$. When $\mathbf{R} > 0$ is also required, the termination $\eta(z)$ has to satisfy an additional norm constraint. For instance, this constraint becomes $|\eta(z)| \leq 1$ for all $z \leq 1$ in the context of positive-definite extensions of Toeplitz matrices. Further constraints may be imposed on $\mathbf{R}$ (and, consequently, on $\eta(z)$) in order to characterize a specific extension, such as:

(i) Maximum-entropy, which leads to $\det \mathbf{R} \to \max$.

(ii) Minimum-norm, which leads to $\lambda_{\min}(\mathbf{R}) \to \max$.

(iii) Minimal-order, which leads to *rank* $\mathbf{R} \to \min$.

Such constraints translate into specific requirements on the termination $\eta(z)$.

One convenient way to describe structured matrices is via the bivariate Z-transform (= generating function) of $\mathbf{R}$, viz.,

$$R(z,w) = [1 \; z \; z^2 \ldots] \; \mathbf{R}[1 \; w \; w^2 \ldots]^* \tag{1.5}$$

The matrix $\mathbf{R}$ is said to have a *displacement structure* when its generating function has the form

$$R(z,w) = \frac{G(z)J \; G^*(w)}{d(z,w)} \quad , \qquad G(z) := \sum_{t=0}^{\infty} g_t z^t \tag{1.6}$$

where the *displacement kernel* $d(z,w)$ is a bivariate function with the Hermitian property $d^*(z,w) = d(w,z)$, the constant matrix J is Hermitian, i.e., $J = J^*$, and $G(z)$ is a power series with (row) vector coefficients $\{g_t\}$. The triple $\{d(z,w), G(z), J\}$ has been called a *generator* of the Hermitian matrix $\mathbf{R}$, since it uniquely determines $R(z,w)$ (and hence also $\mathbf{R}$) via (1.6) [see Lev-Ari and Kailath (1986) for more details]. For instance, the displacement structure exhibited by *Toeplitz* matrices (with $c_0 = 1$) is

$$\mathbf{R} = \{c_{t-s}; 0 \leq t, s < \infty\} \quad , \quad J = J_I \; := \begin{pmatrix} 0 & 1 \\ 1 & 0 \end{pmatrix}$$

$$R(z,w) = \frac{c(z) + c^*(w)}{1 - zw^*} = \frac{[1 \quad c(z)] J_I [1 \quad c(w)]^*}{1 - zw^*}$$

while for *Hankel* matrices

$$\mathbf{R} = \{h_{t+s} : 0 \leq t, s < \infty\} \quad , \quad J = J_R := \begin{pmatrix} 0 & -j \\ j & 0 \end{pmatrix}$$
$$R(z,w) = \frac{zh(z) - w^*h^*(w)}{z - w^*} = \frac{[1 \quad zh(z)]J_R[1 \quad wh(w)]^*}{j(z - w^*)}$$

The importance of the notion of displacement structure lies in the fact that by matching a suitable displacement structure to a particular family of matrices, the *displacement rank*, i.e., the size of the matrix J, can be reduced to a small constant, characteristic of the family. Thus, for instance, the displacement rank of Toeplitz and Hankel matrices is equal to 2.

We have shown that efficient factorization of structured matrices is possible whenever the displacement kernel $d(z,w)$ has the special form [Lev-Ari and Kailath (1986)]

$$d(z,w) = \alpha(z)\alpha^*(w) - \beta(z)\beta^*(w) \ . \tag{1.7}$$

For instance, $d(z,w) = 1 - zw^*$ for Toeplitz matrices, while for Hankel matrices $d(z,w) = j(z - w^*) = 1/2[(1+jz)(1+jw)^* - (1-jz)(1-jw)^*]$. The factorization procedure for such matrices produces a cascade model whose layers are characterized by the chain-scattering matrix

$$\Theta_i(z) = [I - \frac{d(z,\tau_i)}{d(z,0)d(0,\tau_i)} JM_i]W_i \tag{1.8a}$$

where the matrix W_i is arbitrary except for the *J-unitarity* requirement

$$W_iJ\ W_i^* = J \ , \tag{1.8b}$$

the complex coefficient τ_i can be any point such that

$$d(\tau_i, \tau_i) = 0 \tag{1.8c}$$

and M_i is a rank-one matrix constructed from the input to the i-th layer, i.e., from $G_i(z) = G(z)\Theta_0(z)\Theta_1(z)\ldots\Theta_{i-1}(z)$, viz.,

$$M_i := \lim_{z \to 0} \left\{ \frac{G_i^*(z)d(z,z)G_i(z)}{G_i(z)J\ G_i^*(z)} \right\} \tag{1.8d}$$

We establish this factorization procedure in Section 2, while in Section 3 we discuss the one-to-one correspondence (1.4), viz., $\mathbf{R}_{0:m} \longleftrightarrow K_{0:m}$. Finally, in Section 4 we show how this correspondence can be used to solve the structured matrix extension problem, and we study the effect of the termination $\eta(z)$ on the properties of the extension $\mathbf{R}$.

2. Factorization of Structured Matrices. In this section we provide a brief overview of the *Jacobi procedure* for triangular (LDU) factorization of *strongly regular* matrices, i.e., matrices with non-vanishing leading principal minors. In particular, this procedure applies to (infinite) strongly-regular Hermitian matrices, resulting in a factorization of the form

$$\mathbf{R} = \mathbf{L}\mathbf{D}\mathbf{L}^* \tag{2.1}$$

where $\mathbf{L}$ is a lower-triangular matrix with unit diagonal elements and $\mathbf{D}$ is a diagonal matrix. Notice that this triangular factorization is *nested*, i.e.,

$$\mathbf{R}_{0:m} = \mathbf{L}_{0:m}\mathbf{D}_{0:m}\mathbf{L}^*_{0:m}$$

where $\mathbf{R}_{0:m}$ (resp. $\mathbf{L}_{0:m}, \mathbf{D}_{0:m}$) denotes the $(m+1)\times(m+1)$ leading principal submatrix of $\mathbf{R}$ (resp. $\mathbf{L}$,$\mathbf{D}$). Since the actual size of $\mathbf{R}$ does not affect the factorization of its leading principal submatrices we may assume, without loss of generality, that $\mathbf{R}$ (resp. $\mathbf{L}$,$\mathbf{D}$) is, in fact, semi-infinite, viz.

$$\mathbf{R} = \{r_{t,s}; 0 \le t, s < \infty\}. \tag{2.2}$$

When the matrix $\mathbf{R}$ is not strongly-regular one can still pursue block-triangular factorizations, but we shall not do so here.

Denoting the columns of the matrix $\mathbf{L}$ by $\{\mathbf{l}_i; i = 0,1,2,\dots\}$ and the diagonal elements of the matrix $\mathbf{D}$ by $\{d_i; i = 0,1,2,\dots\}$ we observe that the triangular decomposition (2.1) implies that $\mathbf{R}$ admits an additive decomposition, viz.,

$$\mathbf{R} = \sum_{i=0}^{\infty} d_i \mathbf{l}_i \mathbf{l}_i^* \tag{2.3}$$

This produces the following recursive formulation of the *Jacobi procedure*:

$$\mathbf{R}_{i+1} := \mathbf{R}_i - d_i\ \mathbf{l}_i\mathbf{l}_i^*, \qquad \mathbf{R}_0 := \mathbf{R}\ . \tag{2.4a}$$

Since the first $i+1$ rows and columns of $\mathbf{R}_{i+1}$ contain only zeros, we conclude that

$$\mathbf{R}_{i+1}\mathbf{e}_i^* = 0 = \mathbf{R}_i\mathbf{e}_i^* - d_i\ \mathbf{l}_i\ (\mathbf{e}_i\mathbf{l}_i)^*$$

where $\mathbf{e}_i$ is the unit vector

$$\mathbf{e}_i := [\ \underbrace{0\dots0}_{i}\ \ 1\,0\dots\,]\ . \tag{2.4b}$$

Also, since $\mathbf{L}$ has unity diagonal elements, we have $\mathbf{e}_i\mathbf{l}_i = 1$ and, consequently,

$$d_i := \mathbf{e}_i\mathbf{R}_i\mathbf{e}_i^*\quad,\quad \mathbf{l}_i := \mathbf{R}_i\mathbf{e}_i^* d_i^{-1}\ . \tag{2.4c}$$

The strong regularity of **R** implies that $d_i \neq 0$ for all i , so that the division in (2.4c) can always be carried out.

The Jacobi transformation (2.4) can be expressed also in terms of generating functions by defining

$$(2.5a) \qquad (zw^*)^i R_i(z,w) \ := [1\ z\ z^2 \ldots]\mathbf{R}_i[1\ w\ w^2 \ldots]^*$$

and

$$(2.5b) \qquad z^i l_i(z) \ := [1\ z\ z^2 \ldots]\mathbf{l}_i \quad ,$$

which translates (2.4) into the transform-domain recursion

$$(2.6) \quad zw^*\ R_{i+1}(z,w) = R_i(z,w) - R_i(z,0)R_i^{-1}(0,0)R_i(0,w) \ := \mathcal{X}\{R_i(z,w)\} \ ,$$

with the components of the triangular factorization of **R** obtained via the expressions

$$(2.7) \qquad d_i = R_i(0,0 \quad , \qquad l_i(z) = R_i(z,0)d_i^{-1} \ .$$

Consequently, the matrix **R** can be efficiently factored if we can establish an efficient procedure for evaluating the map $\mathcal{X}\{R(z,w)\}$ or the more general map

$$(2.8) \qquad \mathcal{X}_\zeta\{R(z,w)\} \ := R(z,w) - R(z,\zeta)R^{-1}(\zeta,\zeta)R(\zeta,w) \ ,$$

which is useful in certain circuit and network theoretic applications [see, e.g., Dewilde et al. (1978), Dewilde and Dym (1981,1984)]. In this case the recursion (2.6) is replaced by

$$(2.9) \qquad (z-\zeta_i)(w-\zeta_i)^* R_{i+1}(z,w) = \mathcal{X}_{\zeta_i}\{R_i(z,w)\} \ .$$

where some (or all) of the elements of the sequence $\{\zeta_i\ ; 0 \le i < \infty\}$ may coincide.

Now, for matrices with a displacement structure, combining (2.8) with (1.6) results in

$$(2.10a) \qquad \mathcal{X}_\zeta\left\{\frac{G(z)J\ G^*(w)}{d(z,w)}\right\} = \frac{G(z)\left\{J - \dfrac{d(z,w)}{d(z,\zeta)d(\zeta,w)}\ J\ M\ J\right\}G^*(w)}{d(z,w)}$$

where we have defined

$$(2.10b) \qquad M \ := G^*(\zeta)R^{-1}(\zeta,\zeta)G(\zeta) = M^* \ .$$

Note that $R(\zeta,\zeta) \neq 0$ is required and that

$$(2.10c) \qquad M\ J\ M^* = d(\zeta,\zeta)M \ .$$

A significant reduction in the computation of $\mathbf{X}\{\cdot\}$ can be achieved for such matrices if there exists a matrix function $\Theta(z)$, such that

$$J - \frac{d(z,w)}{d(z,\zeta)d(\zeta,w)}\; J\; M\; J = \Theta(z) J\; \Theta^*(w)\;. \tag{2.11}$$

If (2.11) can be solved for $\Theta(z)$ then

$$\mathbf{X}_\zeta\{R(z,w)\} = \frac{\{G(z)\Theta(z)\}J\{G(w)\Theta(w)\}^*}{d(z,w)}$$

and a single step of the Jacobi transformation will be translated into the linear univariate map

$$(z-\zeta_i)G_{i+1}(z) = G_i(z)\Theta_i(z)\;, \tag{2.12}$$

whose computational complexity is significantly lower than that of the (nonlinear) bivariate map (2.9).

If (2.11) admits a solution $\Theta(z)$, it also admits the solution $\Theta(z)W$ where W is any constant *J-unitary* matrix, i.e., $WJW^* = J$. It can be shown [see, e.g. Potapov (1960)] that the general solution of this equation is of the form $\Theta(z) = \overline{\Theta}(z)W$, where $\overline{\Theta}(z)$ is any specific solution and W is any J-unitary matrix. An explicit expression for $\overline{\Theta}(z)$ follows from the observation that for every solution $\Theta(z)$ the matrix $\Theta(\tau)$ is J-unitary for every $\tau \in \Omega$, where

$$\Omega\; := \{z; d(z,z) = 0\}\;, \tag{2.13a}$$

Consequently, if $\Theta(z)$ satisfies (2.11) so does $\Theta(z)\Theta^{-1}(\tau)$ for every $\tau \in \Omega$. In other words, for every $\tau \in \Omega$ equation (2.11) has a solution $\Theta(z)$ with the property $\Theta(\tau) = I$. Therefore, the general solution of (2.11) is given by the explicit expression

$$\Theta(z) = [I - \frac{d(z,\tau)}{d(z,\zeta)d(\zeta,\tau)}\; JM]W \tag{2.13b}$$

where the scalar τ and the matrix W need only to satisfy the constraints

$$\tau \in \Omega, \quad W\; J\; W^* = J\;. \tag{2.13c}$$

Notice that, in general, a different pair $\{\tau_i, W_i\}$ can be used for each layer, i.e., for each $\Theta_i(z)$.

It now follows that (2.11) has a solution if, and only if, there exists a $\tau \in \Omega$ such that $d(z,w)$ satisfies the equation

$$d(z,w) = [d(z,\zeta)\quad \frac{d(z,\tau)}{d(\zeta,\tau)}] \begin{pmatrix} 0 & 1 \\ 1 & -d(\zeta,\zeta) \end{pmatrix} [d(w,\zeta)\quad \frac{d(w,\tau)}{d(\zeta,\tau)}]^*\;, \tag{2.14}$$

which is obtained by substituting (2.13b) into (2.11). Furthermore, this rather implicit condition is equivalent to the more explicit requirement that there exist power series $\alpha(z), \beta(z)$ such that

$$d(z,w) = \alpha(z)\alpha^*(w) - \beta(z)\beta^*(w) \ . \tag{2.15}$$

The equivalence of (2.14) and (2.15) is established by the following observations:

(i) If (2.14) holds then, applying the identity

$$\begin{pmatrix} 0 & 1 \\ 1 & -d(\zeta,\zeta) \end{pmatrix} = T \begin{pmatrix} 1 & 0 \\ 0 & -1 \end{pmatrix} T^* \ , \tag{2.16a}$$

where

$$T = \frac{1}{2} \begin{pmatrix} 2 & -2 \\ 1-d(\zeta,\zeta) & 1+d(\zeta,\zeta) \end{pmatrix} \ , \tag{2.16b}$$

we conclude that (2.15) holds with

$$\alpha(z) = 1 + \frac{1-d(\zeta,\zeta)}{2} \, \frac{d(z,\tau)}{d(z,\zeta)d(\zeta,\tau)}$$

$$\beta(z) = 1 - \frac{1+d(\zeta,\zeta)}{2} \, \frac{d(z,\tau)}{d(z,\zeta)d(\zeta,\tau)} \ .$$

(ii) Conversely, if (2.15) holds then it follows by direct calculation that for every ζ, and for every $\tau \in \Omega$

$$Q^* \begin{pmatrix} 0 & 1 \\ 1 & 1-d(\zeta,\zeta) \end{pmatrix} Q = \begin{pmatrix} 1 & 0 \\ 0 & -1 \end{pmatrix} \tag{2.17a}$$

where

$$Q := \begin{pmatrix} 1 & 0 \\ 0 & d(\tau,\zeta) \end{pmatrix}^{-1} \begin{pmatrix} \alpha(\zeta) & \beta(\zeta) \\ \alpha(\tau) & \beta(\tau) \end{pmatrix} \ . \tag{2.17b}$$

Consequently,

$$\begin{aligned} d(z,w) &= [\alpha(z) \quad -\beta(z)] \begin{pmatrix} 1 & 0 \\ 0 & -1 \end{pmatrix} [\alpha(w) \quad -\beta(w)]^* \\ &= [\alpha(z) \quad -\beta(z)] Q^* \begin{pmatrix} 0 & 1 \\ 1 & -d(\zeta,\zeta) \end{pmatrix} Q [\alpha(w) \quad -\beta(w)]^* \\ &= [d(z,\zeta) \quad \frac{d(z,\tau)}{d(\zeta,\tau)}] \begin{pmatrix} 0 & 1 \\ 1 & -d(\zeta,\zeta) \end{pmatrix} [d(w,\zeta) \quad \frac{d(w,\tau)}{d(\zeta,\tau)}]^* \end{aligned}$$

so that (2.14) holds.

This establishes the equivalence of (2.14) and (2.15).

We shall now illustrate the application of the explicit formula (2.13) by considering two frequently encountered families of structured matrices, and their corresponding $\Theta_i(z)$:

(i) *Quasi-Toeplitz Matrices*: these are matrices whose generating function is

$$R(z,w) = \frac{G(z) J_T G^*(w)}{I - zw^*}, \quad J_T := \begin{pmatrix} 1 & 0 \\ 0 & -1 \end{pmatrix}. \tag{2.18a}$$

This includes Toeplitz matrices, for which $G(z) = 1/2[c(z)+1 \quad c(z)-1]$. By making the following choices

$$\zeta_i = 0, \quad \tau_i = 1, \quad W_i = \frac{1}{\sqrt{1-|k_i|^2}} \begin{pmatrix} 1 & -k_i \\ -k_i^* & 1 \end{pmatrix} \tag{2.18b}$$

and adjusting k_i to get $G_i(0) W_i \sim [1 \quad 0]$ we obtain the chain-scattering matrix

$$\Theta_i(z) = \frac{1}{\sqrt{1-|k_i|^2}} \begin{pmatrix} 1 & -k_i \\ -k_i^* & 1 \end{pmatrix} \begin{pmatrix} z & 0 \\ 0 & 1 \end{pmatrix}. \tag{2.18c}$$

It turns out that k_i is the ratio of the elements of $G_i(0)$, i.e., $G_i(0) \sim [1 \quad k_i]$.

(ii) *Quasi-Hankel Matrices*: these are matrices whose generating function is

$$R(z,w) = \frac{G(z) J_R G^*(w)}{j(z - w^*)}, \quad J_R := \begin{pmatrix} 0 & -j \\ j & 0 \end{pmatrix}. \tag{2.19a}$$

This includes Hankel matrices, for which $G(z) = [1 \quad zh(z)]$. By making the following choices

$$\zeta_i = 0, \quad \tau_i = j\infty, \quad W_i = \begin{pmatrix} \alpha_i^{-1} & -\beta_i \\ 0 & \alpha_i \end{pmatrix} \tag{2.19b}$$

and adjusting α_i, β_i to get $G_i(0) W_i \sim [1 \quad 0]$ we obtain the chain-scattering matrix

$$\Theta_i(z) = \begin{pmatrix} \alpha_i^{-1} & -\beta_i \\ 0 & \alpha_i \end{pmatrix} \begin{pmatrix} 1 & 0 \\ \epsilon_i z^{-1} & 1 \end{pmatrix}. \tag{2.19c}$$

It turns out that $G_i(0) = [\alpha_i \quad \beta_i]\sigma_i$, where σ_i^2 and ϵ_i are the magnitude and the sign of $d_i = R_i(0,0)$, respectively.

3. Cascade Models for Structured Matrices. The factorization procedure (2.12)–(2.13) maps every structured matrix $\mathbf{R}$ into an (infinite) sequence of chain-scattering-matrices $\{\Theta_i(z); i = 0, 1, 2, \dots\}$, which together form a *cascade model*. We shall use this mapping to establish a one-to-one correspondence between the elements of the structured matrix $\mathbf{R}$ and the coefficients of the cascade model, which we have denoted by $\{K_i; i = 0, 1, 2, \dots\}$. Furthermore, we shall show that this correspondence is nested, namely, that the leading principal $(m+1) \times (m+1)$ submatrix $\mathbf{R}_{0:m}$ corresponds to the leading segment of cascade coefficients $K_0, K_1, \dots, K_m$. This makes it possible to obtain a convenient characterization of all solutions to the structured matrix extension problem introduced in Section 1.

Our first observation is that in order to determine the cascade model that corresponds to a given structured matrix $\mathbf{R}$ we need to select a *generator* $\{d(z,w), G(z), J\}$ for this matrix. Since this can be done non-uniquely (i.e., each structured matrix has many generators) it follows that every structured matrix gives rise, in general, to a whole *family of cascade models*. Nevertheless, it is possible in many cases to select a unique representative for each family; in other words we can associate a unique *canonical* generator with certain types of structured matrices. This is true, in particular, for *moment matrices*, i.e., for structured matrices with generating functions $R(z,w)$ of the form

$$R(z,w) = \frac{f(z) + f^*(w)}{d(z,w)} , \tag{3.1a}$$

with

$$f(0) = \frac{1}{2}\, d(0,0) \tag{3.1b}$$

for which we *define* a canonical generator as

$$G(z) = [1 \quad f(z)] , \quad J = \begin{pmatrix} 0 & 1 \\ 1 & 0 \end{pmatrix} \tag{3.2a}$$

where

$$f(z) := d(z,0)R(z,0) - \frac{1}{2}d(0,0) . \tag{3.2b}$$

Thus (3.1) and (3.2) determine a one-to-one correspondence between moment matrices and their generators. Notice that $R(0,0) = 1$ as a consequence of our definition (3.1).

The name 'moment matrix' originates from the observation that such matrices are related to moments of (normalized) measures on curves in the complex plane. In particular, Toeplitz matrices are associated with the unit circle, which is characterized by the displacement kernel $d(z,w) = 1 - zw^*$, while Hankel matrices are associated with the real line, which is characterized by $d(z,w) = j(z - w^*)$. Thus, the characterization (3.1) includes the classical extension problems, as well as a large number of non-classical ones. Moreover, the results that we establish in the sequel for moment matrices can be extended to all structured matrices: for instance, a complete analysis of the case $d(z,w) = 1 - zw^*$ has already been carried out [see Lev-Ari (1983), Lev-Ari and Kailath (1984)].

A closer observation of (3.1)-(3.2) reveals that the one-to-one correspondence between $R(z,w)$ and $f(z)$ extends, in fact, to *nested* sets of coefficients of the power series expansions of these functions. In particular, if $d(0,0) \neq 0$ then this correspondence takes the form

$$(3.3) \qquad \mathbf{R}_{0:m} := \{r_{t,s}; 0 \le t,s \le m\} \quad \longleftrightarrow \quad \mathbf{f}_{0:m} := \{f_t; 0 \le t \le m\}$$

Indeed, the coefficient-domain equivalent of (3.1a) is the difference equation

$$\sum_{i,j=0}^{m} d_{i,j} r_{t-i,s-j} = f_t + f_s^*$$

which implies that, for every (t,s),

$$d_{0,0} r_{t,s} = f_t + f_s^* - \sum_{(i,j)\in D} d_{i,j} r_{t-i,s-j}$$

where $D := \{(i,j); 0 \le i,j \le \min(t,s), (i,j) \neq (0,0)\}$. Consequently, $\mathbf{R}_{0:m}$ can be computed from $\mathbf{f}_{0:m}$ in the following order: $r_{0,0}, r_{1,0}, r_{1,1}, r_{2,0}, r_{2,1}, r_{2,2}, \ldots, r_{m,0}, r_{m,1}, \ldots, r_{m,m}$. Conversely, given $\mathbf{R}_{0:m}$ it follows from (3.2b) that

$$f_t = \sum_{i=0}^{t} d_{i,0} r_{t-i,0} - 1/2d(0,0)$$

so that $\mathbf{f}_{0:m}$ can be computed from the given elements of $\mathbf{R}_{0:m}$.

It should be observed that (3.3) does not hold when $d(0,0) = 0$. For instance, for Hankel matrices (i.e., $d(z,w) = j(z - w^*)$) it can be shown that (3.3) is replaced by a correspondence $\mathbf{R}_{0:m} \longleftrightarrow \mathbf{f}_{0:2m}$. It is not clear whether this modified form of nested correspondence is also valid for other choices of $d(z,w)$ with $d(0,0) = 0$.

Given a canonical generator of a moment matrix $\mathbf{R}$ we can now determine a cascade model for $\mathbf{R}$ via the factorization procedure described in Section 2 (with $\zeta_i = 0$). Notice that this still involves selecting the parameters $\{\tau_i, W_i\}$, and that every specific choice serves to determine a unique mapping from $\mathbf{R}$ to the parameters of the cascade model. For instance, (2.18b) and (2.19b) present convenient choices for Toeplitz and Hankel matrices, respectively. Conversely, given a cascade model we can uniquely determine a canonical generator of the form (3.2) and, consequently, a moment matrix $\mathbf{R}$. This is accomplished by 'reversing' the factorization procedure that was used to obtain the cascade model from $f(z)$. Since this procedure can be reformulated as a linear fractional map similar to the one occurring in the classical Schur algorithm, the reversed procedure can also be easily characterized as a linear fractional map.

Indeed, the basic recursion $zG_{i+1}(z) = G_i(z)\Theta_i(z)$ implies that

$$f_{i+1}(z) = \frac{\theta_{i,22}(z)f_i(z) + \theta_{i,12}(z)}{\theta_{i,11}(z) + \theta_{i,21}(z)f_i(z)} \tag{3.4a}$$

where

$$G_i(z) := [u_i(z) \quad v_i(z)], \quad f_i(z) := \frac{v_i(z)}{u_i(z)} \tag{3.4b}$$

and

$$\Theta_i(z) := \begin{pmatrix} \theta_{i,11}(z) & \theta_{i,12}(z) \\ \theta_{i,21}(z) & \theta_{i,22}(z) \end{pmatrix}. \tag{3.4c}$$

Furthermore, the chain-scattering matrix $\Theta_i(z)$ can be expressed in terms of $f_i(z)$ rather than in terms of the individual functions $u_i(z), v_i(z)$ that make up $G_i(z)$. Indeed, it follows from (2.13) that $\Theta_i(z)$ depends upon $G_i(z)$ only via the rank 1 matrix $M_i := G_i^*(0)R_i^{-1}(0,0)G_i(0)$, which can be rewritten as

$$M_i = \delta_i \begin{bmatrix} 1 \\ f_i^*(0) \end{bmatrix} [1 \quad f_i(0)] \tag{3.5a}$$

where

$$\delta_i := \lim_{z\to 0} \frac{d(z,0)}{f_i(z) + f_i^*(0)}. \tag{3.5b}$$

If we adopt the convention that the J-unitary matrix W_i is also determined in terms of $f_i(z)$ alone then $f_{i+1}(z)$ can be determined from (3.4a) once $f_i(z)$ is known. Finally, we observe that the assumption of strong regularity (i.e., $R_i(0,0) \neq 0$) implies that $G_i(0) := [u_i(0) \quad v_i(0)]$ can not vanish. Consequently, either $u_i(0) \neq 0$ or $v_i(0) \neq 0$ (or both). If we agree to replace $G_i(z)$ by the equivalent generator $G_i(z)J$ whenever the first element of $G_i(z)$ vanishes, the we shall always have $u_i(0) \neq 0$. Consequently, both, δ_i and $f_i(0)$ are well-defined, which establishes the fact that the functions $f_i(z)$ can be recursively computed via (3.4a), starting with $f_0(z) \equiv f(z)$.

The foregoing analysis shows that only $f_i(0)$ and δ_i are required at each step in order to determine the chain-scattering matrix $\Theta_i(z)$. Moreover, when $d(0,0) \neq 0$, then δ_i is determined by $f_i(0)$, viz.,

$$d(0,0) \neq 0 \Longrightarrow \delta_i = \frac{d(0,0)}{2Re\ f_i(0)}.$$

Thus, the coefficient vector K_i, which we introduced in Section 1, consists of *two parameters* when $d(0,0) = 0$, while for $d(0,0) \neq 0$ it consists of a *single parameter* viz.,

$$K_i = \begin{cases} f_i(0) & d(0,0) \neq 0 \\ [f_i(0) \quad \delta_i] & d(0,0) = 0 \end{cases} \tag{3.6}$$

The parameter δ_i is always real while $f_i(0)$ may be either real or complex, depending on the nature of the moment matrix $\mathbf{R}$. Moreover, when $d(0,0) \neq 0$ the parameters $K_{0:m}$ depend only upon the coefficients $\mathbf{f}_{0:m}$ of the power series representation of $f(z)$. This follows from the observation that (with $d(0,0) \neq 0$) the matrix function $\Theta_i(z)$ is analytic at the origin and, as a consequence, the recursion $zG_{i+1}(z) = G_i(z)\Theta_i(z)$ implies that:

(i) If $G_0(z) := [1 \quad f(z)]$ is analytic at the origin then the same property is inherited by all subsequent $G_i(z)$, which can therefore be expanded into power series, viz.,

$$G_i(z) = \sum_{t=0}^{\infty} g_{i,t} z^t$$

(ii) For every fixed m, and for every $0 \leq i \leq m$ the coefficients $\{g_{i,t}; 0 \leq t \leq m-i\}$ of $G_i(z)$ depend only upon the coefficients $\{g_{i-1,t}; 0 \leq t \leq m-i+1\}$ of $G_{i-1}(z)$. It follows, by induction on i, that for every $0 \leq i \leq m$ the coefficients $\{g_{i,t}; 0 \leq t \leq m-i\}$ depend only upon $\mathbf{f}_{0:m} := \{f_t; 0 \leq t \leq m\}$.

In particular, the coefficients $\{g_{i,0}; 0 \leq i \leq m\}$ depend only upon $\mathbf{f}_{0:m}$. Since $f_i(0)$ is the ratio of the two elements of $G_i(0) \equiv g_{i,0}$, it follows that, for $d(0,0) \neq 0$, the parameters $K_{0:m}$ are determined only by $\mathbf{f}_{0:m}$.

We now turn to investigate the mapping from cascade model parameters to the coefficients of the function $f(z)$. In order to recover $f(z)$ from its cascade model we need to reverse the recursion (3.4a), resulting in

$$f_i(z) = \frac{\theta_{i,11}(z) f_{i+1}(z) - \theta_{i,12}(z)}{\theta_{i,22}(z) - \theta_{i,21}(z) f_{i+1}(z)} \tag{3.7}$$

Notice that (3.7) is related to the elements of $\Theta_i^{-1}(z)$ in precisely the same manner as (3.4a) is related to the elements of $\Theta_i(z)$. A repeated application of this recursive inversion produces an explicit *loading formula* which expresses $f(z) \equiv f_0(z)$ in terms of $\Theta_0(z)$, $\Theta_1(z), \ldots, \Theta_m(z)$ and $f_{m+1}(z)$. Indeed, let

$$\Pi_m(z) := \Theta_0(z)\,\Theta_1(z)\ldots\Theta_m(z) = \begin{bmatrix} \pi_{m,11}(z) & \pi_{m,12}(z) \\ \pi_{m,21}(z) & \pi_{m,22}(z) \end{bmatrix} \tag{3.8a}$$

then, by the same reasoning that produced (3.7), we find that

$$f(z) = \frac{\pi_{m,11}(z) f_{m+1}(z) - \pi_{m,12}(z)}{\pi_{m,22}(z) - \pi_{m,21}(z) f_{m+1}(z)} \tag{3.8b}$$

Notice that since $f(z)$ is assumed to be analytic at the origin, it follows that

$$\pi_{m,22}(0) - \pi_{m,21}(0) f_{m+1}(0) \neq 0\,. \tag{3.8c}$$

Now assume that we replace $f_{m+1}(z)$ in (3.8b) by some other $\tilde{f}_{m+1}(z)$ (which still satisfies the constraint (3.8c)), resulting in a corresponding replacement of $f(z)$ in (3.8b) by $\tilde{f}(z)$, which is also analytic at the origin. It follows that

$$f(z) - \tilde{f}(z) = \frac{\det \Pi_m(z)[f_{m+1}(z) - \tilde{f}_{m+1}(z)]}{[\pi_{m,22}(z) - \pi_{m,21}(z) f_{m+1}(z)][\pi_{m,22}(z) - \pi_{m,21}(z)\tilde{f}_{m+1}(z)]}$$

and since $\det \Pi_m(z)$ has a multiple zero at the origin, i.e., $\det \Pi_m(z) = z^{m+1}\phi(z)$, where $\phi(z)$ is analytic at the origin, we conclude that $f(z) - \tilde{f}(z) = O(z^{m+1})$. This means that the coefficients $\mathbf{f}_{0:m}$ of $f(z)$ depend only upon $\Pi_m(z)$ (i.e., upon the parameters $K_{0:m}$) and are independent of the 'load' $f_{m+1}(z)$. The multiple zero of $\det \Pi_{m(z)}$ is a direct consequence of the identity

$$\det \Theta_i(z) = 1 - \frac{d(0,0)d(z,\tau_i)}{d(z,0)d(0,\tau_i)} \tag{3.9}$$

which shows that $\det \Theta_i(z)$ has (at least) one zero at the origin.

In summary, we have established in this section the following nested one-to-one correspondence:

$$\mathbf{R}_{0:m} \longleftrightarrow \mathbf{f}_{0:m} \longleftrightarrow K_{0:m} \qquad \text{when} \quad d(0,0) \neq 0 \tag{3.10}$$

Even though the individual correspondences $\mathbf{R}_{0:m} \longleftrightarrow \mathbf{f}_{0:m}$ and $\mathbf{f}_{0:m} \longleftrightarrow K_{0:m}$ do not hold when $d(0,0) = 0$, it appears that the correspondence $\mathbf{R}_{0:m} \longleftrightarrow K_{0:m}$ may nevertheless hold; this happens, for instance, for $d(z,w) = j(z - w^*)$.

4. Extension of Moment Matrices. The nested one-to-one correspondence

$$\mathbf{R}_{0:m} \longleftrightarrow \mathbf{f}_{0:m} \longleftrightarrow K_{0:m} \ ,$$

which we established in Section 3 (for $d(0,0) \neq 0$), makes it possible to characterize all structured extensions of a given submatrix $\mathbf{R}_{0:m}$ in terms of the unspecified cascade coefficients $K_{m+1}, K_{m+2}, \dots$ or, equivalently, in terms of the function $f_{m+1}(z)$ that is associated with these coefficients. Indeed, the loading formula (3.8) provides an explicit expression for all such extensions: every choice of $f_{m+1}(z)$ results in a function $f(z)$ that is an extension of $\mathbf{f}_{0:m}$, and every extension of $\mathbf{f}_{0:m}$ must be of the form (3.8). Also, by the uniqueness of the mapping between moment matrices and their generators, the generator $G(z) = [1 \quad f(z)]$ (with $f(z)$ given by (3.8)) determines all structured extensions of the given submatrix $\mathbf{R}_{0:m}$.

In order to impose further constraints on the extension we need to study the relation between the properties of the termination $\eta(z) := [1 \quad f_{m+1}(z)]$ and the properties of the corresponding matrix extension $\mathbf{R}$. For instance, in order to get a minimal rank extension, i.e., one with $rank\ \mathbf{R} = rank\ \mathbf{R}_{0:m}$, we should choose a *constant isotropic termination*, namely $\eta(z) = \eta$ and $\eta J \eta^* = 0$. With this choice, the Schur-complement of $\mathbf{R}$ with respect to $\mathbf{R}_{0:m}$ vanishes and, consequently, $rank\ \mathbf{R} = rank\ \mathbf{R}_{0:m}$.

Constant terminations often serve to make the function extension $f(z)$ rational. In particular this is so for $d(z,w) = \alpha + \beta z + (\beta w)^* + \delta z w^*$, which includes the two classical extension problems mentioned in the introduction. A constant termination also imposes a constraint on the values attained by $f(z)$ on the boundary curve Ω: since $d(\tau,\tau) = 0$ for every $\tau \in \Omega$, it follows from (2.13) that $\Theta(\tau)$ is J-unitary and, therefore, that for all $i \geq 1$ and for all $\tau \in \Omega$

$$G_i(\tau) J\ G_i^*(\tau) = G_0(\tau) J\ G_0^*(\tau) \ . \tag{3.11a}$$

In particular, since $G_{m+1}(z) \sim [1 \quad f_{m+1}(z)] \ := \ \eta$, it follows that

$$\text{(3.11b)} \qquad sgn\ \eta J\ \eta^* = \ sgn\ G_0(\tau) J\ G_0^*(\tau) \qquad \text{for all} \ \ \tau = \Omega$$

The significance of this constraint is best demonstrated by a specific example. Consider the following variant of the Schur extension problem: given $\mathbf{f}_{0:m}$, find an extension $f(z) := \sum f_t z^t$ such that $|f(e^{j\theta})| \leq \mu$ for all θ. This problem is solved by constructing a cascade model for the generator $G(z) = [\mu \quad f(z)]$, $J = diag\{1, -1\}$, $d(z,w) \ = \ 1 - zw^*$. The constraint (3.11) implies that in this case

$$sgn\{\mu^2 - |f(e^{j\theta})|^2\} = sgn\ \eta J \eta^*$$

so by selecting $\eta = [1 \quad 1]$ we obtain $\eta J \eta^* \ = 0$ and, consequently, $|f(e^{j\theta})| \ = \mu$ for all θ. This means that the function $f(z)$ is a *rational allpass function* and can be expressed in the form

$$f(z) = \mu \frac{p^{\#}(z)}{p(z)}$$

where $p(z)$ is a polynomial in z, and $p^{\#}(z)$ denotes the conjugate reversal of this polynomial, i.e.,

$$p^{\#}(z) \ := z^{deg\ p(z)} [p(z^{-*})]^* \ .$$

In particular, when $\mu \ = \ 1$, the function $f(z)$ can be interpreted as a *scattering function of a lossless medium.* The reader can easily verify that the structured matrix $\mathbf{R}$, which corresponds to such a lossless scattering function $f(z)$, is congruent to a Schur-Cohn Bezoutian. Consequently, the *inertia* of $\mathbf{R}$ (i.e., the number of its positive, zero and negative eigenvalues) serves to characterize the root-distribution of the polynomial $p(z)$ with respect to the unit circle.

The same is true for the more general case $d(z,w) \ = \ \alpha + \beta z + (\beta w)^* + \delta z w^*$. We have shown that in this case the function $f(z)$ that is obtained by selecting a lossless termination for the cascade model (i.e., $\eta J \eta^* \ = 0$) can still be expressed in the form $f(z) \ = \ p^{\#}(z)/p(z)$, but with a suitably modified definition of $p^{\#}(z)$ [Lev-Ari, Bistritz and Kailath (1988)]. Moreover, the matrix generated by $G(z) = [p(z) \ \ p^{\#}(z)]$, and $J \ = \ diag\{1, -1\}$ is a *generalized Bezoutian* in the sense that its inertia characterizes the distribution of the roots of $p(z)$ with respect to the curve Ω.

REFERENCES

[1] AKHIEZER, N.I., *The Classical Moment Problem*, Hafner Publishing Co., New York (1965) (Russian original, 1961).

[2] BAREISS, E.H., *Numerical Solution of Linear Equations with Toeplitz and Vector Toeplitz Matrices*, Numer. Math., 13 (Oct. 1969), pp. 404-424.

[3] BRENT, R.P., F.G. GUSTAVSON AND D.Y.Y. YUN, *Fast Solution of Toeplitz Systems of Equations and Computation of Pade Approximants*, Journal of Algorithms, 1, 3 (1980), pp. 259-295.

[4] BRUCKSTEIN, A.M. AND T. KAILATH, *An Inverse Scattering Framework for Several Problems in Signal Processing*, IEEE ASSP Magazine, 4, 1 (Jan. 1987), pp. 6-20.

[5] Citron, T.K., A.M. Bruckstein and T. Kailath, *An Inverse Scattering Interpretation of the Partial Realization Problem*, Proc. 23rd IEEE Conf. Dec. Contr., Las Vegas, NV (Dec. 1984).

[6] Dewilde, P., and H. Dym, *Schur Recursions, Error Formulas, and Convergence of Rational Estimators for Stationary Stochastic Processes*, IEEE Trans. Inform. Theory, IT-27, 4 (July 1981), pp. 446-461.

[7] Dewilde, P., and H. Dym, *Lossless Inverse Scattering, Digital Filters, and Estimation Theory*, IEEE Trans. Inform. Theory, IT-30, 4 (July 1984), pp. 644-662.

[8] Dewilde, P., A. Vieira and T. Kailath, *On A Generalized Szego-Levinson Realization Algorithm for Optimal Linear Predictors Based on a Network Synthesis Approach*, IEEE Trans. Circ. Systs., CAS-25 (Sept. 1978), pp. 663-675.

[9] Kailath, T., A.M. Bruckstein and D. Morgan, *Fast Matrix Factorization via Discrete Transmission Lines*, Linear Algebra and Its Applications, 75 (March 1986) , pp. 1-25.

[10] Kailath, T., S.-Y. Kung and M. Morf, *Displacement Ranks of Matrices and Linear Equations*, J. Math. Anal. and Appl., 68, 2 (1979), pp. 395-407. See also Bull. Amer. Math. Soc., 1 (Sept. 1979), pp. 769-773.

Kailath, T., and H. Lev-Ari, *On Mappings Between Covariance Matrices and Physical Systems*, Proc. 1984 AMS Summer Research Conference on Linear Algebra and Its Role in Systems Theory, Maine, (July 1984).

Kalman, R.E., *On Partial Realizations, Transfer Functions and Canonical Forms*, Acta Polytech. Scand., MA31 (1979), pp. 9-32.

[13] Kalman, R.E., *Realization of Covariance Sequences*, Proc. Toeplitz Memorial Conference, Tel-Aviv, Israel (May 1981).

[14] Krein, M.G., and A.A. Nudelman, *The Markov Moment Problem and Extremal Problems*, American Mathematical Society, Providence (1977).

[15] Kung, S.Y., Ph.D. Dissertation, Dept. of Electrical Engineering, Stanford University, Stanford, CA (1977).

[16] Le Roux, J., and C. Gueguen, *A Fixed Point Computation of Partial Correlation Coefficients*, IEEE Trans. Acoust. Speech and Signal Processing, ASSP-25 (1977), pp. 257-259.

[17] Lev-Ari, H., *Nonstationary Lattice-Filter Modeling*, Ph.D. Dissertation, Dept. of Electrical Engineering, Stanford University, Stanford, CA (December 1983).

[18] Lev-Ari, H., Y. Bistritz and T. Kailath, *Generalized Bezoutians and Efficient Matrix Factorization*, Proc. 1987 Int. Symp. Mathematical Theory of Networks and Systems, Phoenix, AZ (June 1987). Reprinted in C.I. Byrnes, C.F. Martin and R.E. Saeks (eds.), *Linear Circuits, Systems and Signal Processing: Theory and Application*, North Holland, New York (1988).

[19] Lev-Ari, H., and T. Kailath, *Lattice-Filter Parametrization and Modeling of Nonstationary Processes*, IEEE Trans. Inform. Thy., IT-30 (January 1984).

[20] Lev-Ari, H. and T. Kailath, *Triangular Factorization of Structured Hermitian Matrices*, in *I. Schur Methods in Operator Theory and Signal Processing, Operator Theory: Advances and Applications*, I. Gohberg (ed.), Vol. 18, pp. 301-324, Birkhaüser Verlag, Basel (1986).

[21] Morf, M., *Fast Algorithms for Multivariable Systems*, Ph.D. Dissertation, Dept. of Electrical Engineering, Stanford University, Stanford, CA (1974).

[22] Pták, V., *Lyapunov, Bezout and Hankel*, Lin. Alg. Appl., 58 (Apr. 1984), pp. 363-390.

[23] Rissanen, J., *Recursive Identification of Linear Systems*, SIAM J. Contr., 9 (1971), pp. 420-430.

[24] Rissanen, J., *Algorithms for Triangular Decomposition of Block Hankel and Toeplitz Matrices with Application to Factoring Positive Matrix Polynomials*, Math. Comput., 27 (Jan. 1973), pp. 147-154.

[25] Schur, I., *Über Potenzreihen, die in Innern des Einheitskreises Beschränkt Sind*, Journal für die Reine und Angewandte Mathematik, 147, Berlin (1917), pp. 205-232, and 148, Berlin, (1918) , pp. 122-145. English translation in I. Schur Methods in Operator Theory and Signal Processing, Operator Theory: Advances and Applications, I. Gohberg (ed.), Vol. 18, pp. 31-88, Birkhäuser Verlag, Basel (1986).

RECENT EXTENSION OF THE SAMPLING THEOREM

GILBERT G. WALTER*

Abstract. A number of recent extensions of the Shannon sampling theorem

$$f(t) = \Sigma f(nT)\frac{\sin\sigma(t-nT)}{\sigma(t-nT)}$$

are discussed. In particular extension to other orthogonal systems and to bandlimited signals with infinite energy are presented in greater detail.

Key words. distributions, orthogonal polynomials, band limited signals.

AMS(MOS) subject classifications. 46F12

Introduction. The classical Shannon sampling theorem for bandlimited signals is the name given to the formula

$$f(t) = \sum_{n=-\infty}^{\infty} f(nT)\frac{\sin\sigma(t-nT)}{\sigma(T-nT)}\ , \ t\varepsilon\mathbf{R}^1\ . \tag{1.1}$$

By bandlimited signal we mean the function $f(t)$ whose Fourier transform $F(w)$ vanishes for $|w| > \sigma$. Here $T = \frac{\pi}{\sigma}$ is the sampling period.

The theory is easy to prove for $f\varepsilon L^2$ by taking the inverse Fourier transform of $F(w)$ and its Fourier series. It goes back to Whittaker [26] but was applied to communication theory first by Kotelnikov and then Shannon [20]. Subsequently it was generalized in a number of different directions as outlined in the survey papers of Jerri [10], Higgins [8] and Butzer, et al [2], [3]. In this work we shall explore two of these generalizations involving

(i) orthogonal systems other than the trigonometric system;

(ii) infinite energy signals (i.e. functions not in L^2).

There are many other extensions which may be found in the survey papers. These include:

(iii) derivative sampling,

(iv) generalized sampling,

(v) non-harmonic sampling,

(vi) sampling stochastic processes.

*Department of Mathematical Sciences, University of Wisconsin - Milwaukee, Milwaukee, WI 53201, USA

Derivative sampling refers to sampling theorems in which the functions and its derivatives are sampled. This enables one to sample less frequently and still recover the function.

In generalized sampling the function $\sin \sigma t/\sigma t$ is replaced by another function $\phi(t)$ whose Fourier transform is smoother. If the Fourier transform $\Phi(w)$ is equal to 1 on the support of $F(w)$ and vanishes outside of $[-\sigma, \sigma]$ then the theorem becomes

$$f(t) = \sum_{n=-\infty}^{\infty} f(nT)\phi(t - nT) \tag{1.2}$$

with the same proof as (1.1). This series however will converge more rapidly particularly if $\Phi(w) \varepsilon C^{\infty}$ with compact support.

Non-harmonic sampling involves sampling at points $\{t_n\}$ that are not multiples of T. One particular result arises from a classical theorem in complex analysis, Kadec's $\frac{1}{4}$-theorem. [27, p.42]. If $|t_n - n| \leq \alpha < \frac{1}{4}$ then there exists functions $\{s_k(t)\}$ such that $s_k(t_n) = \delta_{kn}$ and

$$f(t) = \sum_{k=-\infty}^{\infty} f(t_k)s_k(t) \tag{1.3}$$

for bandlimited functions. Other results arise in conjunction with sampling other orthogonal systems.

The extensions (vi) and (vii) are self-explanatory.

Much of the recent research related to the sampling theorem involves not so much extensions of the theorem but rather analysis of the error involved. Errors may arise from a number of different sources which are

(a) truncation error, which arises when the series is replaced by a finite sum;

(b) aliasing error, which arises when the signal is not σ bandlimited;

(c) amplitude error, which arises when the exact value of $f(nT)$ is unknown or rounded off;

(d) jitter error, which arises from inexact sampling points.

We shall not explore these error sources further except as they may relate to (i) and (ii). Again the interested reader is referred to the survey articles mentioned above.

2. Other Orthogonal Systems and Transforms. The classical sampling theorem (1.1), is closely related to an orthogonal system $\{e^{inw}\}$ and an integral transform, the Fourier transform. They are related by the fact that the kernel of the latter when restricted to the integers gives the former. For $F(w)$ with support in $[-\pi, \pi]$, the (conjugate) Fourier transform is

$$f(t) = \frac{1}{2\pi} \int_{-\pi}^{\pi} e^{-iwt} F(w)dw \tag{2.1}$$

Its values on the integers are just the coefficients of the Fourier series of $F(w)$,

$$F(w) = \sum_{n=-\infty}^{\infty} f(n)e^{iwn} . \tag{2.2}$$

If (2.2) is now substituted in (2.1) and the summation and integral interchanged, the sampling theory (1.1) results, at least formally.

Kramer [14] observed that this procedure can be initiated for any orthogonal system and integral transform related the same way. Let $K(t,x)$ be the kernel of the transform

$$f(t) = \int_I K(t,x)F(x)dx \tag{2.3}$$

which maps $L^2(I)$ into $L^2(\mathbf{R}^1)$; let $\{t_n\}$ be a sequence such that $\{K(t_n,x)\}$ is a complete orthogonal system in $L^2(I)$. Then the expansion of $F(x)$ is

$$F(x) = \sum_n f(t_n)K(t_n,x)/\|K(t_n,\cdot)\|^2 \tag{2.4}$$

which when substituted into (2.3) gives

$$f(t) = \sum_n f(t_n)s_n(t) \tag{2.5}$$

where

$$s_n(t) = \int_I K(t,x)\overline{K(t_n,x)}dx/\|K(t_n,\cdot)\|^2 .$$

Most orthogonal systems arising as eigenfunctions of a differential operator can be put into this framework. The kernel $K(t,x)$ is a solution of the differential equation which may satisfy some but not all of the boundary conditions.

In particular the theorem has been applied to Bessel function series [14], to Sturm-Liouville series [4], to Legendre polynomial series [4], to Gegenbauer polynomial series [12], to prolate spheroidal function series [12], to Laguerre function series [11], [19], and to Jacobi polynomial series [25], [13].

However most of these results are incomplete, since the functions $f(t)$ given by (2.3) are not usually characterized in a precise way. In the classical case (2.1), the function may be characterized by means of the Paley-Wiener theorem [17]. This says that each f in the range of the integral transform is an entire function of exponential type $\leq \pi$ in L^2 on the real axis. In some other cases, similar theorem can be obtained with no difficulty, e.g. in the regular Sturm-Liouville series, but in others, e.g. the Jacobi polynomials, more work is required.

Another approach [22] involves beginning with an orthogonal system $\{\phi_n\}$ on I and then extending it to continue values of n by using a sampling sequence $\{s_n(t)\}$ such that $s_n(k) = \delta_{nk}$. We define $\phi(t,x)$ by

$$\phi(t,x) = \sum_n \phi_n(x)s_n(t) \tag{2.6}$$

provided $\Sigma\|\phi_n\|^2 s_n^2(t)$ converges uniformly to a bounded function on $\mathbf{R}^1$. Any $\{s_n(t)\}$ may be used, but if $\{\phi_n\}$ are the eigenfunctions of a differential operator A, we may wish $\phi(t,x)$ to satisfy the same differential equation. Then we have

$$\lambda_t \phi(t,x) = A\phi(t,x) = \sum_n \lambda_n \phi_n(x) s_n(t) \tag{2.7}$$

and hence formally

$$\Sigma(\lambda_t - \lambda_n)\phi_n(x)s_n(t) = 0\ ,\ x \epsilon I\ ,\ t \epsilon \mathbf{R}^1\ . \tag{2.8}$$

Fortunately the series (2.8) does not converge in the sense of $L^2(I)$. If it did the coefficients would have to be zero. It does converge in the sense of distributions on I to a distribution which is zero in the interior of I.

Example 2.1. In the classical case, the orthogonal system is $\{e^{inw}\}$ on $[-\pi,\pi]$. The extension given by

$$\Sigma e^{inw} s_n(t)$$

converges for $|w| < \pi$ to e^{itw}. The e^{inw} are eigenfunctions of the operator $-iD$ with eigenvalues n. Hence (2.8) becomes

$$\Sigma(t-n)e^{inw}s_n(t) = \Sigma(t-n)e^{inw}\frac{\sin \pi(t-n)}{\pi(t-n)}$$

$$= \frac{\sin \pi t}{\pi}\Sigma e^{inw}(-1)^n = \frac{\sin \pi t}{\pi}\delta^*(w-\pi)$$

which is zero in $(-\pi,\pi)$.

Example 2.2. Laguerre transform. [6], [7], [11], [19]. The Laguerre polynomials $L_n^\alpha(x)$ are orthogonal on $(0,\infty)$ with respect to the weight $x^\alpha e^{-x}, \alpha > -1$. They satisfy the equation

$$xy'' + (\alpha + 1 - x)y' + ny = 0\ .$$

An extension $L_t^\alpha(x)$ would be expected to satisfy the same equation with n replaced by t. This can be given in terms of confluent hypergeometric functions, ${}_1F_1(\alpha;\nu;x)$, [21, p.103]

$$L_n^\alpha(x) = {}_1F_1(-n;\alpha+1;x)$$

which may be extended to non-integral values of n. Our extension is by

$$L_t^\alpha(x) = \Sigma s_n(t) L_n^\alpha(x)$$

where s_n may be taken as

$$s_n(t) = \frac{\sin 2\pi(t-n)}{2\pi(t-n)}$$

Then

$$2\pi\Sigma(n-t)s_n(t)L_n^\alpha(x) = \Sigma \sin 2\pi t L_n^\alpha(x)$$

which converges in the sense of distributions on $(0,\infty)$. But we see from the generating function [21, p. 101]

$$\sum_{n=0}^{\infty} w^n L_n^\alpha(x) = (1-w)^{\alpha-1} e^{-xw/(1-w)},$$

that this limit is zero as $w \to 1, x > 0$, and hence the series is Abel summable to zero. Notice that $L_t^2(0) = \infty, t \neq n$ in our extension and hence is not ${}_1F_1$.

However our L_t^α may be given as a linear combination of the hypergeometric functions of the first and second kind since it satisfies the same differential equations.

The appropriate sampling theorem in this case would be for f given by

$$f(t) = \int_0^\infty L_t^\alpha(x)\, F(x) x^\alpha e^{-x} dx$$

where $F \varepsilon L^2((0,\infty);\ e^{-x}x^\alpha)$. Then

$$f(t) = \sum_{n=0}^{\infty} f(n) \frac{\sin 2\pi(t-n)}{2\pi(t-n)} .$$

Example 2.3. Sturm-Liouville Transforms. [14], [28]

Let $\{\phi_n\}$ be a regular Sturm-Liouville system on $(0,\pi)$ satisfying a differential equation

$$(q - D^2)\phi_n \ = \ \lambda_n U_n .$$

Let $\phi(t^2, x)$ and $\psi(t^2, x)$ be solutions to the same equation with λ_n replaced by t^2 and each satisfying one boundary condition. Let $W(t^2)$ be their Wronskian. If

$$f(t) = \int_0^\pi \phi_t(x) F(x) dx$$

than the appropriate sampling theorem is

$$f(t) = \sum_{n=0}^{\infty} f(t_n^2) W(t^2)/W'(t_n^2)(t^2 - t_n^2) .$$

Example 2.4. Jacobi Transform. [21], [13]

The Jacobi polynomials $P_n^{(\alpha,\beta)}$ are orthogonal on (-1,1) with weight $w(x) = (1-x)^\alpha(1+x)^\beta$. We use a normalized version $R_n^{(\alpha,\beta)}$ which satisfy $R_n^{(\alpha,\beta)}(1) = 1$. They satisfy a hypergeometric equation and may be given by

$$R_n^{(\alpha,\beta)}(x) = {}_2F_1(-n, n+\alpha+\beta+1; \alpha+1, \frac{1-x}{2})$$

where $2F_1$ is the hypergeometric function. This same function may be used to extend these to all real values of n, i.e. to the Jacobi functions $R_t^{(\alpha,\beta)}$. The finite Jacobi transform is given by

$$f(t) = 2^{-2\lambda} \int_{-1}^{1} F(x) R_{t-\lambda}^{(\alpha,\beta)}(x) w(x) dx \ , \ \alpha, \beta > -1 \ , \ t \varepsilon \mathbf{R}^1$$

where $\lambda = \frac{\alpha+\beta+1}{2}$. Here the index has been shifted to $t-\lambda$ in order to make f even.

The function F may not be taken to be in $L^2(w(x))$ necessarily since the transform does not exist for all such functions. However if F is bounded on [-1,1], the transform exists and the sampling theorem is given by

$$f(t) = \sum_{n=0}^{\infty} f(n+\lambda)s_n^{\lambda}(t-\lambda)$$

where

$$s_n^{\lambda}(t-\lambda) = \frac{(2n+2\lambda)\Gamma(n+2\lambda)\Gamma(t-\lambda+1)\sin\pi(t-\lambda-n)}{\Gamma(n+1)\Gamma(t+\lambda)\pi(t^2-(\lambda+n)^2)} .$$

Those transforms are characterized to some extent in [13] where is shown that $f(t)$ is an even, entire function of exponential type satisfying

$$|f(t)| \leq (|t|+1)^{\mu} e^{\pi|\sin t|} \quad , \quad t\varepsilon C .$$

provided $F(x)$ is sufficiently restricted. A converse is also true; if f satisfies these conditions for μ sufficiently large, then f has such a sampling expansion.

3. Sampling Infinite Energy Signals. Two difficulties arise when the signal $f(t)$ does not have finite energy ($\not\in L^2(\mathbf{R}^1)$. The series in (1.1) may not converge and even if it does may converge to some thing other than $f(t)$.

Example 3.1. Let $f(t) = t$; then

$$\sum_{n=-\infty}^{\infty} nT\frac{\sin\sigma(t-nT)}{\sigma(t-nT)}$$

does not converge pointwise for all t.

Example 3.2. Let $f(t) = \sin\pi t$; then $f(n) = 0$ for all n and the series converges but not to $f(t)$.

Example 3.3. Let $f(t) = \cos\pi t$; then the series is

$$\sum_{n=-\infty}^{\infty} (-1)^n \frac{\sin\pi(t-n)}{\pi(t-n)}$$

which converges to $\cos\pi t$ for all t, but $f(t)$ has infinite energy.

There is no difficulty in defining bandlimited signals provided $f(t)$ has polynomial growth at most. In that case f is a tempered distribution in S' for which a Fourier transform always exist [1]. We say that such a function is σ bandlimited if its Fourier transform is a distribution with support in $[-\sigma,\sigma]$. However as Example 3.2 shows, not all σ-bandlimited functions of polynomial growth have sampling series.

A number of approaches have been proposed. The earliest is due to Campbell [4] who obtained a generalized sampling theorem for functions whose Fourier transform had support in some interior subinterval, say $[-\sigma', \sigma'] \subset (-\sigma, \sigma)$. Let $\phi \varepsilon C^\infty$ such that $\phi(w) = 1$ on $[-\sigma', \sigma']$ and $\phi(w) = 0$ on $\mathbf{R}^1 - (-\sigma, \sigma)$. Then the distribution $F(w)$ satisfies

$$F(w) = \phi(w)F(w) = \sum_{n=-\infty}^{\infty} f(nT)e^{iwnT}\phi(w)$$

with convergence in the sense of S'. By taking inverse Fourier transforms, he obtained the sampling theorem

$$f(t) = \sum_{n=-\infty}^{\infty} f(nT)\tilde{\phi}(t - nT) \tag{3.1}$$

where the series converges uniformly on bounded sets. By choosing ϕ judiciously, he obtained an expression for (3.1) close to the classical form of the sampling theorem (1.1).

3.1. Polynomial Correction Methods. Unfortunately Campbell's results did not generalize (1.1) since the support was not all of $[-\sigma, \sigma]$. Pfaffelhuber [18] obtained a result which held for all σ-bandlimited functions.

He did so by correcting $f(t)$; he subtracted a certain function and then obtained a sampling theorem for the difference. It had the form

$$f(t) = q_N(t)\cos\sigma t + \sum_{n=-\infty}^{\infty} [f(nT) - (-1)^n q_N(nT)] \left(\frac{t}{nT}\right)^N \frac{\sin\sigma(t - nT)}{\sigma(t - nT)} \tag{3.2}$$

where the series converged absolutely. The function $q_N(t)$ is a polynomial of degree N given by the Taylor polynomial of $f(t)/\cos\sigma t$ whose N is an integer (the order of F) such that $f(t)/(|t| + 1))^N \varepsilon L^2(\mathbf{R}^1)$. Similar results were obtained by Lee [16] and by Hoskins and de Sousa Pinto [9]. Lee replaced $\cos\sigma t$ by an exponential function and thus obtained a simpler expression for $q_N(t)$. Hoskins and deSousa Pinto replaced $\cos\sigma t$ by an arbitrary σ-bandlimited function which did not vanish at $t = 0$. These results were extended further in [24] to obtain a basic lemma involving an arbitrary polynomial

$$P(t) = \prod_{j=1}^{k}(t - a_j)^{m_j} \quad , \quad \sum_{j=1}^{k} m_j = N$$

of the same order as F. It is

LEMMA 3.1. *Let $f(t)$ be a σ-bandlimited function of polynomial growth which satisfies $f^{(i)}(a_j) = 0$, $i = 0, 1, \cdots, m_j - 1; j = 1, \cdots, k$; then*

$$f(t) = \sum_{n=-\infty}^{\infty} f(nT)\frac{P(t)}{P(nT)} \frac{\sin\sigma(t - nT)}{\sigma(t - nT)} \tag{3.3}$$

uniformly on bounded sets in **C**.

This enables us to obtain a number of different versions of the sampling theorem depending on the choice of $P(t)$. The most interesting perhaps is the interpolation formula given in

Corollary 3.2. Yet $P(t)$ have N distinct zeros and $L_N(t)$ be the polynomial which interpolates $f(t)$ as those zeros; then

$$f(t) = L_N(t) + \sum_{n=-\infty}^{\infty} [f(nT) - L_N(nT)] \frac{P(t)}{P(nT)} \frac{\sin \sigma(t-nT)}{\sigma(t-nT)} \tag{3.4}$$

The points nT are not excluded as possible zeros of $P(t)$; the modification needed in the formula are obvious.

3.2. Summability methods. The formulae (3.2)-(3.4), while generalizing (1.1) and applying to σ-bandlimited functions lack the simplicity and elegance of the original sampling theorem. Fortunately as was shown in [23] and [24], certain summability methods enable us to avoid both generalized sampling as in (3.2) and the polynomial correction of the other methods.

We first give a result applicable to σ'-bandlimited functions of polynomial growth where $\sigma' < \sigma$. It is

$$f(t) = \sum_{n=-\infty}^{\infty} f(nT) \frac{\sin \sigma(t-nT)}{\sigma(t-nT)} \tag{3.5}$$

where, however, the convergence is taken in the sense of (C, α) summability for any $\alpha > N + 1$ where N is the order of F. The (C, α) summability can be replaced by Abel summability which holds for F of all orders.

A series $\sum_{n=-\infty}^{\infty} u_n$ is (C, α) summable to s if

$$s = \lim_{N \to \infty} \sum_{n=-N}^{N} C_{N,n}^{\alpha} w_n \ , \ C_{N,n}^{\alpha} = \frac{\binom{N-|n|+\alpha}{\alpha}}{\binom{N+\alpha}{\alpha}}$$

and is Abel summable to s if

$$\lim_{r \to 1} \sum_{n=-\infty}^{\infty} r^{|n|} w_n = s \ , \ 0 < r < 1 \, .$$

These results are based on the following lemma and some summability results for Fourier series.

LEMMA 3.2. *Let $F(w)$ be a distribution with support in $[-\sigma, \sigma]$; then there exists a function $G(w)$ in L^2 with support in $[-\sigma, \sigma]$, an integer p and constants $c_0, \cdots, c_{p-1}$ such that*

$$F = D^p G + \sum_{i=1}^{p-1} c_i \delta^{(i)} \tag{3.6}$$

in the sense of distributions.

This same lemma allows us to get stronger results by looking at σ more carefully. The counterexample 3.2 still applies to these summability methods so we cannot expect to get a theorem including all σ- bandlimited functions of polynomial growth. However, we can get one which is applicable to σ-bandlimited functions in L^2 (rather than σ') and hence generalizes (1.1). It involves the concept of strong integration of distributions over closed intervals [1].

Definition 3.1. A distribution $F(w)$ is strongly integrable over [-σ, σ] if its anti-derivative $F^{(-1)}$ has a value at σ and $-\sigma$.

The concept of value [1] of a distribution generalizes the concept of value of a function at a point. The distribution $G(w)$ has value γ at $w = 0$ if there is a p and anti-derivative $G^{(-p)}(w)$ such that

$$\lim_{w \to 0} \frac{G^{(-p)}(w)}{w^p} = \frac{\gamma}{p!} . \tag{3.7}$$

The integral of $F(w)$ over $[-\sigma, \sigma]$ is defined as

$$\int_{-\sigma}^{\sigma} F = F^{(-1)}(\sigma) - F^{(-1)}(-\sigma) .$$

THEOREM 3.3. *[23] Let $f(t)$ be σ-bandlimited function of polynomial growth whose Fourier transform $F(w)$ is strongly integrable over $[-\sigma, \sigma]$. Then (3.5) holds in the sense of Abel summability uniformly on bounded sets.*

The proof involves using the characterization (3.6) together with the fact that the family $\{K_r\}$ where

$$K_r(w) = \frac{(-w)^k P_r^{(k)}(w)}{k!}$$

is a quasi-positive delta-family (kernel) as $r \to 1$. Here $P_r(w)$ is the Poisson kernel associated with Abel summability. (See [23] for details).

Many questions associated with those summability methods remain to be answered. In particular the error estimates have only begun to be explored. Krezner [15] has obtained a bound for the truncation error for (C, α) - summability, but no other errors have been studied.

REFERENCES

[1] P. ANTOSIK, J. MIKUSINSKI, R. SIKORSKI, *Theory of distributions. The sequential approach*, Polish Scientific Publishers Warsaw, 1973.

[2] P. BUTZER, *A survey of the Whittaker-Shannon sampling theorem and some of its extensions*, J. Math. Res. Exposition, 3 (1983), pp. 185–212.

[3] P. BUTZER, W. SPLETTSTÖßER AND R. STENS, *The sampling theorem and linear predictions in signal analysis*, Jahresber. Deutsch. Math.-Verein., 90 (1988), pp. 1-70.

[4] L. Campbell, *A comparison of the sampling theorems of Kramer and Whittaker*, J. SIAM, 12 (1964), pp. 117–130.

[5] A. Erdelyi et al. (Bateman Manuscript project), *Higher Transcendental Functions*, vol I, McGraw-Hill, New York, 1953.

[6] H. Glaeske, *The Laguerre transform of some elementary functions*, Zeitschrift fúr Analysis, 3, 3 (1984), pp. 237–244.

[7] H. Glaeske, *Die Laguerre-Pinney transformation*, Aequationes Math., 22 (1981), pp. 73–85.

[8] J. Higgins, *Five short stories about the cardinal series*, Bull. Amer. Math. Soc., 12, 1 (1985), pp. 45–89.

[9] R.F. Hoskins and J. deSousa Pinto, *Sampling expansions for functions bandlimited in the distributional sense*, SIAM J. Appl. Math., 44 (1984), pp. 6050–610.

[10] A. Jerri, *The Shannon sampling theorem - its various extensions and applications: A tutorial review*, Proc. IEEE 65, 11 (1977), pp. 1565–1596.

[11] A. Jerri, *Sampling expansion for Laguerrre - L_ν^α transforms*, J. Res. Nat. Bur. Standards, sec. B(80),(B) 3 (1976), pp. 415–418.

[12] A. Jerri, *On the application of some interpolating functions in physics*, J. Res. Nat. Bur. Standards, sec. B(80),(B) 3 (1969), pp. 241–245.

[13] T. Koornwinder and G. Walter, *The finite continuous Jacobi transform and its inverse*, preprint.

[14] H. Kramer, *A generalized sampling theorem*, J. Math. Phys., 38 (1959), pp. 68–72.

[15] R.J. Kreczner, *Generalized cardinal series*, Ph.D. Thesis University of Wisconsin-Milwaukee, (1988).

[16] A. Lee, *A note on the Campbell sampling theorem*, SIAM J. Appl. Math., 41 (1981), pp. 553–557.

[17] R. Paley, N. Wiener, *Fourier transforms in the complex domain*, Colloq. Publ., vol. 19, Amer. Math. Soc., Providence, R.I. (1934).

[18] E. Pfaffelhuber, *Sampling series for bandlimited generalized functions*, IEEE Trans. Inform. Theory IT-17 (1971), pp. 650–654.

[19] Sri Dharma Selvaratnam, *Shannon-Wittaker sampling theorems*, Ph.D. Thesis, University of Wisconsin-Milwaukee (1987).

[20] C. Shannon, *Communication in the presence of noise*, Proc. IRE, 37 (1949), pp. 10–21.

[21] G. Szegö, *Orthogonal Polynomials*, Colloq. Pub., 23, Amer. Math. Soc., Providence, R.I. (1939).

[22] G. Walter, *A finite continuous Gegenbauer transform and its inverse*, SIAM J. Appl. Math. (to appear).

[23] G.G. Walter, *Abel summability for a distributional sampling theorem*, Proceeding of GFCA-87, 28 (1988), pp. 680–688.

[24] G.G. Walter, *Sampling bandlimited functions of polynomial growth*, SIAM J. Anal. Math., 19 (1988), pp. 1198–1201.

[25] G. Walter and A. Zayed, *The continuous (α, β) Jacobi transform and its inverse when $\alpha + \beta + 1$ is a positive integer*, Trans. Amer. Math. Soc., 305, 2 (1988), pp. 653–664.

[26] E. Whittaker, *On the functions which are represented by the expansion of the interpolation theory*, Proc. Roy. Soc. Edinburgh, 35 (1915), pp. 181–194.

[27] R.M. Young, *An introduction to non-harmonic Fourier series*, Academic Press, New York, 1980.

[28] A. Zayed, G. Hinsen and P. Butzer, *On Lagrange interpolation and Kramer-type sampling theorems associated with Sturm-Liouville problems*, preprint.

[29] A. Zygmund, *Trigonometric series*, Cambridge University Press, New York, 1957.

GENERALIZED SPLIT LEVINSON, SCHUR, AND LATTICE ALGORITHMS FOR ESTIMATION AND INVERSE SCATTERING

ANDREW E. YAGLE*

Abstract. A unified treatment of fast algorithms for estimation and inverse scattering in one and three dimensions is given. The classical Levinson, Schur, and lattice algorithms can be used to compute the reflection coefficients of the optimal linear least-squares prediction filter for a stationary random process; they can also be used to compute the reflection coefficients at the interfaces of a one-dimensional layered scattering medium probed with an impulsive plane wave. The split versions of these algorithms compute the potentials associated with the prediction filter and the scattering medium; the split algorithms operate by propagating a discrete Schrodinger equation.

The split algorithms are then generalized to three dimensions. This allows fast algorithms for three-dimensional random field estimation problems, and for three-dimensional inverse scattering problems, to be developed. The covariance of the random field is required to have a Toeplitz-plus-Hankel structure in the double Radon transform domain; this is a generalization of the stationarity required in the one-dimensional problem. The connection between random field estimation and inverse scattering in three dimensions illuminates both problems.

Key words. Estimation, fast algorithms, random fields

AMS(MOS) subject classifications. 93E10, 93E11, 35R30

I. INTRODUCTION

The linear least-squares estimation problem is to compute the optimal filter for estimating a stationary random process from noisy observations of the process, given knowledge of its covariance. The inverse scattering problem is to reconstruct a scattering medium from knowledge of its reflection or transmission response to an impulsive plane wave.

In the one-dimensional case, it is well known that these two problems are equivalent. This is commonly thought to be due to the fact that wave propagation in a one-dimensional scattering medium is identical to signal propagation in the lattice form of the linear prediction filter. However, the connection between the two problems is much deeper than this. Both problems have orthogonality conditions at their base–the well known orthogonality condition of linear prediction, and the orthogonality of the scattering solutions in scattering problems. This allows the connection between estimation and inverse scattering to be extended from one to three dimensions, even though the latter case admits a lattice formulation of neither scattering wave propagation nor filtering signal propagation.

The connection in one dimension allows fast algorithms for solving one problem to be applied to the other problem as well. The well known Levinson and Schur algorithms of linear prediction theory compute the reflection coefficients associated with the lattice form of the optimal linear prediction filter from the covariance of the random process. These same algorithms can also be used to compute the reflection coefficients of a layered medium from its reflection response to an impulsive plane

*Department of Electrical Engineering and Computer Science, The University of Michigan, Ann Arbor, Michigan 48109-2122

wave. We briefly review these algorithms and their applications to both problems below; for a more detailed treatment see [1]. Recently it has been shown [2] that the lattice algorithm of linear prediction can also be used to compute the reflection coefficients of a scattering medium from its *transmission* response to an impulsive plane wave.

Alternatives to the classical Levinson, Schur, and lattice algorithms are the *split* versions of each of these algorithms [3]. The split algorithms replace the lattice computations with three-term recurrences that require only half as many multiplications. The reflection coefficients computed by the classical algorithms are replaced by potentials; the reflection coefficients for the lattice filter can easily be computed from these potentials (see (2-9) below). The three-term recurrence propagated by the split algorithms has been related to three-term recurrences for orthogonal polynomials [3]; it can also be interpreted as the *discrete Schrodinger equation* for the scattering medium.

The significance of the split algorithms is that the split Levinson, Schur, and lattice algorithms generalize to three dimensions, while the classical algorithms do not. This follows since the wave propagation described by the one-dimensional algorithms does not generalize to three dimensions; wave propagation in three dimensions is far too complicated to be described by a lattice structure. However, the Schrodinger equation easily generalizes to three dimensions; hence the split algorithms based on it do also. Furthermore, these three-dimensional split Levinson, Schur, and lattice algorithms solve *both* estimation problems *and* inverse scattering problems, just as they do in one dimension.

This paper is organized as follow. Section 2 quickly reviews the one-dimensional classical and split Levinson, Schur, and lattice algorithms, their application to linear prediction and inverse scattering, and their continuous-parameter forms. Section 3 formulates the three-dimensional estimation and inverse scattering problems in detail, noting the integral equation that solves both problems. Section 4 gives the differential form of this integral equation, presents the generalized split Levinson, Schur, and lattice algorithms, and explains how these algorithms are used to solve the problems of Section 3. Section 5 concludes by summarizing the paper and noting directions for possible future research. This paper summarizes the results contained in [1]-[5], including background ([1],[3]) and recent results by the author ([2],[4],[5]).

II. THE ONE-DIMENSIONAL CLASSICAL AND SPLIT ALGORITHMS

We quickly summarize the one-dimensional split Levinson, Schur, and lattice algorithms of [3], discuss briefly their application to linear prediction and inverse scattering, and their continuous-parameter versions.

2.1 Classical Levinson algorithm. Consider the one-dimensional *linear prediction problem* of estimating the present value of a zero-mean, stationary, discrete-time random process $x(i)$ from observations $\{x(j), i-n \leq j \leq i-1\}$ of its past n values. It is well known that the optimal linear prediction filter coefficients can be obtained using the Levinson algorithm. Let $R(z)$ be the z-transform of one side of the autocorrelation sequence of $x(i)$. Then the nth-order prediction error filter

$A_n(z)$ can be recursively computed as follows:

$$(2-1a) \qquad \begin{bmatrix} A_n(z) \\ B_n(z) \end{bmatrix} = \begin{bmatrix} 1 & zk_n \\ k_n & z \end{bmatrix} \begin{bmatrix} A_{n-1}(z) \\ B_{n-1}(z) \end{bmatrix}$$

$$(2-1b) \qquad k_n = -A_{n-1}(z)R(z)/P_{n-1}|_{z=0}$$

$$(2-1c) \qquad P_n = (1-k_n^2)P_{n-1}$$

$$(2-1d) \qquad A_0(z) = 1; B_0(z) = 0$$

Equations (2-1) also recursively compute the *backwards prediction* error filter $B_n(z)$. This is the error filter for estimating $x(i-n-1)$ from its future n values $\{x(j), i-n \leq j \leq i-1\}$.

The $\{k_i, i = 1 \dots n\}$ characterize the optimal prediction filters of all orders up to n. They are called *reflection coefficients*, since equations (2-1) can be implemented on a lattice filter in which signals in one rail are scattered into the other rail, with gain k_i in the ith section of the lattice.

Now consider the one-dimensional *inverse scattering problem* of reconstructing a layered medium from its reflection response to an impulsive plane wave. The medium is characterized by the reflection coefficients at its interfaces. The goal is to compute these reflection coefficients from the impulse reflection response.

It is not difficult to show [1] that this inverse scattering problem can also be solved using the Levinson algorithm. In fact, if $R(z)$ is the z-transform of the reflection response, then the Levinson algorithm (2-1) will compute the layer interface reflection coefficients as the $\{k_i, i = 1 \dots n\}$.

Superficially this seems due to the fact that signal propagation in a lattice filter mimics wave propagation in a layered medium (see [1] for details). However, the true connection between these problems is that both involve *orthogonality conditions*– the well known orthogonality condition of linear prediction, and the orthogonality of the scattering solutions to the Schrodinger equation (see [6]). It is this deeper connection that allows the connection between these problems to be generalized to three dimensions.

In the Levinson algorithm the reflection coefficients k_i are computed using (2-1b), which is called the "inner product" computation. This is a non-parallelizable computational bottleneck, and it would be desirable to avoid this computation if possible. This motivates the next two algorithms.

2.2 Classical Schur algorithm. If the recursions (2-1a) are initialized using $R(z)$ instead of (2-1d), the result is the Schur algorithm:

$$(2-2a) \qquad \begin{bmatrix} U_n(z) \\ D_n(z) \end{bmatrix} = \begin{bmatrix} 1 & zk_n \\ k_n & z \end{bmatrix} \begin{bmatrix} U_{n-1}(z) \\ D_{n-1}(z) \end{bmatrix}$$

$$(2-2b) \qquad k_n = -U_{n-1}(z)/zD_{n-1}(z)|_{z=0}$$

$$(2-2c) \qquad U_0(z) = D_0(z) = R(z)$$

The Schur algorithm also computes the reflection coefficients k_i for the lattice prediction filter from the z-transform $R(z)$ of one side of the covariance. It propagates the expectation of the residuals times the process; (2-2b) then follows from the orthogonality condition of linear prediction.

The Schur algorithm can also be used to solve the inverse scattering problem of computing the reflection coefficients k_i of a layered medium from its impulse reflection response $R(z)$. For this application the quantities $D_n(z)$ and $U_n(z)$ have the physical interpretation of downgoing and upgoing waves in the scattering medium, resulting from the impulsive excitation. Equation (2-2b) then follows directly from time causality.

Note that in the Schur algorithm the reflection coefficients are computed using (2-2b), which is not an "inner product" computation. Hence the Schur algorithm can be propagated in parallel with the Levinson algorithm, solely for the purpose of computing the reflection coefficients k_i, and thus avoiding the inner product (2-1b) required by the Levinson algorithm alone [7].

2.3 Classical lattice algorithm. If the recursions (2-1a) are initialized using $X(z)$ instead of (2-1d), where

$$(2-3) \qquad 1 + R(z) + R(1/z) = X(z)X(1/z)$$

the result is the lattice algorithm:

$$(2-4a) \qquad \begin{bmatrix} U_n(z) \\ D_n(z) \end{bmatrix} = \begin{bmatrix} 1 & zk_n \\ k_n & z \end{bmatrix} \begin{bmatrix} U_{n-1}(z) \\ D_{n-1}(z) \end{bmatrix}$$

$$(2-4b) \qquad k_n = -X_{n-1}(z)D_{n-1}(1/z)/P_{n-1}|_{z=0}$$

$$(2-4c) \qquad P_n = (1-k_n^2)P_{n-1}$$

$$(2-4d) \qquad U_0(z) = D_0(z) = X(z)$$

The lattice algorithm thus computes the reflection coefficients k_i from the spectral factor of the covariance, rather than the covariance itself.

Equation (2-3) is well known in the seismics literature as the relation between the reflection response $R(z)$ and the transmission response $X(z)$ of a layered medium to an impulsive plane wave. Hence the lattice algorithm can be used to compute the reflection coefficients k_i of a layered medium from its *transmission response* $X(z)$ [2].

2.4 Split Levinson algorithm. There is some redundancy in the above algorithms. Defining $h_n(z)$ as

$$h_n(z) = A_n(z) + B_n(z) \tag{2-5}$$

it may be shown [3] that $h_n(z)$ can be computed using the *three-term recurrence*

$$h_{n+1}(z) = (z+1)h_n(z) - za_n h_{n-1}(z) \tag{2-6}$$

and that a_n may be computed using

$$a_n = v_n(0)/v_{n-1}(0); v_n(0) = 2R(z)h_n(z)|_{z=0} \tag{2-7}$$

Equations (2-6) and (2-7) constitute the *split Levinson algorithm.* The point is that the two coupled recursions (2-1) are replaced by the three-term recurrence (2-6). Note that an "inner product" computation (2-7) is still required at each recursion.

Since the classical Levinson algorithm can be applied to an inverse scattering problem, the split Levinson algorithm can also. In fact, equation (2-6) corresponds to the *discrete Schrodinger equation* associated with with the layered medium. The scattering potential of this medium is then (see [3],[8])

$$V_n = 1 - a_n \tag{2-8}$$

Applied to scattering problems, the split algorithms merely propagate the field quantities (voltage, pressure, etc.) associated with the scattering medium, while the classical algorithms propagate waves in the medium. Applied to linear prediction problems, the $\{a_i\}$ characterize the optimal filters of all orders just as the reflection coefficients do; indeed we have [3]

$$a_n = (1 + k_n)(1 - k_{n-1}). \tag{2-9}$$

Since the decomposition of the field quantity into forward and backward travelling waves (2-5), (2-10), and (2-12) is not possible in three dimensions, this explains why only the split algorithms can be generalized to three dimensions. The potential V_n defined in (2-8) generalizes to three dimensions, but there are additional dependencies in it.

2.5 Split Schur algorithm. Defining

$$v_n(z) = U_n(z) + D_n(z) \tag{2-10}$$

from (2-2), $v_n(z)$ can be computed using the split Schur algorithm [3]:

$$v_{n+1}(z) = (z+1)v_n(z) - za_n v_{n-1}(z) \tag{2-11a}$$

$$a_n = v_n(0)/v_{n-1}(0) \tag{2-11b}$$

Applied to linear prediction, the split Schur algorithm computes the filter potentials a_n from the covariance $R(z)$; the reflection coefficients of the lattice filter can then be computed using (2-9). Applied to inverse scattering, the split Schur algorithm propagates the voltage or pressure in the scattering medium resulting from the impulsive excitation, and computes the scattering potential from the reflection response $R(z)$.

Note that there is no "inner product" computation. Equation (2-11) can be derived using causality arguments.

2.6 Split lattice algorithm. Defining

$$u_n(z) = U_n(z) + D_n(z) \tag{2-12}$$

from (2-4), $u_n(z)$ can be computed using the split lattice algorithm [3]:

$$u_{n+1}(z) = (z+1)u_n(z) - za_n u_{n-1}(z) \tag{2-13a}$$

$$a_n = v_n(0)/v_{n-1}(0); v_n(0) = u_n(z)u_n(1/z)|_{z=0} \tag{2-13b}$$

Applied to linear prediction, the split lattice algorithm computes the filter potentials a_n from the spectral factor $X(z)$; the reflection coefficients of the lattice filter can then be computed using (2-9). Applied to inverse scattering, the split Schur algorithm propagates the voltage or pressure in the scattering medium resulting from the impulsive excitation, and computes the scattering potential from the transmission response $X(z)$.

2.7 Continuous parameter forms. The continuous-parameter form of the three-term recurrence (2-6) (and also (2-11a) and (2-13a)) is determined by noting that (2-6) is a discrete Schrodinger equation. The continuous-parameter Schrodinger equation in the time domain is

$$\left(\frac{\partial^2}{\partial x^2} - \frac{\partial^2}{\partial y^2}\right)h(x,y) = V(x)h(x,y) \tag{2-14}$$

where $h(x,y)$ is the continuous-parameter version of $h_n(z)$ (y is time). Equation (2-14) describes a continuous one-dimensional scattering medium with continuous scattering potential $V(x)$ (the continuous version of (2-8)). It is also the equation for a vibrating, elastically-based string used in [9] for one-dimensional linear estimation problems.

In the following sections the three-dimensional version of (2-14) is used for three-dimensional linear estimation and inverse scattering problems. It should be clear why this can be construed as a three-dimensional, continuous-parameter generalization of the three-term recurrences that constitute the one-dimensional split algorithms.

III. THREE-DIMENSIONAL PROBLEM SPECIFICATIONS

3.1 Estimation Problem Specification. The basic three-dimensional linear least-squares estimation problem is as follows. Let

$$w(x) = s(x) + v(x), \quad x \in R^3 \tag{3-1}$$

be some noisy observations of a zero-mean real-valued random field $s(x)$ having covariance

$$E[s(x)s(y)] = k(x,y). \tag{3-2}$$

$v(x)$ is a zero-mean real-valued white noise field with covariance

$$(3-3) \qquad E[v(x)v(y)] = \delta(x-y)$$

and $v(x)$ is uncorrelated with $s(x)$.

We wish to compute the linear least-squares estimate $\hat{s}(x)$ of $s(x)$ given the noisy observations $\{w(y), |y| \leq |x|\}$. This is the *filtering* problem of estimating $s(x)$ on the *surface* of the sphere of observations. It can also be viewed as the three-dimensional analogue of the linear prediction problem solved by the Levinson algorithm in one dimension. The forward and backward predictors for either end of the segment of observations generalize to the predictors for all points on the surface of the sphere of radius $|x|$.

The optimal filter for this problem is $h(x,y)$, which in turn yields $\hat{s}(x)$ by

$$(3-4) \qquad \hat{s}(x) = \int h(x,y)w(y)dy = \int_0^{|x|} \int_S h(x,|y|e)w(|y|e)|y|^2 de\, d|y|$$

Here S is the unit sphere and $y = |y|e$, where e is a unit vector. By the orthogonality principle, $h(x,y)$ solves the three-dimensional Wiener-Hopf integral equation

$$(3-5) \qquad k(x,y) = h(x,y) + \int_{|z| \leq |x|} h(x,z)k(z,y)dz, \qquad |y| \leq |x|.$$

Without loss of generality, we define $h(x,y) = 0$ for $|y| > |x|$. $h(x,y)$ can be viewed as analogous to a continuous-parameter quarter-plane AR filter, except that the causality is defined in terms of $|x|$ and $|y|$, so that there is no "corner" and no ambiguity over in which direction to proceed.

The covariance function $k(x,y)$ is positive definite, and it is assumed to have the *generalized displacement property*

$$(3-6) \qquad (\Delta_x - \Delta_y)k(x,y) = 0,$$

which is a direct generalization of the Toeplitz-plus-Hankel structure exploited by the one-dimensional Levinson, Schur, and lattice algorithms. Indeed, the double Radon transform of $k(x,y)$ has the one-dimensional Toeplitz-plus-Hankel structure in the Radon transform variables. The structure (3-6) of the covariance makes possible fast algorithms for solving the integral equation (3-5).

The structure of $k(x,y)$ implied by (3-6) reduces the number of degrees of freedom in the function $k(x,y)$ from six to five. This is still far more general than the case of a *homogeneous* random field having covariance $k(x-y)$ (three degrees of freedom), or the case of an *isotropic* random field having covariance $k(|x-y|)$ (one degree of freedom) treated in [10]. Note that both homogeneous and isotropic random fields are included as special cases of the property (3-6).

3.2 Inverse Scattering Problem Specification. The basic three-dimensional inverse scattering problem is as follows. The wave field $u(x,k)$ satisfies the Schrodinger equation

$$(\Delta + k^2 - V(x))u(x,k) = 0 \tag{3-7}$$

where the scattering potential $V(x)$ is real-valued, smooth, and has compact support. Two different solutions to (3-7) are considered.

The *scattering solution* $\psi(x,k,e)$ has the boundary condition

$$\psi(x,k,e_i) \simeq e^{-ike_i\cdot x} + (e^{-ik|x|}/4\pi|x|)A(k,e_s,e_i) + O(|x|^{-2}) \tag{3-8}$$

This corresponds to a scattering experiment in which an impulsive plane wave $e^{-ike\cdot x}$ is used to probe the scattering medium, resulting in a scattered field that spreads out to the scattering amplitude $A(k,e_s,e_i)$ in the direction e_s. The goal is to reconstruct the potential $V(x)$ from the scattering amplitude $A(k,e_s,e_i)$.

The *regular solution* $\phi(x,k,e)$ to (3-7) is the solution that is an entire analytic function of k and is of exponential order $|x|$. Subject to mild assumptions [11], this solution exists generically, and its inverse Fourier transform $\phi(x,t,e)$ has support in t on the interval $[-|x|,|x|]$.

The scattering and regular solutions are related by

$$\psi(x,k,e_1) = \int_S \phi(x,k,e_2)F(k,e_1,e_2)de_2 \tag{3-9a}$$

$$\phi(x,k,e_1) = \int_S \psi(x,k,e_2)F^{-1}(k,e_1,e_2)de_2 \tag{3-9b}$$

where $F^{-1}(k,e_1,e_2)$ is the inverse kernel to $F(k,e_1,e_2)$, which is analytic in the lower half-plane, and can be determined using [12]

$$\delta(e_1-e_2) + \frac{k}{2\pi i}A(k,e_1,e_2) = \int_S F^{-1}(k,e_1,e_3)F(-k,e_3,e_2)de_3. \tag{3-10}$$

The inverse scattering problem may be solved as follows [12]. Let

$$M(k,e_1,e_2) = \int_S F(k,e_1,e_3)F(k,e_3,e_2)^* de_3 \tag{3-11}$$

and let $k(x,y)$ be its double inverse Fourier transform. Then the inverse Fourier transform of the regular solution $\phi(x,y) = \Delta(x-y) - h(x,y)$ is determined by solving an integral equation *identical to the Wiener-Hopf equation (3-5) above!* ((8.4) of [12]). The scattering potential is then computed from the regular solution using

$$V(x) = -\frac{2}{|x|^2}\frac{d}{d|x|}|x|^2 h(x,x) \tag{3-12}$$

which is comparable to (2-2b).

The reason that the inverse scattering and estimation problems both involve the same integral equation (3-5), in both one and three dimensions, is that this integral equation expresses an orthogonality condition for both problems–the orthogonality principle in estimation theory, and the orthogonality of the $\psi(x,k,e)$ in scattering theory (see [6]).

IV. THREE-DIMENSIONAL SPLIT ALGORITHMS

In this section we present fast, differential algorithms that are generalizations of the one-dimensional split algorithms of Section II. These algorithms solve the estimation problem of computing the optimal filter $h(x,y)$ from the covariance $k(x,y)$, and they also solve the inverse scattering problem of computing the potential $V(x)$ from the scattering amplitude $A(k, e_s, e_i)$. These algorithms are derived and presented in more detail in [4]; here we merely summarize them for comparison with the algorithms of Section II.

4.1 Differential form of the Wiener-Hopf integral equation. Applying the operator $(\Delta_x - \Delta_y)$ to the integral equation (3-5) and using the generalized displacement property (3-6), Green's theorem, and the unicity of solution to (3-5) when $k(x,y)$ is positive definite and both $k(x,y)$ and $h(x,y)$ are L^2 yields, after some algebra (see [4])

$$(4-1) \qquad (\Delta_x - \Delta_y)h(x,y) = \int_S V(x,e)h(|x|e,y)|x|^2 de$$

where the *non-local filter potential* $V(x,e)$ is defined as

$$(4-2) \qquad V(x,e) = -\frac{2}{|x|^2}\frac{d}{d|x|}|x|^2 h(x,|x|e).$$

Note that although the Wiener-Hopf equation (3-5) is only valid for $|y| \leq |x|$, the differential form (4-1) is valid for *all* x and y, since for $|y| > |x|$ we have trivially $0 = 0$. Equation (4-1) is a direct generalization of (2-14) to three dimensions; the only difference is the extra dependence in the potential $V(x,e)$. Even this is not surprising; since $k(x,y)$ has five degrees of freedom, $h(x,y)$ does also, and thus the potential function characterizing the $h(x,y)$ must also have five degrees of freedom.

4.2 Generalized Split Levinson Algorithm. Since (4-1) holds for all x and y, we can perform a Radon transform of (4-1) taking y into t and e_i, yielding

$$(4-3) \qquad (\Delta - \frac{\partial^2}{\partial t^2})\hat{h}(x,t,e_i) = \int_S V(x,e)\hat{h}(|x|e,t,e_i)|x|^2 de$$

Equation (4-3) describes a continuous three-dimensional scattering medium with *non-local* scattering potential $V(x,e)$. Equation (4-3) is the inverse Fourier transform of a three-dimensional Schrodinger equation with a non-local potential. In this form, the applicability to three-dimensional inverse scattering problems becomes apparent.

The three-dimensional split Levinson algorithm consists of propagating (4-3) in increasing $|x|$ and t, with the potential $V(x,e)$ computed using

$$(4-4) \qquad V(x,e) = -\frac{2}{|x|^2}\frac{d}{d|x|}|x|^2\left(k(x,|x|e) - \int_{|z|\leq|x|} h(x,z)k(z,|x|e)dz\right)$$

Equation (4-4) follows directly from (4-2) and the integral equation (3-5); it is the three-dimensional version of (2-7). Note the "inner product" in (4-4) takes the form of a three-dimensional integral. Equation (4-3) can be discretized in $|x|$ and t into a three-term recurrence like (2-6); there are three major differences:

1. A separate set of independent, parallelizable recurrences is required for each e_i;
2. The simple multiplication by the potential in (2-6) becomes an integration over the unit sphere;
3. The transverse Laplacian on the surface of the sphere of radius $|x|$ must be computed at each recursion; this involves only values of $h(x,t,e_i)$ on the surface of this sphere.

The estimation problem can be solved using the three-dimensional split Levinson algorithm as follows. Propagate (4-3) and (4-4) in increasing $|x|$ and $-|x| < t < |x|$, using the given covariance $k(x,y)$ in the boundary condition (4-4). Then compute the optimal filter $h(x,y)$ by taking the inverse Radon transform of $\hat{h}(x,t,e_i)$. See [4] for more details.

The inverse scattering problem can be solved using the three-dimensional split Levinson algorithm as follows. First, use (3-10) to compute the spectral factor $F(k,e_1,e_2)$ from the scattering amplitude $A(k,e_s,e_i)$; this amounts to solving an integral equation (see [5] and [12]). Next, compute $M(k,e_1,e_2)$ from (3-11), and take the double inverse Fourier transform $k(x,y) = \mathfrak{F}^{-1}\mathfrak{F}^{-1}\{M(k,e_1,e_2)\}$. Finally, compute the regular solution

$$(4-5) \qquad \hat{\phi}(x,t,e_i) = \delta(t - e_i \cdot x) - \hat{h}(x,t,e_i)$$

by propagating (4-3) and (4-4) as above; note that the scattering data $A(k,e_s,e_i)$ enters into the algorithm via $k(x,y)$ and (4-4). Note that unlike the estimation problem, the regular solution is of no interest; the scattering potential computed using (4-4) is the desired quantity.

For a real-world inverse scattering problem, the non-local potential $V(x,e)$ will be *local*: $V(x,e) = V(x)\delta(x/|x| - e)$. It is not at all obvious that (4-4) should yield a $V(x,e)$ of this form; in inverse scattering theory this is called the *miracle* [11],[12] (see (3-12)). For proper scattering data $A(k,e_s,e_i)$ the miracle will happen; this phenomenon is still not well understood in inverse scattering theory.

4.3 Generalized Split Lattice Algorithm. The split lattice algorithm propagates the scattering solution $\psi(x,k,e_i)$, as opposed to the regular solution $\phi(x,k,e_i)$ propagated by the split Levinson algorithm. The two solutions are related by (3-9).

The *estimation* problem can be solved using the split lattice algorithm as follows. We assume the spectral factor $F(k,e_1,e_2)$ of $k(x,y)$ is given. Propagate

$$(4-6) \qquad (\Delta - \frac{\partial^2}{\partial t^2})u(x,t,e_i) = \int_S V(x,e)u(|x|e,t,e_i)|x|^2 de$$

in increasing $|x|$ and t; note that (4-6) has the same form as (4-3), but propagates the non-impulsive part of the scattering solution (compare to (4-5))

$$(4-7) \qquad \hat{\psi}(x,t,e_i) = \delta(t - e_i \cdot x) - u(x,t,e_i).$$

Equation (4-6) is initialized using the inverse Fourier transform of

$$(4-8) \qquad u(0,k,e_1) = \int_S F(k,e_1,e_2)de_2$$

which follows from setting $x = 0$ in (3-9a). The potential $V(x,e)$ is computed using (3-9b), (4-5), and (4-4). Note that the "inner product" (4-4) is again needed, and that the spectral factor $F(k,e_1,e_2)$ enters into the computation both at the initial condition (4-8) and the boundary condition (4-4).

The above quantities all have illuminating stochastic interpretations. $\hat{\phi}(x,y)$ is the *residual* filter that converts the observation field $w(x)$ into the residual field $w(x) - s(x|w(y), |y| \leq |x|)$. This residual field can then be decorrelated on the circle $|y| = |x|$ to give an innovations field. $F(k,e_1,e_2)$ is then the transfer function of a modelling filter that transforms the white innovations field into $w(x)$, while $F^{-1}(k,e_1,e_2)$ is the transfer function of a (non-causal) whitening filter that whitens $w(x)$. This whitened field can be filtered into the residuals field using $\psi(x,k,e_1)$; this follows from (3-9b). Further details are given in [5].

Note that the output of the split lattice algorithm is the potential $V(x,e)$. This can then be plugged into the split Levinson algorithm, eliminating the "inner product" computation (4-4), but replacing it with another. As in the one-dimensional case, the main use of the lattice algorithm is to avoid the computation of $k(x,y)$ from its spectral factor $F(k,e_1,e_2)$.

The *inverse scattering* problem is solved using the split lattice algorithm as follows. Recall that the one-dimensional split lattice algorithm solves the inverse scattering problem given *transmission* data, which is related to *reflection* data by (2-3). Similarly, the three-dimensional split lattice algorithm solves the three-dimensional inverse scattering problem given the *transmission data* (4-8). Note that in *all* problems the transmission data is the same as the spectral factor (integrated over a variable in (4-8)); compare (2-3) with (3-10) and (4-8).

For the inverse scattering problem we are given the transmission data $u(0,t,e_i)$. Initialize (4-6) with this, and propagate it as above. Note that we don't explicitly know $F(k,e_1,e_2)$ from $u(0,k,e_i)$ (see (4-8)); however, the *local* potential $V(x)$ may be computed using (3-12) and (3-9b).

4.4 Generalized Split Schur Algorithm. The split Schur algorithm must be propagated in x and y, rather than in x and t. The reason for this is that the Schur algorithm propagates the convolution of the prediction error filter and the observation field covariance, which is zero for $|y| < |x|$ by the orthogonality principle. However, the triangularity property of being zero for $|y| < |x|$ does NOT map to the Radon transform domain. This is unlike the Levinson algorithm, in which $h(x,y) = 0$ for $|y| > |x|$ implies $\hat{h}(x,t,e) = 0$ for $t > |x|$. Since the triangularity property is the essential structure of the Schur algorithm (in one or three dimensions), we are forced back to the $x-y$ domain.

Define $\chi(x,y)$ as the convolution of the prediction error filter and the observation field covariance, just as in the one-dimensional case:

$$(4-9) \qquad \chi(x,y) = \int \phi(x,z)(\delta(z-y) + k(z,y))dz$$

Then $\chi(x,y)$ satisfies the differential form (4-1), just as $\phi(x,y)$ does. Furthermore, by the orthogonality principle we have $\chi(x,y) = 0$ for $|y| < |x|$. Then, from the form (4-5) of $\phi(x,y)$, $\chi(x,y)$ must have the form

$$\chi(x,y) = \delta(x-y) + v(x,y)1(|y| - |x|) \tag{4-10}$$

Inserting (4-10) into the differential form (4-1) results in

$$(\Delta_x - \Delta_y)h(x,y) = \int_S V(x,e)h(|x|e,y)|x|^2 de \tag{4-11a}$$

$$V(x,e) = -2\left(\frac{\partial}{\partial x} + \frac{\partial}{\partial y}\right) v(x, y = |x|e) \tag{4-11b}$$

The split Schur algorithm consists of propagating (4-11) in increasing $|x|$ and $|y| > |x|$ using (4-11a), with $V(x,e)$ reconstructed using (4-11b). The algorithm is initialized using [4]

$$v(|x|e, y) = k(|x|e, y),\ |x| \to 0 \tag{4-12}$$

Note that the dependence of $k(x,y)$ on $e = x/|x|$ for small $|x|$ is needed. This ensures the five degrees of freedom in the data necessary to compute $V(x,e)$, which also has five degrees of freedom.

For the estimation problem, the split Schur algorithm allows the potentials $V(x,e)$ to be computed without an "inner product" computation. The $V(x,e)$ can then be plugged into the split Levinson algorithm, replacing (4-4). This is the three-dimensional generalization of the approach proposed in [7].

For the inverse scattering problem, the split Schur algorithm allows the scattering potential to be computed from reflection data. It is first necessary to compute $k(x,y)$ from the scattering data, as discussed for the split Levinson algorithm, but again the "inner product" computation (4-4) is avoided.

V. CONCLUSION

Generalized split Levinson, Schur, and lattice algorithms for the three-dimensional random field linear least-squares estimation problem, and for the three-dimensional inverse scattering problem, have been obtained. These algorithms directly solve the three-dimensional integral equation associated with both problems. The algorithms are fast since they exploit the Toeplitz structure of the double Radon transform of the quantity $k(x,y)$ to reduce the amount of computation necessary to solve the integral equation.

The one-dimensional split Levinson and split fast Cholesky algorithms are three-term recurrences that are equivalent to a discretization of a one-dimensional Schrodinger equation in the time domain. The three-dimensional algorithms of Section 4 are equivalent to three-dimensional Schrodinger equations in the time domain, which is why these algorithms are referred to as generalized three-dimensional split algorithms.

Some issues that need further research are as follows. The non-local potential $V(x,e)$ complicates matters enormously, since it has no one-dimensional analogue and introduces non-causality. It would be very desirable to be able to characterize the set of covariance functions $k(x,y)$, or spectral factors $F(k,e_1,e_2)$, associated with a *local* potential $V(x)$. This set would have three degrees of freedom, like the set of covariance functions associated with homogeneous random fields. We note here that this is a major unsolved problem in inverse scattering theory; an estimation viewpoint may well be more appropriate for solving this problem. Another issue is the numerical performance of these algorithms, which is currently being investigated.

ACKNOWLEDGMENT

The support and hospitality of the Institute for Mathematics and its Applications, Dept. of Mathematics, University of Minnesota, is gratefully acknowledged.

REFERENCES

1. A.M. BRUCKSTEIN AND T. KAILATH, *Inverse Scattering for Discrete Transmission Line Models*, SIAM Review 29(3), 359-389 (1987).
2. A.E. YAGLE, *The Lattice Algorithm of Linear Prediction Applied to the Inverse Scattering Problem Given Transmission Data*, Proc. Int'l Conf. on Acoust., Speech, Sig. Proc., New York, April 1988, pp. 1655-1658.
3. P. DELSARTE AND Y. GENIN, *On the Splitting of Classical Algorithms in Linear Prediction Theory*, IEEE Trans. Acoust., Speech, Sig. Proc. ASSP-35(5), 645-653 (1987).
4. A.E. YAGLE, *Generalized Split Levinson, Schur, and Lattice Algorithms for Three-Dimensional Random Field Estimation Problems*, submitted to SIAM J. Appl. Math.
5. A.E. YAGLE, *Connections between Three-Dimensional Inverse Scattering and the Linear Least-Squares Estimation of Random Fields*, to appear in Acta Appl. Mathe.
6. A.E. YAGLE, *Multidimensional Inverse Scattering: An Orthogonalization Formulation*, J. Math. Phys. 28(7), 1481-1491 (1987).
7. S.Y. KUNG AND Y.H. HU, *A Highly Concurrent Algorithm and Pipelined Architecture for Solving Toeplitz Systems*, IEEE Trans. Acoust., Speech, Sig. Proc. ASSP-31(1), 66-75 (1983).
8. A.M. BRUCKSTEIN AND T. KAILATH, *On Discrete Schrodinger Equations and their Two-Component Wave Equation Equivalents*, J. Math. Phys. 28(12), 2914-2924 (1987).
9. B.C. LEVY AND J.N. TSITSIKLIS, *Linear Estimation of Stationary Stochastic Processes, Vibrating Strings, and Inverse Scattering*, Tech. Report #LIDS-P-1155, Laboratory for Information and Decision Systems, M.I.T., (1982).
10. B.C. LEVY AND J.N. TSITSIKLIS, *A Fast Algorithm for Linear Estimation of Two-Dimensional Isotropic Random Fields*, IEEE Trans. Info. Th. IT-31(5), 635-644 (1985).
11. R.G. NEWTON, *Inverse Scattering. IV. Three Dimensions: Generalized Marchenko Construction with Bound States, and Generalized Gel'fand-Levitan Equations*, J. Math. Phys. 23(4), 594-604 (1982).
12. R.G. NEWTON, *Inverse Scattering. II. Three Dimensions*, J. Math. Phys. 21(7), 1698-1715 (1980).

MIX
Papier aus verantwortungsvollen Quellen
Paper from responsible sources
FSC® C105338

If you have any concerns about our products,
you can contact us on
ProductSafety@springernature.com

In case Publisher is established outside the EU,
the EU authorized representative is:
Springer Nature Customer Service Center GmbH
Europaplatz 3, 69115 Heidelberg, Germany

Printed by Libri Plureos GmbH
in Hamburg, Germany